Australia

300 Years of Botanical Illustration

The research for this book
was generously supported by
Dame Elisabeth Murdoch AC DBE
and the
Centre for Plant Biodiversity Research, Canberra

(opposite)
Barringtonia asiatica (L.) Kurz (= *Barringtonia*)
Copper engraving drawn by George Forster from
J.R. Forster & J.G.A. Forster, *Characteres Generum Plantarum*, published in 1776.
[NLA 19.8 x 12.0]

(cover)
Rhododendron lochiae F.Muell.
Original watercolour painting by Margaret Stones, 1973.
This rare rhododendron was named by Ferdinand von Mueller in 1887 in honour of Lady Loch, wife of Victoria's Governor, for her patronage of horticulture.
[IPMA 44.0 x 35.3]

Australia

300 Years of Botanical Illustration

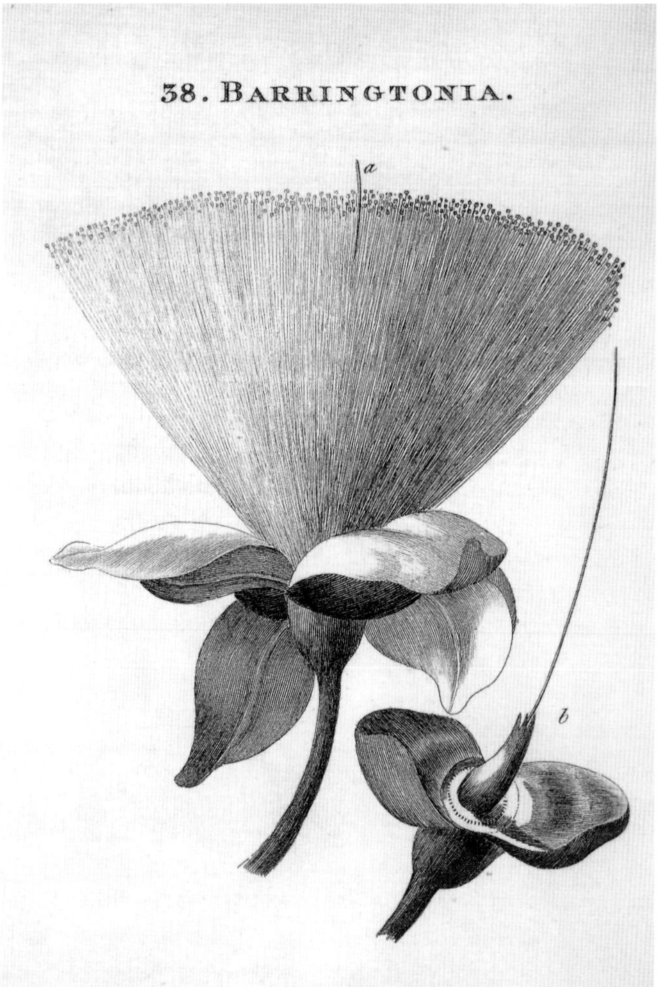

Helen Hewson

CSIRO
PUBLISHING

National Library of Australia
Cataloguing-in-Publication entry

Hewson, Helen.
Australia : 300 years of botanical illustration.

Bibliography.
Includes index.
ISBN 0 643 06365 X (limited ed.).
ISBN 0 643 06366 8.

1. Botanical illustration – Australia. 2. Botanical illustration – Australia – History. I. Title. II. Title : 300 years of botanical illustration.

743.93409

Editor: Jane Holth
Designer: Jo Waite
Printer: Craft Print Pte Ltd

First published in 1999 by
CSIRO PUBLISHING
PO Box 1139
(150 Oxford Street)
Collingwood VIC 3066
Australia

Tel: (03) 9662 7666 Int: +61 3 9662 7666
Fax: (03) 9662 7555 Int: +61 3 9662 7555
www.publish.csiro.au

10 9 8 7 6 5 4 3 2 1

09 08 07 06 05 04 03 02 01 00 99

International Edition:
ISBN 1 85149 334 4

Published throughout the world (excluding Australia) by
ANTIQUE COLLECTORS' CLUB Ltd.

UK OFFICE
5 Church Street, Woodbridge
Suffolk IP12 1DS
Tel: (01394) 385501 Fax: (01394) 384434
Email: sales@antique-acc.com
Website: www.antique-acc.com

US OFFICE
Market Street Industrial Park
Wappingers' Falls, NY 12590
Tel: (914) 297 0003 Fax: (914) 297 0068
Email: info@antiquecc.com
Website: www.antiquecc.com

Contents

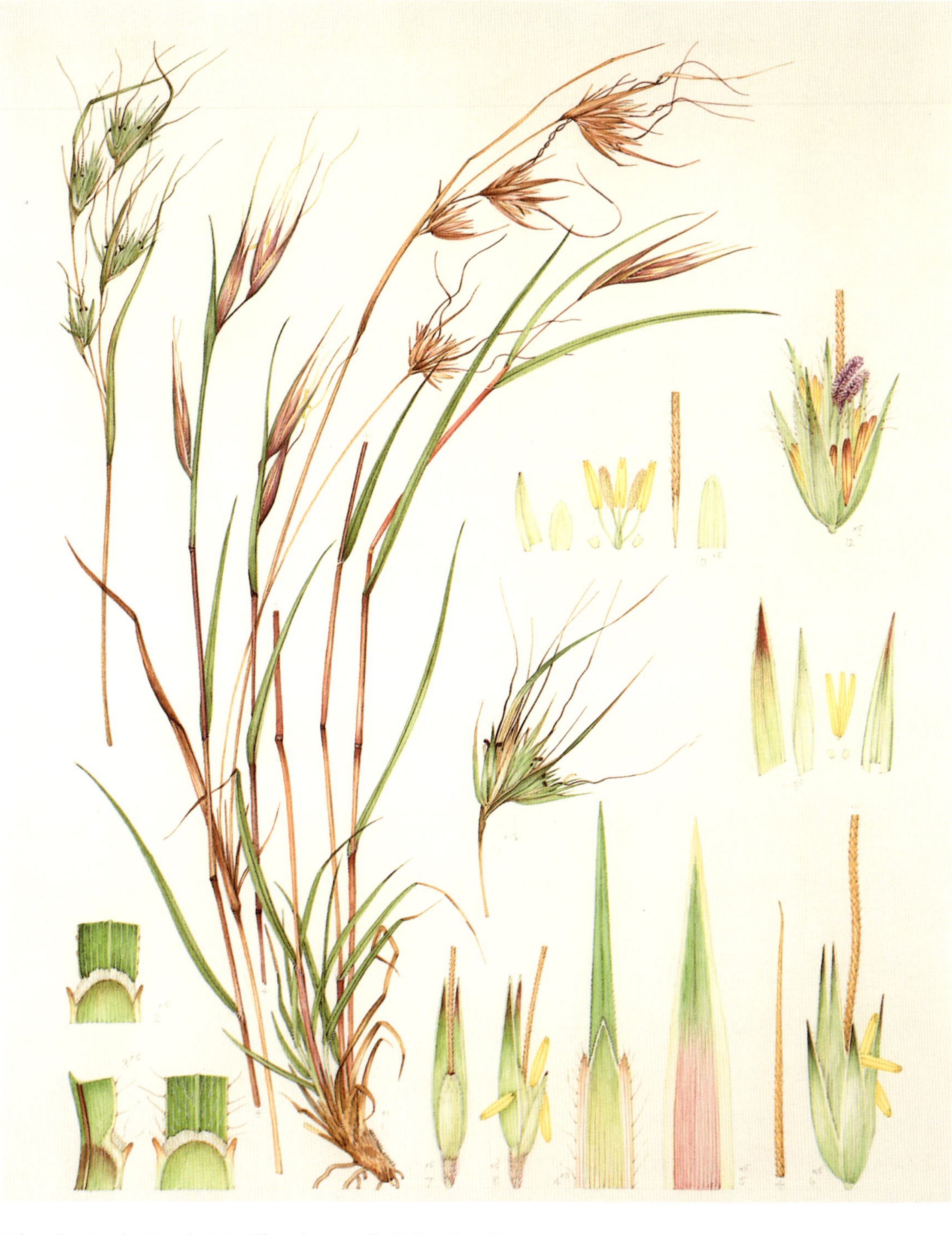

Themeda triandra Forsskal. (= *Themeda australis* (R.Br.) Stapf.)
Watercolour by Margaret Stones, 1975.
{MEL 54.9 x 38.3}

Preface

The importance of illustration by means of engravings, whether on wood or (later) on metal, did not escape {post-Renaissance botanists}. Naturalistic and empirical plant illustration developed rapidly in the course of the sixteenth century and was a major factor in spreading information about plants. Obviously one of the great disadvantages which faced the pre-Renaissance botanists was the difficulty of comparing species from different localities. There are three ways to solve this problem: illustration, growing plants in botanic gardens, and herbaria.

Frans Stafleu[1]

The European Renaissance which began in the fourteenth century saw the re-invigoration of art, literature, culture, science, medicine and technology. Previously scientific inquiry was stultified by myth and superstition. The Renaissance led to an explosion of new observations and interpretations of nature, and to such intrepid investigation and exploration as had never before been envisaged.

In natural history, old observations and images had been hand-copied for centuries. Now original observations were made, and new recording methods invented. Woodcut printing was gradually replaced by copper engraving. The printing press was invented. The *Herbals* (books devoted to describing herbs) were replaced with texts describing more than the useful plants. The Flat Earth myth was disproved. Voyages of discovery to lands in other parts of the world introduced a wealth of new plants for food, medicine, and horticultural delight. Soon a worship of the exotic began. Botanic gardens and private gardens flourished; herbaria and museums were established and expanded; and horticultural expertise and the nursery trade grew. For science, new classification, naming and description methods developed. The resultant record of the newly found exotic wonders was published in grandly illustrated books and journals.

In the midst of this enlightenment Australia was 'discovered' and the recording of its plant and animal wealth began.

For Australia much of the early history of botanical illustration is that of collecting plants during voyages of exploration in the southern hemisphere. Explorers, natural historians and artists had to carry out their respective activities under exceptionally difficult conditions. Many of the voyages suffered tragedies of varying kinds, sometimes for the ship, and sometimes for the individuals and materials on board. The history of botanical illustration is a colourful and dramatic story.

When the story shifted to Australia, geographic and scientific exploration continued to go hand in hand. The explorers were confronted with an inland that was much harsher than the coastal fringe, and many succumbed to attacks by hostile Aborigines, to dehydration and starvation, or to poisoning through ignorance of the fruits and seeds they harvested. At the same time, scientific exploration continued with the aim of a complete census and documentation of the plant heritage.

By the twentieth century settlement and agriculture had expanded into the hinterlands. The needs of the time were very practical and tools with which to identify plants became a fundamental requirement. Primary industries and economic botany placed the greatest demand on the resources of the herbaria and their botanists. This was a time of consolidation for taxonomic botany, hence diagnostic and pseudo-diagrammatic representation was much more in demand than were grand works of art. By then the quantity and quality of botanical illustration throughout the world had diminished. At the same time, methods of reproduction of artwork had mechanised, resulting in poorer images.

Not until around the 1970s and the bicentennial of James Cook's first landing at Botany Bay did the trend obviously begin to reverse and botanical illustration and botanical art begin to flourish again. At the end of the century Australia now has numerous botanical artists, some of whom are rivalling the greats of the eighteenth and early-nineteenth centuries.

1. The Art of Illustration

A painter, an engraver and a botanist are all equally necessary to produce a good illustration; if one of them goes wrong, the illustration will be wrong in some respect. Hence botanists who have practised the arts of painting and engraving along with botany have left us the most outstanding illustrations.

Carl Linnaeus[1]

The beauty of flowers has inspired artists for centuries. For scientists, however, the purpose of illustrating flowers is more to capture their fundamental essence than to portray their beauty. Over the centuries, these two objectives have merged. While botanical illustration is the subject of this story it will soon be evident that there is no clear difference between botanical illustration and botanical art. Frequently the distinction lies with the reproduction, not with the original artwork.

The illustrations in this book have predominantly been prepared and published for science. Through them is revealed the progression from the original artwork to the published image of it in a scientific text or journal.

Description of plants is done by a combination of words and drawings. For over two millennia plants have been described by drawings to make reliable identification easier. The *Herbals* of the Middle Ages, in particular, catalogued useful plants, especially medicinal herbs, in this way. They represent the beginnings of modern pharmacology and descriptive botany. The development of these sciences went hand-in-hand with that of the art of botanical illustration.

In the Middle Ages repeated illustrations for the production of books were achieved either by hand-copying or by wood-block printing. In the latter, an impression of the plant is carved into a smooth block of wood. The high surfaces are then inked and an imprint effected by pressure onto a sheet of paper (technically, engraving in *relief*).

The illustration of Australian plants came after the next method of printing was developed. Accordingly, there are no early wood-block prints of Australian plants for scientific purposes. A pinnacle of perfection in wood-block printing was achieved by Thomas Bewick in the late-eighteenth century. He perfected the method of engraving on the end-grain, rather than the long-grain, of the wooden block and in this way depicted some Australian animals in his *A History of Quadrupeds* (1792).

For book illustration, metal engraving superseded wood-block printing. The impression of the plant is incised with a hard metal tool into the surface of a polished soft-metal sheet (usually copper). The grooves are then filled with ink and an imprint effected by pressure, in a press, onto a sheet of paper (technically, engraving in *intaglio*). The very first published drawings of Australian plants were produced by this method in William Dampier's *A Voyage to New Holland* in 1703. Indeed most of the published drawings of Australian plants for about one hundred and thirty years were produced by this means.

Etching (acid incision into a metal surface) seems to have been insignificant for botanical illustration, or restricted to the achievement of particular effects. Charming exceptions are seen in the works of Johann Wendland, his *Collectio plantarum* (1805–1819) being an example.

Illustrations of Australian plants in the latter decades of the eighteenth century serve to highlight another aspect of illustration for book production. Typically, three or more people were involved in the production of the final plate. First there was the artist, who provided the original artwork. The artist was the observer and capturer of the essence and morphology of the plant (with or without the supervision of a scientist) and was acknowledged on the plate with *del.*, or *delin.*, or *pinxt.* In Australian publications sometimes two people were involved at this stage, for example the Sydney Parkinson and Frederick Nodder combination where Parkinson did the sketches and Nodder made paintings based on these. The scientist, for whom the drawing was being done, was sometimes acknowledged with *direxit.* Ferdinand von Mueller in his *Fragmenta* (1858–1882) was acknowledged in this way. The engraver — an artisan or technician — converted the image onto the block for printing and was acknowledged with *sc.*, or *sculp.*, or *gravé* (or *lith.*). The printer, the artisan who inked the block and created the print on paper, was rarely acknowledged. Finally the colourist, the artist who hand-coloured the resultant print, was not acknowledged.

The artist usually received the accolades for published illustrations. Certainly draftsmanship is fundamental to the successful portrayal of an image. However, over the years, some artists must have been either over-rated or under-rated depending on the expertise of the engraver, in particular. Obviously the image which one sees in the book is executed by the engraver (or lithographer) and colourist, if hand-coloured. Ferdinand Bauer resolved this in his *Illustrationes Florae Novae Hollandiae* (1813) by taking all roles. Bauer is reputed to have done the engraving himself because he could not find a competent engraver. A comparison of

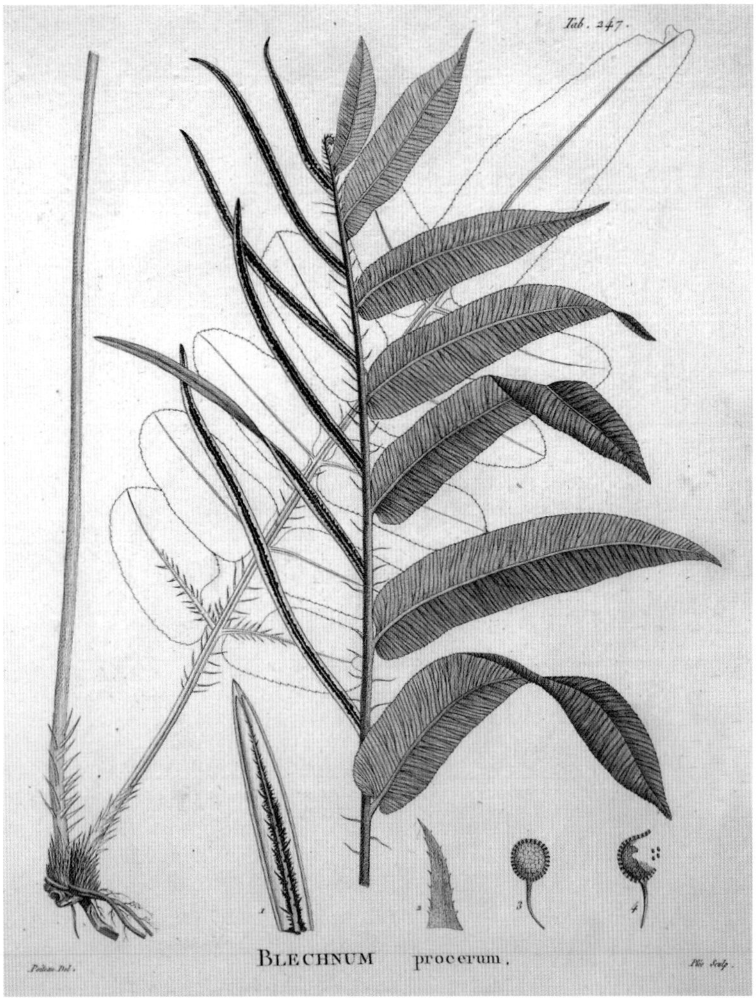

Blechnum wattsii Tindale (= *Blechnum procerum* (G.Forst.) Sw.)
Copper engraving drawn by Pierre-Antoine Poiteau, engraved by Auguste Plée and published in Labillardière's *Novae Hollandiae Plantarum Specimen*, 1804–1806. The central frond is vegetative to the right and spore-bearing to the left. These two states do not occur together in this way in nature, neither do the spore-bearing pinnae attach naturally to the frond. The scientist and artist have cooperated to make best use of space by using 'artistic licence'.
[NLA 28.3 x 21.7]

his own engraving with that of others (based on Bauer drawings) in Matthew Flinders' *A Voyage to Terra Australis* (1814) exposes competency in engraving as a problem for artists.

Accuracy in botanical illustration is another issue with which scientists and artists battle routinely — the scientist because accuracy is part of the work ethic and fundamental to the scientific record, and the artist because accuracy is required for the scientific record. In the arts, however, the phenomenon of artistic licence is common. Perhaps botanical illustration is the only discipline in science which may, from time to time, indulge in artistic licence. An obvious example of this is seen in the Pierre-Antoine Poiteau plate of *Blechnum* in Jaques Labillardière's *Novae Hollandiae Plantarum Specimen* (1804–1806). The depiction of the half-fertile and half-vegetative frond on the one frond is not true to nature. An occasional anomaly

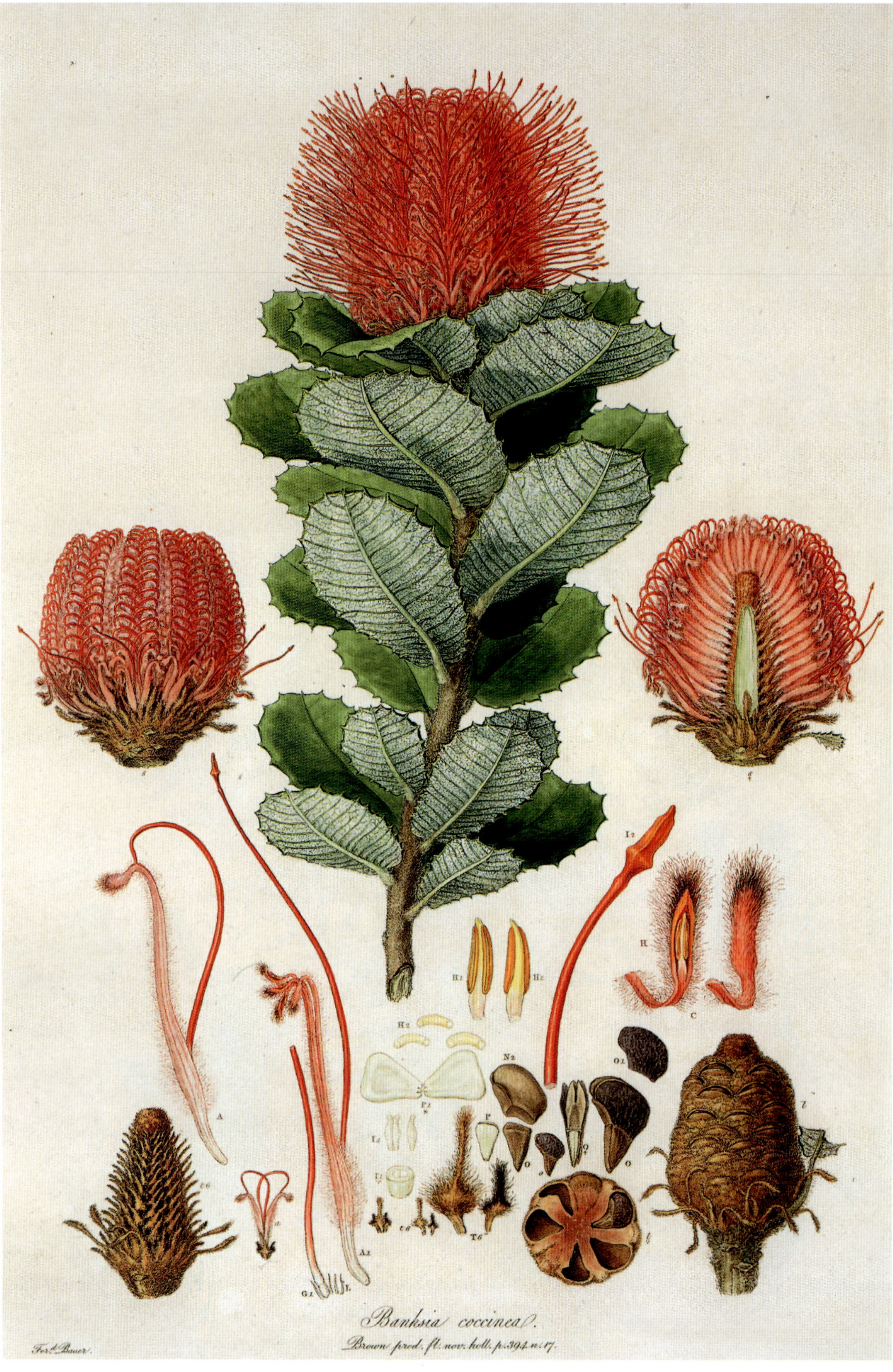
Banksia coccinea.
Brown prod. fl. nov. holl. p. 394. n. 17.
Ferd. Bauer

Hibbertia scandens (Willd.) Dryander
(= *Hibbertia volubilis* Andrews)
Copper engraving drawn by Pierre-Jean Turpin and engraved by Victor from F.G. Levrault, *Dictionnaire des Sciences Naturelles*, published 1816–1830.
The illustration has relied heavily on the engraving detail to give form. This style is not suitable for hand colouring. Both the artist and the engraver are acknowledged below the image.
[NLA 24.0 x 16.0]

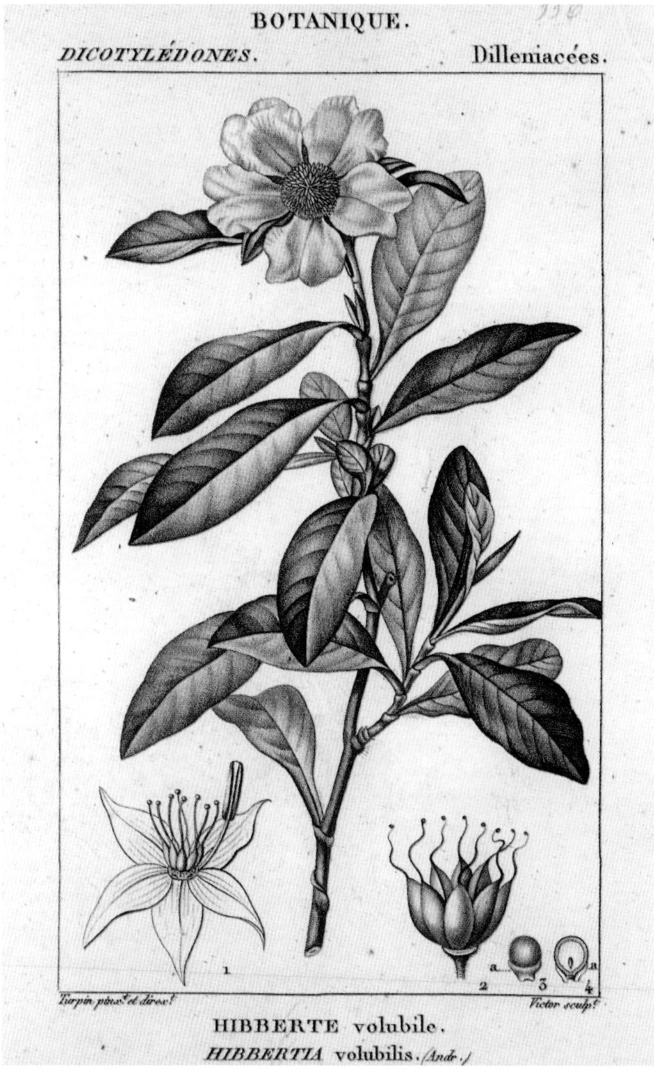

might be seen, but always with a gradual change from the vegetative to the fertile condition, or vice versa. Space limitations within one plate may have induced Poiteau and Labillardière to depict the plant in this way, though usually artists follow nature closely. There is another example of artistic licence in von Mueller's *The Plants Indigenous to the Colony of Victoria* (1860–1865). For the plate depicting *Panax dendroides*, von Mueller supervised the artist to draw apparent leaf-variation on one plant that is now known to represent more than one species.

Refinements to the technique of metal engraving came with the development of sophisticated tools to achieve a better three-dimensional illusion of the subject, its shape and its tonal qualities. Line engraving was the most commonly employed technique for early botanical illustration, graded depth and width of incision being used to achieve the illusion of form and tone. The French engravers, especially Auguste Plée and Poiteau, were masters of this technique as is seen in Labillardière's *Novae Hollandiae Plantarum Specimen*. Generally the French used parallel lines. Another method was cross-hatching, which is seen in the work of Frederick Nodder in James Britten's *Illustrations of the Botany of Captain Cook's Voyage Round the World* (1900–1905). Although not common in scientific botanical illustration, the mezzotint technique was sometimes used in conjunction with other techniques, including line and stipple. An early example appears in Ferdinand Bauer's plates in his *Illustrationes Florae Novae Hollandiae*.

Banksia coccinea R.Br.
Hand-coloured copper engraving published in Ferdinand Bauer's *Illustrationes Florae Novae Hollandiae* in 1813. The drawing, engraving and colouring are all by Bauer. It is not common for the artist to take three roles in the creation of a published image.
[SLV 39.0 x 25.0]

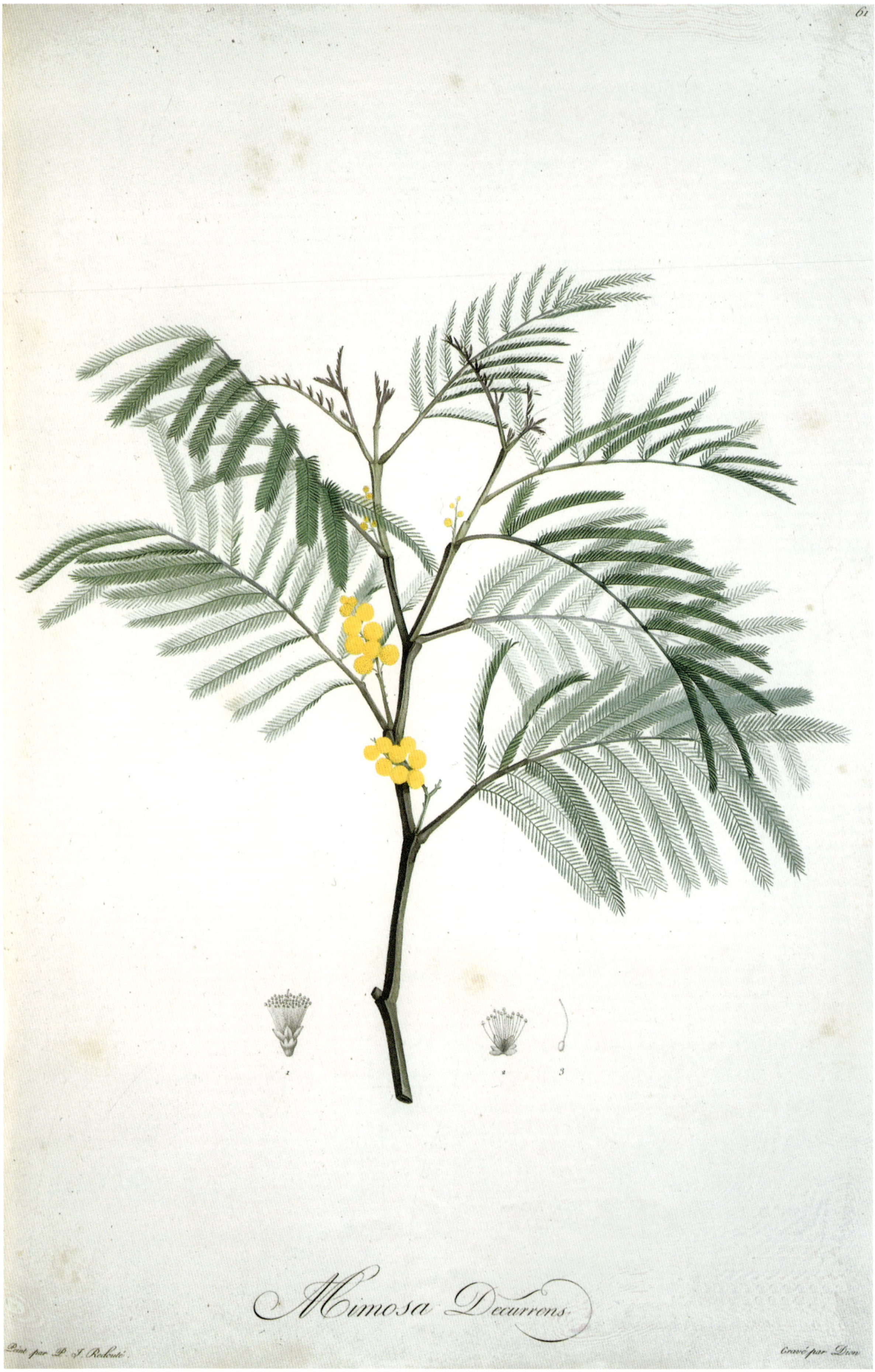
61
Mimosa Decurrens
Peint par P. J. Redouté
Gravé par Dien

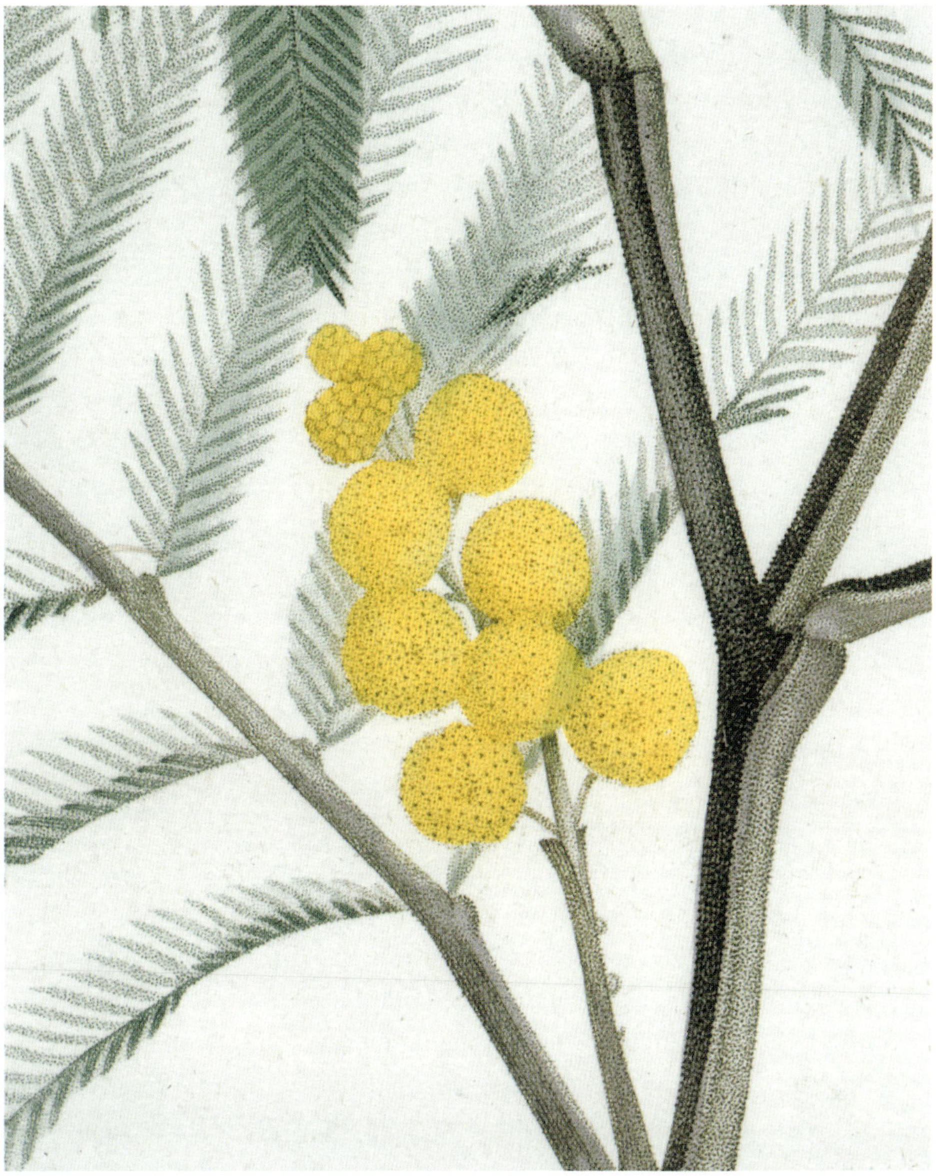

Acacia decurrens (Wendl.) Willd. (= *Mimosa decurrens* Ventenat)
Coloured stipple engraving by Pierre-Joseph Redouté (opposite), published in É.P. Ventenat, *Jardin de la Malmaison,* in 1804–1805. A small section of the engraving has been enlarged (above) to show the technique in greater detail.
[CANB 54.0 x 35.5]

The stipple technique (dots replacing line) was introduced to botanical illustration and developed to an 'art-form' by Pierre-Joseph Redouté as in Étienne Ventenat's *Jardin de la Malmaison* (1803–1805). In this book Redouté employed *à la poupée* – a method of inking the stipple engraving with coloured ink using a fine cloth tool to produce the full-coloured image in one printing only, with minimal hand-finishing. Indeed, the French artists, led by Redouté, became the great exponents of the method. Redouté's famous *Les Liliacées* (1802–1816) and *Les Roses* (1817–1824) (no Australian examples in the latter) represent the pinnacle of his achievement.

Colour printing did not make a significant impact on Australian plant illustration until well after the advent of lithography. A notable exception is the work of Redouté in *Jardin de la Malmaison*. However hand-coloured engravings are relatively common. The great periodicals (e.g. *Curtis's Botanical Magazine*) of the turn of the eighteenth century were all hand-coloured (watercolour painted) by an army of unacknowledged and unknown colourists. James Sowerby and James Edward Smith teamed to create a huge volume of hand-coloured engraving. Smith's *A Specimen of the Botany of New Holland* (1793–1795) is an example of this team's work with Australian plants.

Colourists were of varying talent. A poor example of hand colouring is seen in some plates of Robert Sweet's *Flora Australasica* (1827–1828). The colouring is heavy and opaque and the paint clumsily applied, obscuring the line of the engraver. Comparison between several copies of the same work may reveal varying ability in colouring which, of course, highlights the fact that each copy of the early illustrated books is a work of art in its own right.

This ceased to be so when the process of reproducing images became more mechanical with the advent of lithography. In the late 1790s in Germany Alois Senefelder invented the technique. He drew on a highly polished stone block with a greasy pencil to sensitise the drawing to accept an oil-based ink. The background was de-sensitised and wet, to repel the ink. In this way, the block was inked and the image printed on paper under pressure (technically *planographic* printing).

Lithography was soon adopted for botanical illustration in France but English artists were slower to embrace it. In England Walter Fitch was one of its great early exponents. In his book *Flower and Fruit Prints* (1938), G. Dunthorne claimed that the quality of lithographic printing was on the decline by the time Fitch fully embraced it. One of its virtues is that tone can so easily be achieved. The method allows the use of both chalk and pen. Thus both line and tone are able to be produced with a different technique in the same work. Dunthorne observed that the subtlety of tone was being lost. The exquisiteness of line seems to have suffered with the use of lithography, many uncoloured lithographs appearing more like pencil sketches than finished drawings.

Another talented English lithographer for scientific illustration was Maxim Gauci, a Maltese-born artist who worked in London. His work is acclaimed by W. Blunt and W. Stearn as having the qualities which Dunthorne so highly prized: 'he ranged his tone from the palest of silvery greys to the richest velvet black; his outline is never mechanical or obtrusive.'[2] Maybe Dunthorne is correct in attributing some of the loss in quality to the artisanship of the lithographers. However, the quality of botanical illustration in general seems to have to begun to decline by the mid-1800s, more due to changing technology and printing methods plus early attempts at automated colour printing than to the workmanship of artists and artisans. Friederich Schoenfeld's lithography in von Mueller's *The Plants Indigenous to the Colony of Victoria* is surely amongst the best botanical lithography ever executed.

John Brown's *The Forest Flora of South Australia* (1882–1890) is a very fine early example of colour lithography or chromolithography, using three or more printings of the primary colours (each from separately prepared blocks) to effect the final colour.

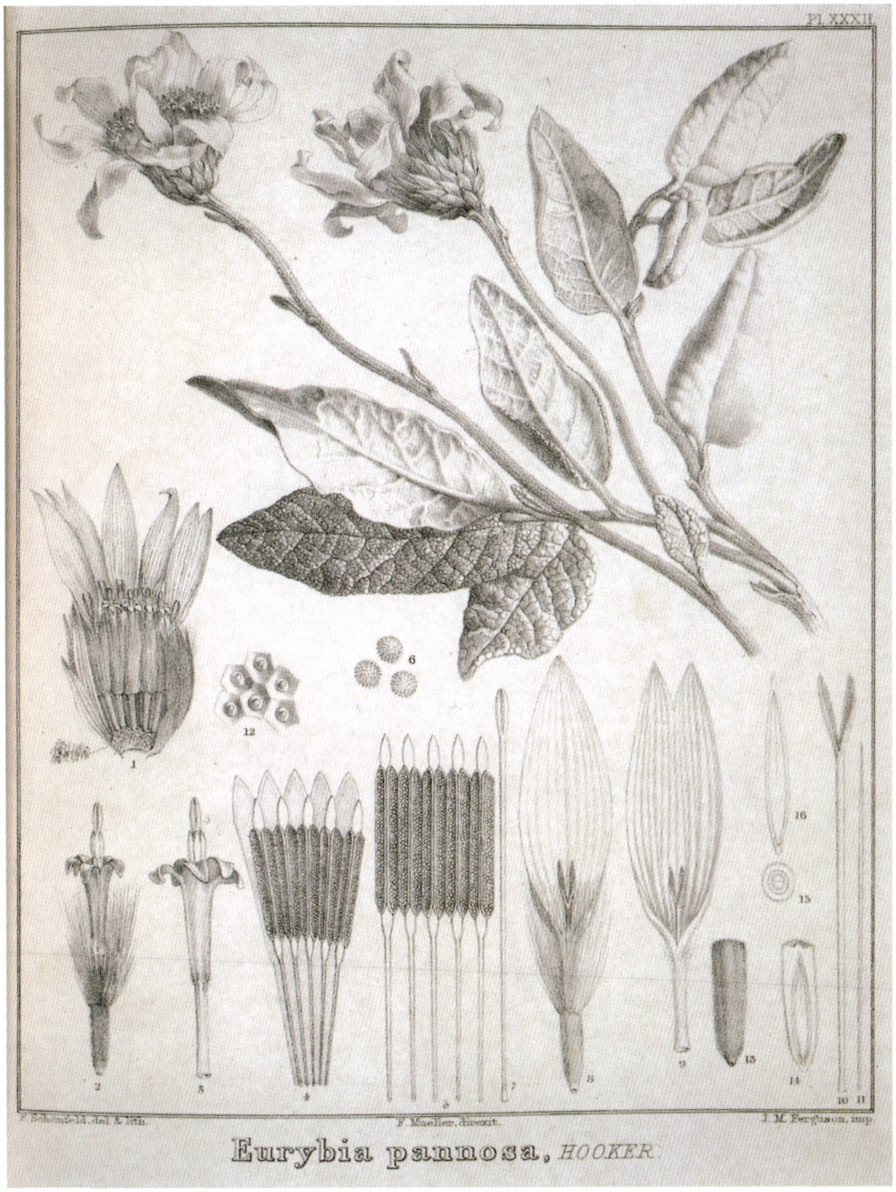

Olearia pannosa Hook. (= *Eurybia pannosa* (Hook.) F.Muell.)
Velvet Daisy-bush drawn and lithographed by Friederich Schoenfeld, published in F. Mueller, *The Plants Indigenous to the Colony of Victoria*, 1864–1865. The technique has been used to great effect to achieve both form and the velvet texture.
[NLA 25.3 x 19.8]

An unusual printing method was nature printing, which was discovered in the fifteenth century and taken up by some German scientists in the eighteenth century. Nature printing was effected by inking a piece of plant, usually a leaf, and subjecting it to pressure as with inked plates and blocks — similar to the method of obtaining finger- and palm-prints for forensic purposes. Leaf veins and other high points form the impression. The method is highly accurate for depicting veins, but has practical limitations. The first example found for Australian plant material is a plate of *Macadamia ternifolia* illustrating von Mueller's description of the species in 1858.[3] Later examples appear in Frederick Bailey's publications during

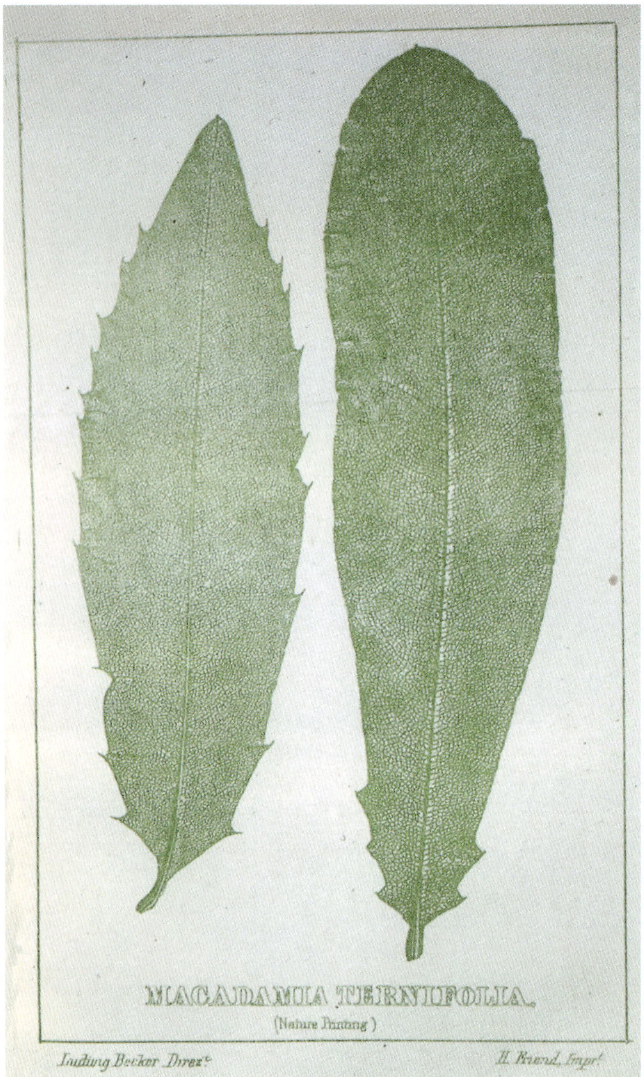

Macadamia ternifolia F.Muell.
In nature printing the leaf is used as the block, inked, and the impression results from the high points provided by the strong venation. This print, supervised by Ludwig Becker, was published in *Transactions of the Philosophical Institute of Victoria*, Volume 2, 1858.
[NLA 18.1 x 10.3]

the 1880s and 1890s. The printer had the specimens inked and the image transferred to a lithographic block. From this image on the block, the lithograph was prepared in the normal way.

Other methods of nature printing have been used, including an electrotype process to prepare the block.

The invention of photography in the first half of the nineteenth century ultimately helped solve a range of problems which had previously made illustrated books expensive and/or limited in print run. Initially, photography was used to transfer the image to a 'lithographic' plate. Photo-sensitisation of plates removed the need for an artisan-lithographer and permitted artists to draw in pencil, or ink, directly onto paper. The printer could then make a photolithograph or photoengraving to produce a print with little obvious loss in quality.

Photoengraving, or process engraving, also permitted artists to make drawings smaller than the size of the printed page, and allowed the printer to incorporate these drawings into text. Previously, printers could only do this with wood-engraved blocks. Wood engraving is a laborious and specialised craft with limited life in the printing process and it could limit the size of print runs. John Lindley used woodcut engraving in his texts during the mid-1800s. There are some very fine examples of woodcut engraving of Australian plants in *Paxton's Flower Garden* (1850–1853). Ferdinand von Mueller also used woodcuts in a couple of his more popular publications, the first being *Introduction to Botanic Teachings* (1877). The use of photoengraving to achieve the same effect was capitalised on around the turn of the century. John Black's *Naturalised Flora of South Australia* (1909) is an early Australian example. In a sense, this development was an emancipation for the art of illustration.

The next major use of photolithography was to print in colour. This was achieved through the development of screen-printing. The technique was to photograph the original artwork through a screen to effect tone or halftone. A series of filters allowed the creation of the individual screens for each primary colour. Colour printing by this method resulted in a

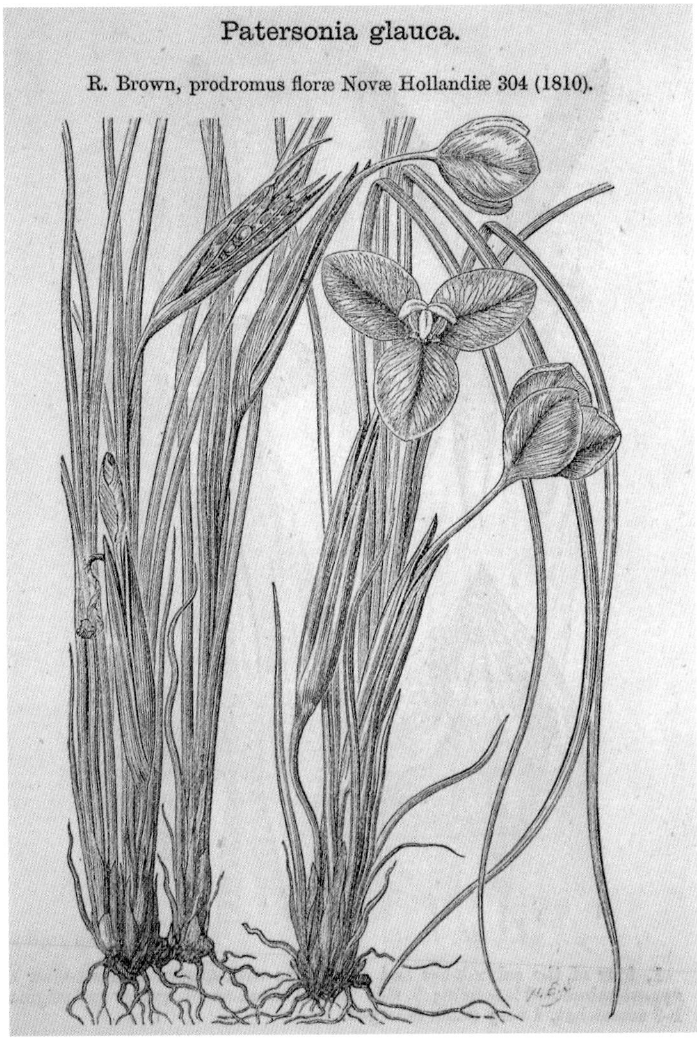

Patersonia fragilis (Labill.) Asch. & Graebner (= *Patersonia glauca* R.Br.)
Woodcut engraved by H.B. Smith and published in Ferdinand von Mueller's *Key to The System of Victorian Plants*, in 1885.
[NLA 12.5 x 7.7]

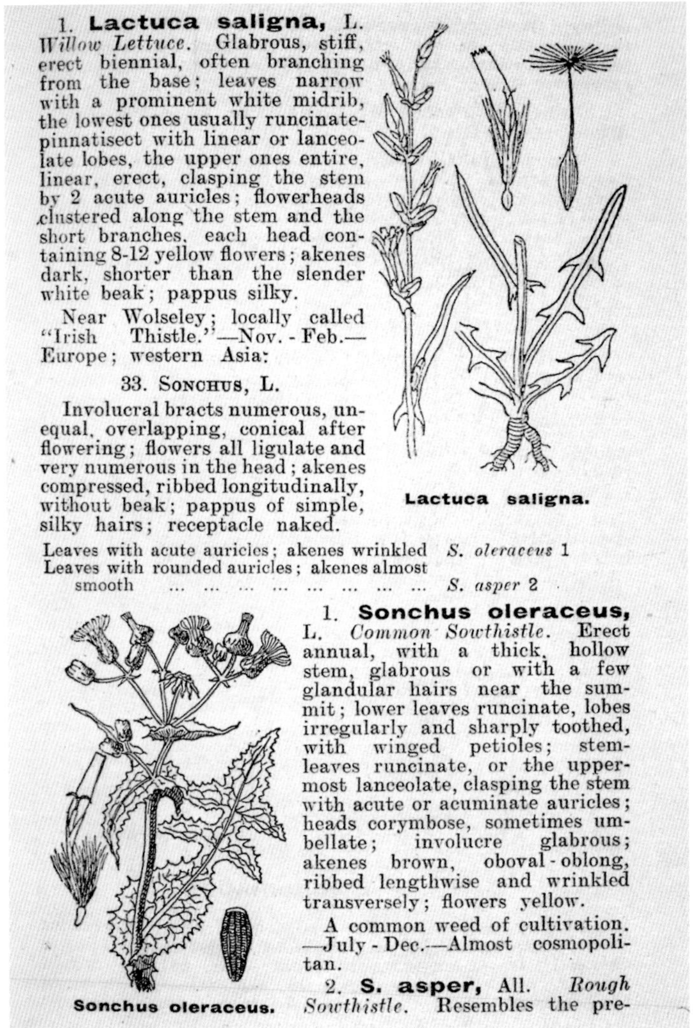

1. **Lactuca saligna,** L. *Willow Lettuce.* Glabrous, stiff, erect biennial, often branching from the base; leaves narrow with a prominent white midrib, the lowest ones usually runcinate-pinnatisect with linear or lanceolate lobes, the upper ones entire, linear, erect, clasping the stem by 2 acute auricles; flowerheads clustered along the stem and the short branches, each head containing 8-12 yellow flowers; akenes dark, shorter than the slender white beak; pappus silky.

Near Wolseley; locally called "Irish Thistle."—Nov. - Feb.—Europe; western Asia.

33. Sonchus, L.

Involucral bracts numerous, unequal, overlapping, conical after flowering; flowers all ligulate and very numerous in the head; akenes compressed, ribbed longitudinally, without beak; pappus of simple, silky hairs; receptacle naked.

Lactuca saligna.

Leaves with acute auricles; akenes wrinkled ... *S. oleraceus* 1
Leaves with rounded auricles; akenes almost smooth ... ... *S. asper* 2

1. **Sonchus oleraceus,** L. *Common Sowthistle.* Erect annual, with a thick, hollow stem, glabrous or with a few glandular hairs near the summit; lower leaves runcinate, lobes irregularly and sharply toothed, with winged petioles; stem-leaves runcinate, or the uppermost lanceolate, clasping the stem with acute or acuminate auricles; heads corymbose, sometimes umbellate; involucre glabrous; akenes brown, oboval-oblong, ribbed lengthwise and wrinkled transversely; flowers yellow.

A common weed of cultivation.—July - Dec.—Almost cosmopolitan.

2. **S. asper,** All. *Rough Sowthistle.* Resembles the pre-

Sonchus oleraceus.

Lactuca saligna L. & *Sonchus oleraceus* L.
An early example of process engraving which replaced wood engraving for the incorporation of illustrations within the letterpress. The pen drawings by John McConnell Black were published in *The Naturalised Flora of South Australia* in 1909.
[NLA 17.9 x 11.9]

significant drop in quality of the image. One of the major problems was the need to have very accurate 'registration' of each colour print, one on top of the other (not slightly offset). Generally early colour work was very poor. The colour plates in Bailey's *Comprehensive Catalogue of Queensland Plants* (1909) demonstrate the inadequacies very well.

Photographic and printing techniques and equipment have improved since the late-nineteenth and early-twentieth centuries, greatly enhancing the quality of reproduction. The computer is now employed to scan images and printers are using finer screens than previously possible. Using a combination of stochastic screening and hand-lithographic processes they can now achieve a quality of colour-image almost the same as the original.

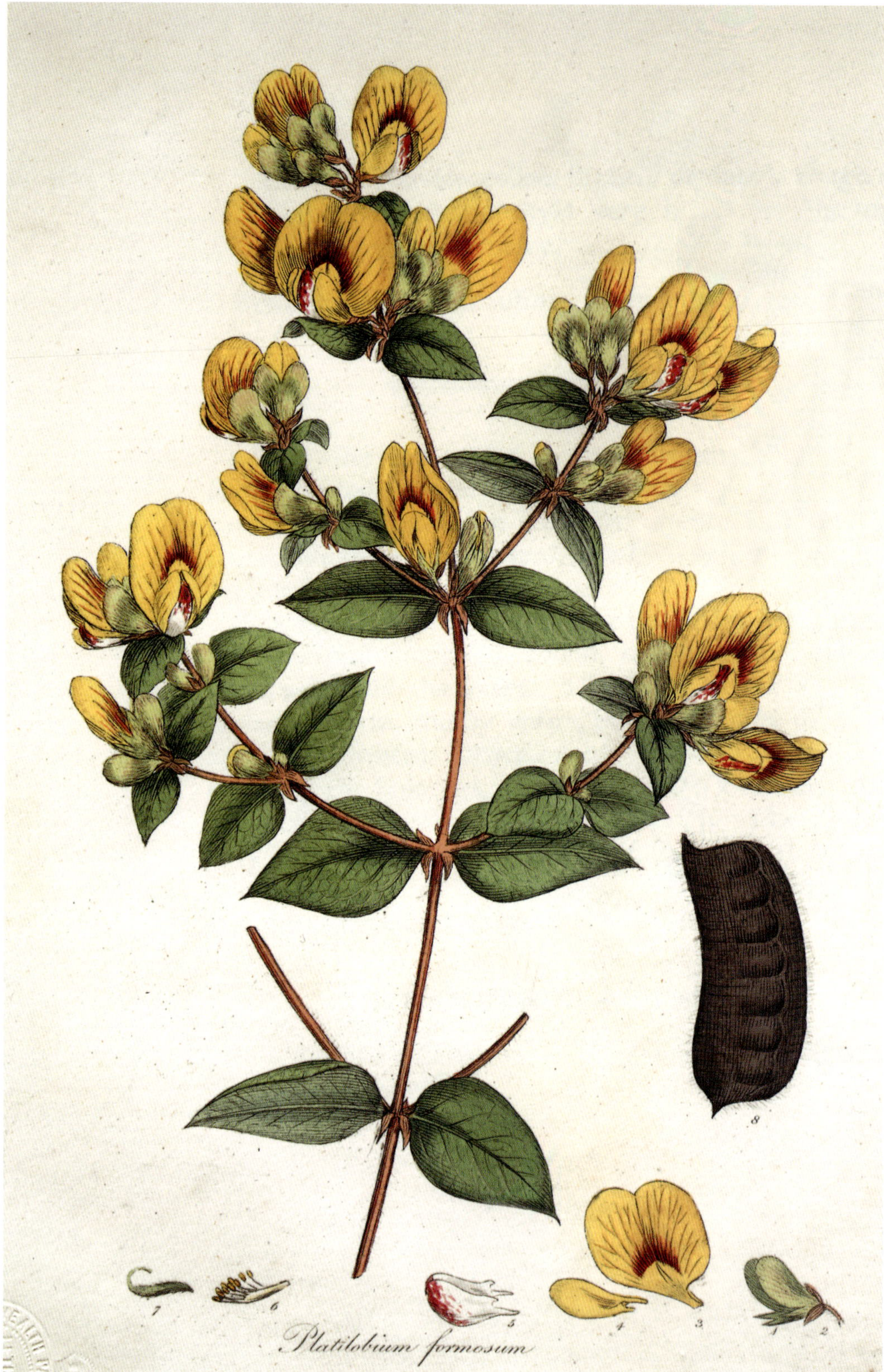

Platylobium formosum Smith
Hand-coloured copper engraving by James Sowerby, published in J.E. Smith, *A Specimen of the Botany of New Holland*, 1793–1795, the first Flora of Australian plants. This plate is part of the original description of this genus and species of 'egg and bacon' plant.
[NLA 22.3 x 15.0]

2. Discovery

Of the special skills and indefatigability demanded of those who took part in the successful voyages of discovery to the South Seas in the eighteenth and early nineteenth centuries, none were greater than those required of the artists who accompanied the navigators.

Sir Rutherford Robertson[1]

During the fifteenth and sixteenth centuries expansion of trade for the fruits of South East Asia (then the East Indies) and beyond was one of the objectives of European exploration in the southern hemisphere; confirming the existence of a great southland, Terra Australis, was another.

Ferdinand Magellan sailed west from Spain in 1519 to discover a western route to the East Indies and so avoid confrontation with the Portuguese who, at the time, controlled the eastern route via the Cape of Good Hope. His flagship, the *Trinidad*, was accompanied by four other ships. Magellan discovered a strait, the Magellan Strait, which separates Tierra del Fuego from the mainland of South America. The discovery enabled him to sail north-west across the Pacific in the direction of the East Indies. Magellan himself did not complete the voyage because he was killed by natives in the Philippines in 1521. Only one of his ships, the *Victoria*, succeeded in circumnavigating the world (1519–1522), the first to do so. While Magellan did not discover Terra Australis, because he sailed north-east of Australia and New Guinea, the voyage did fuel stronger speculation of the existence of a great southland. That speculation lead to further exploration in the Pacific and eventually to charting the eastern coast of Australia.

The existence of the western coast of Australia became well known during the sixteenth century while European traders used it to navigate to Batavia and the spice islands. At that time cartographers interpreted it as a northward extension of Terra Australis.

While Spain and Portugal were active participants, trade in the Indian Ocean and the mid-western Pacific was soon dominated by Holland, and during the seventeenth century the Dutch East India Company was all-powerful. The Dutch named the land on the eastern boundary of the Indian Ocean, New Holland. The dry, somewhat inhospitable coast had no potential for trade and so was not even explored for its natural history. If any traders and explorers did collect plant material it has not made its way into scientific record.

Exploration continued in the southern hemisphere. Abel Tasman[2] was commissioned by the Dutch East India Company to make a specific voyage to discover the great southland and record its geography and inhabitants and their ways of living. Tasman set out from Batavia in 1642–1643 with the *Heemskerk* and *Zeehaen*. He discovered the southern shores of Tasmania, naming it Van Diemen's Land in honour of the then Governor General of the Dutch East Indies, Anthony van Diemen.[3]

Tasman sailed on to New Zealand and other islands of the Pacific including Tonga, Fiji and New Guinea. On this voyage there were few records relating to natural history. As for the great southland, Tasman concluded that New Zealand formed part of its western coast and he named it Great Staten Landt. Staten Landt had been discovered by Jacob le Maire in 1616 and was thought to be part of the great southland. In reality it is an island in the south-western complex of islands near Cape Horn of South America. Tasman's voyage, however, was significant because it proved that New Holland was not part of the great southland.

England was preoccupied by war with France during much of the seventeenth century. During this time the British relied on trade in the West Indies, and on piracy and dealing with other European merchants for their needs from the East Indies. Towards the end of the century, however, their interest in the southern hemisphere was awakened.

In 1698 William Dampier was commissioned by the British Admiralty to explore in the region of New Guinea and northern and western New Holland. Dampier had a reputation for being a successful pirate and enemy of the Dutch East India Company, for being a good navigator, and for having some prior knowledge of the region. In 1688 he had landed at Cygnet Bay on the west coast of New Holland in his ship, the *Cygnet*, to effect repairs and take on fresh water.

Dampier's commissioned voyage set sail in January 1699 in the *Roebuck*. He became the first Englishman known to have landed on Australia's shores. Willem Jansz and crew, on the Dutch vessel *Duyfken*, are reported to be the first Europeans to have landed on the Australian continent. They did so in 1606.

While seeking fresh water at Shark Bay Dampier collected some plants which found their way into science. In August 1699 Dampier observed:[4]

> The Land is of an indifferent Heighth, so that it may be seen 9 or 10 Leagues off. It appears at a distance very even; but as you come nigher you find there

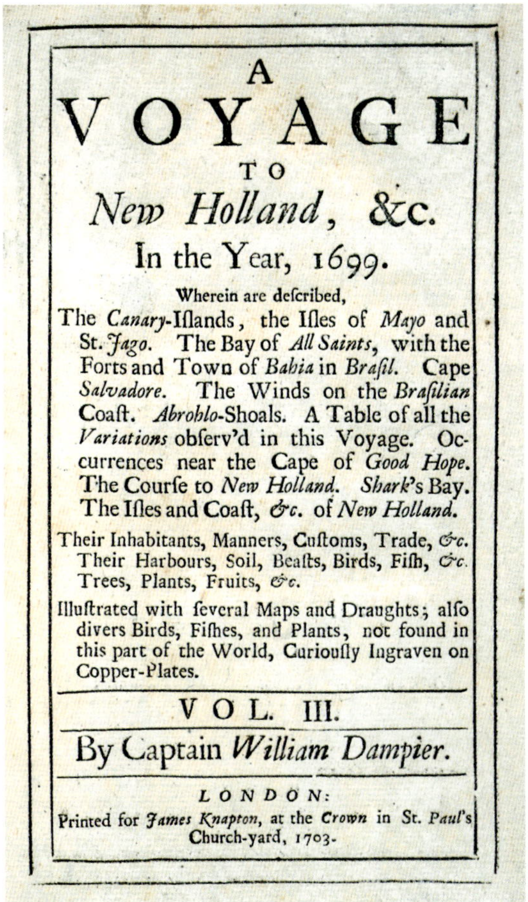

A VOYAGE TO *New Holland*, &c. In the Year, 1699.

Wherein are deſcribed,

The *Canary*-Iſlands, the Iſles of *Mayo* and St. *Jago*. The Bay of *All Saints*, with the Forts and Town of *Bahia* in *Braſil*. Cape *Salvadore*. The Winds on the *Braſilian* Coaſt. *Abrohlo*-Shoals. A Table of all the *Variations* obſerv'd in this Voyage. Occurrences near the Cape of *Good Hope*. The Courſe to *New Holland*. *Shark's* Bay. The Iſles and Coaſt, *&c.* of *New Holland*.

Their Inhabitants, Manners, Cuſtoms, Trade, *&c.* Their Harbours, Soil, Beaſts, Birds, Fiſh, *&c.* Trees, Plants, Fruits, *&c.*

Illuſtrated with ſeveral Maps and Draughts; alſo divers Birds, Fiſhes, and Plants, not found in this part of the World, Curiouſly Ingraven on Copper-Plates.

VOL. III.

By Captain *William Dampier*.

LONDON:

Printed for *James Knapton*, at the *Crown* in St. *Paul's* Church-yard, 1703.

William Dampier's, *A voyage to New Holland*, etc. *in the year, 1699* is the first book known to contain drawings of Australian plants. A map of the world showing the voyage reveals how little was known of the south-western Pacific Ocean at the time. [NLA]

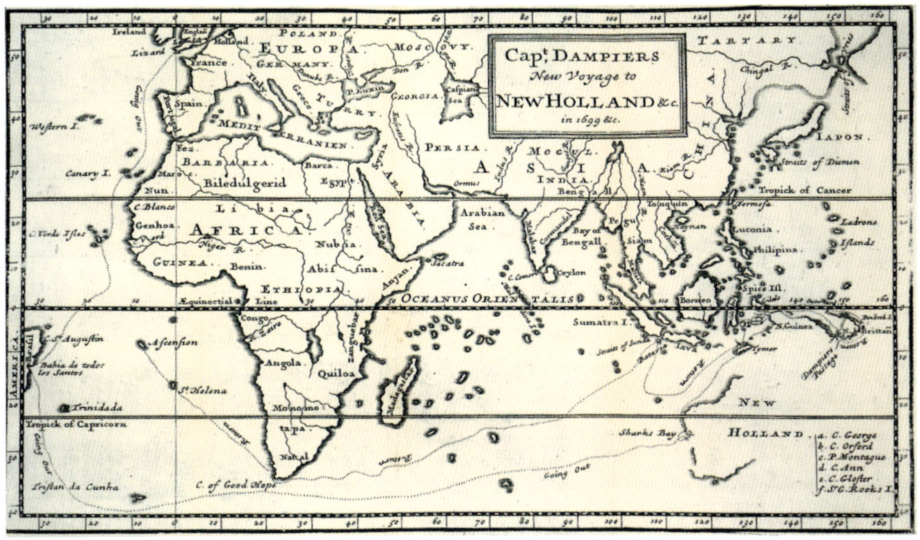

are many gentle Risings, tho' none steep nor high. 'Tis all a steep Shore against the open Sea: But in this Bay or Sound we were now in, the Land is low by the Sea-side, rising gradually in within the Land. The Mould is Sand by the Sea-side, producing a large Sort of Sampier, which bears a white flower. Farther in, the Mould is reddish, a Sort of Sand producing some Grass, Plants, and Shrubs. The Grass grows in great Tufts as big as a Bushel, here and there a Tuft: Being intermixe'd with much Heath, much of the kind we have growing on our Commons in England. Of Trees or Shrubs here are divers Sorts; but none above 10 Foot high: There Bodies about 3 Foot, and 5 or 6 Foot high before you come to the Branches, which are bushy and composed of small Twigs there spreading abroad, tho' thick set, and full of Leaves; which were mostly long and narrow. The Colour of the Leaves was on one Side whitish, and on the other side green; and the Bark of the Trees was generally of the same Colour with the Leaves, of a pale green. Some of the Trees were sweet-scented, and reddish within the bark, like Sassafrass, but redder. Most of the Trees and Shrubs had at this Time either Blossoms or Berries on them. The Blossoms of the different Sort of Trees were of several Colours, as red, white, yellow, &c. but mostly blue: And these generally smelt very sweet and fragrant, as did some also of the rest. There were also beside some Plants, Herbs, and tall Flowers, some

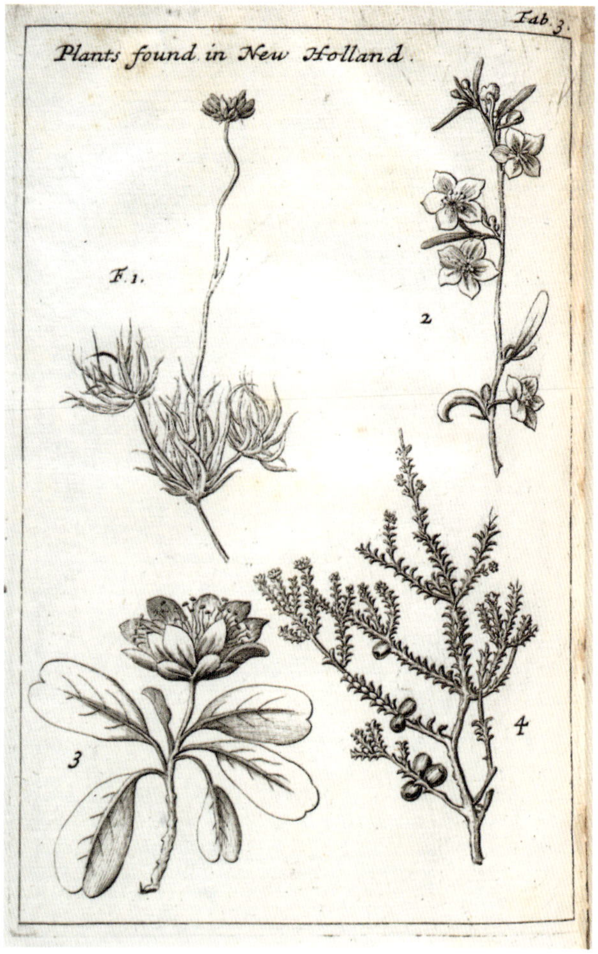

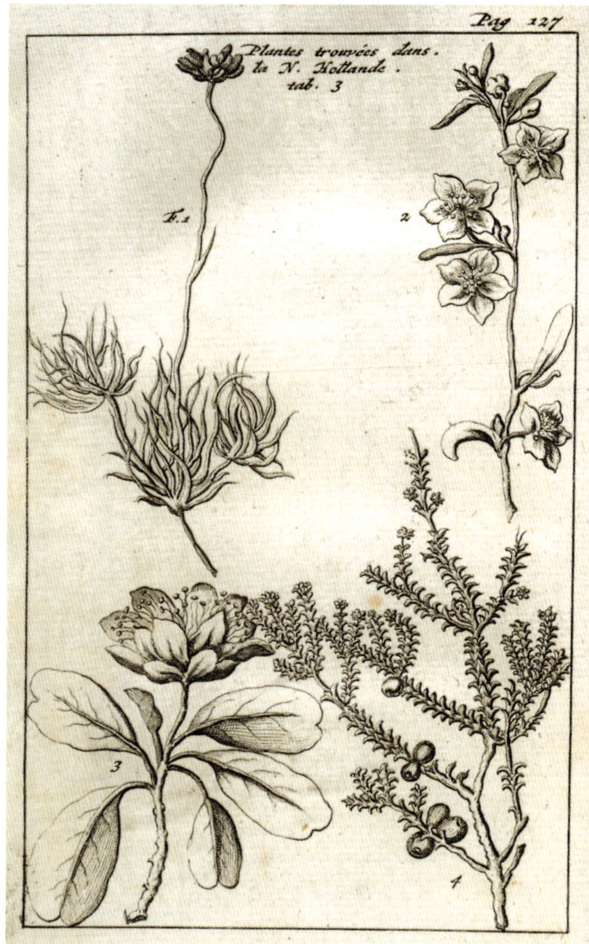

The copper engravings for William Dampier's *A voyage to New Holland,* etc. *in the year, 1699,* published in 1703 (left), were executed by an unknown artist and engraver. The French version of Dampier's book, *Voyage de GUILLAUME DAMPIER aux terres Australes á La Nouvelle Hollande* etc., published in 1712 (right), had the plates re-engraved by L. Duvivier Tardieu. They were drawn from Dampier's original plates, corrected to prevent a mirror-image effect, and show only minor differences. The plants are: f.1. *Conostylis styloides* F.Muell.; f.2. *Sida calyxhymenia* J.Gay; f.3. *Diplolaena grandiflora* Desf.; and, f.4. *Beaufortia sprengelioides* (DC.) Craven.
[NLA 17.0 x 10.5]

> very small Flowers, growing on the Ground, that were sweet and beautiful, and for the most part unlike any I had seen elsewhere.

On the return voyage to England Dampier nearly met disaster when he anchored off Ascension Island. The *Roebuck* sprang a leak. Attempts to caulk her failed and she began to sink. In February 1701 Dampier reported:

> Where having fastned her, I made a Raft to carry the Men's Chests and Bedding ashore. I had sent ashore a Puncheon, and a 36 Gallon Cask of Water, with one Bag of Rice for our common use; But great Part of it was stolen away, before I came ashore; and many of [?] Books and Papers lost.

They were rescued two months later by some English Men-of-War ships.

This dramatic beginning for Australian recorded history — the land not worthy of attention, the first British explorer being a well-known buccaneer, and his record nearly lost when the ship sank off Ascension Island — seemed to set a pattern for many of the voyages of exploration for more than a century.

The plant collection made by Dampier[5] also included specimens from South America and New Guinea. They were passed to John Woodward, a botanist and council member of the Royal Society of London. John Ray, a leading British botanist, regarded as the father of English natural history, studied the specimens and provided brief descriptions of them in his *Historia Plantarum* (1704).

Details of Ray's observations were made available to Dampier in time for publication of his book. Dampier had already begun to write of his adventures. The book about his voyage, *A voyage to New Holland*, etc. *in the year, 1699* (1703), was third in the series. This volume included drawings of some of the plants he had collected around Shark Bay. The drawings were done by an anonymous draftsman on board the *Roebuck* and were published as copper engravings.

Dampier's record of his voyage was of such significance that the French had it translated and published as *Voyage de GUILLAUME DAMPIER aux terres Australes á la Nouvelle Hollande*, etc. *fait en 1699* (1705). The plates were re-engraved, unacknowledged, for this book. A second French edition, published in 1712, had the plates re-engraved again, this time by L. Duvivier Tardieu.

Leonard Plukenet, Queen's Botanist, saw Dampier's specimens too. Plukenet not only described them but also provided some illustrations in an appendix to his *Amaltheum Botanicum* (1705). These are copper engravings based on drawings by John Collins. Jean Simon, born in Normandy, was the major engraver for these plates. An apparently Australian plant appears earlier in the book. This was engraved by Henry Hulsbergh, a Dutch-born engraver. The caption reads 'Amaranthoides parvum. Polygoni montani nivei facie Nov hollandicum'. It is possible that this was collected prior to 1699.

Following Dampier's little flurry of collection, description and illustration of Australian plants ceased for several decades.

In 1768 the Dutch botanist, Nicolaas Laurens Burman, acquired two specimens, reportedly from the East Indies, and described and illustrated them in his *Flora Indica* as ferns. He was well off-beam. His *Polypodium spinulosum* Burm.f. is a *Synaphea* and *Adiantum truncatum* Burm.f. is an *Acacia*, both being flowering plants and both endemic to Australia. From a scientific point of view, Australian endemic plants occur only in Australia and Australian native plants occur naturally in Australia and other countries.

In 1996 Alex George[6] and others pointed out that these plants may have been collected by Dutch explorer Willem de Vlamingh and crew on Rottnest Island when they landed in the *Geelvinck* in December 1696, or along the Swan River early in January 1697. The collection

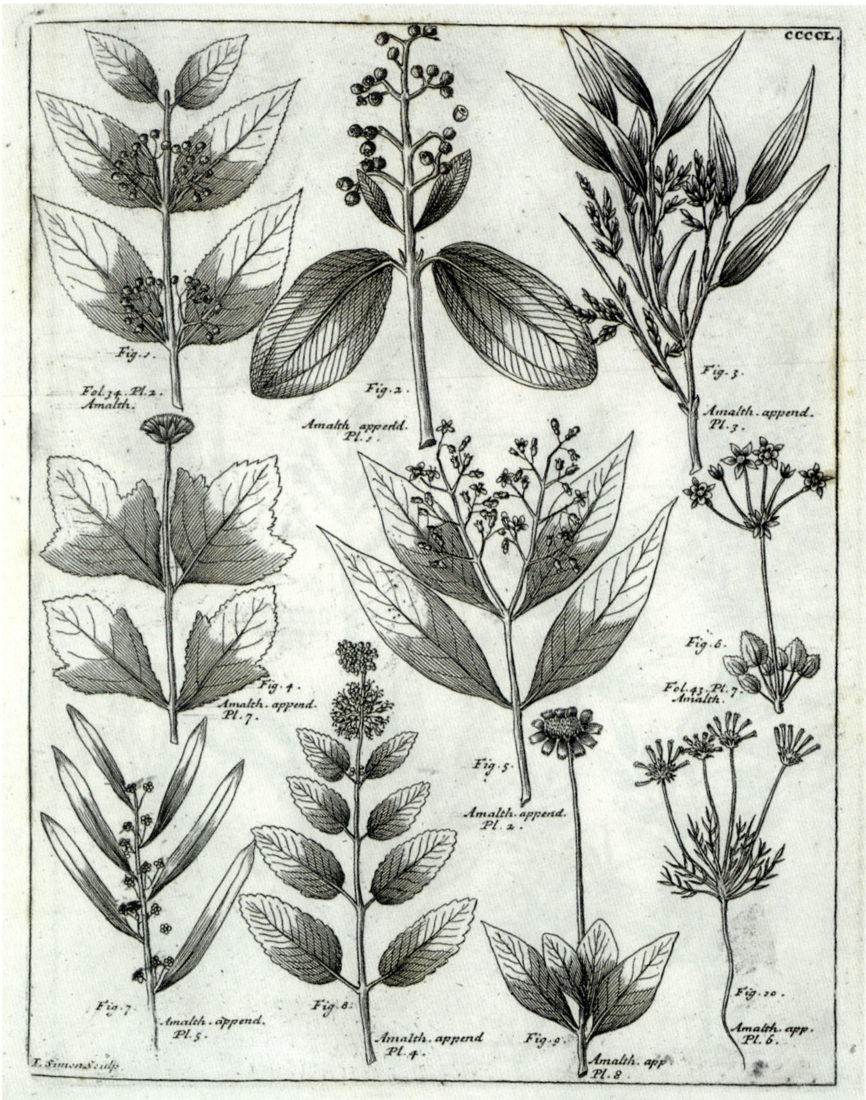

A composite plate with two Australian plants collected by William Dampier. The drawings Fig.7 and Fig.10 are of *Acacia ligulata* A.Cunn. ex Benth. and *Brachycome* aff. *cheilocarpa* F.Muell. Copper engraving by Jean Simon of drawings by John Collins published in L. Plukenet, *Phytographia: Amaltheum Botanicum*, 1705. [KEW 27.3 x 20.3]

of these two specimens will forever remain an enigma. The illustrations — engravings by Adolf van der Laan — are of leaves and are little more than outlines.

Meanwhile, botanical illustration for both art and horticultural purposes blossomed in Europe, particularly in Germany, Holland and France. With some notable exceptions botanical illustration for scientific purposes, however, was relatively minor and slow to develop. Generally, major botanical treatises contained either no or few illustrations. Nevertheless botanical illustration gradually became a significant component of descriptive taxonomy.

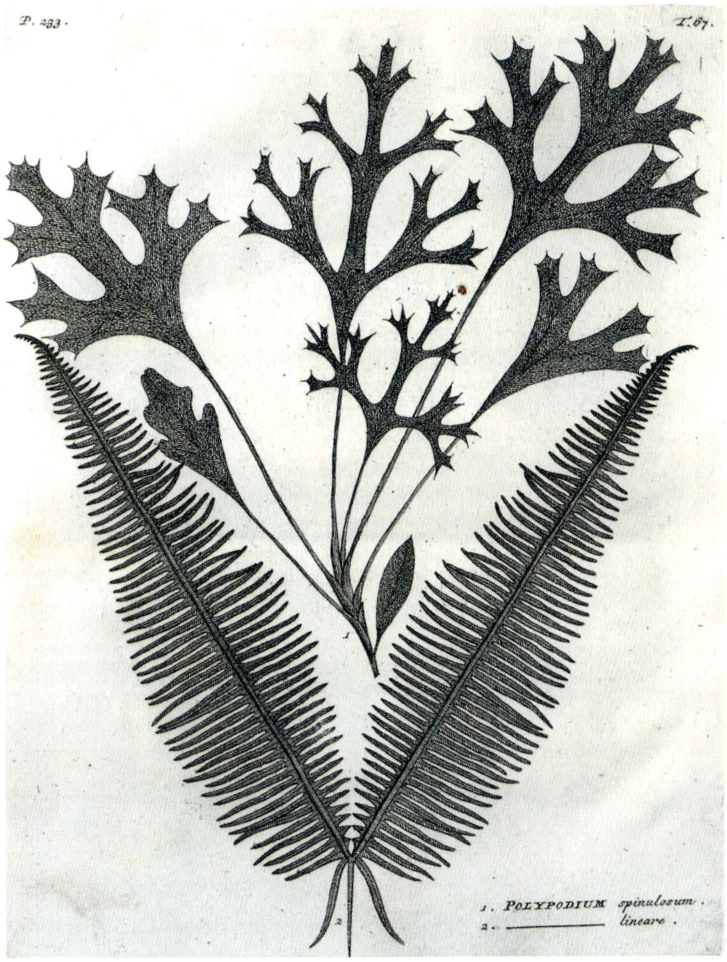

Synaphea spinulosa (Burm.f.) Merrill (= *Polypodium spinulosum* Burm.f.)
Copper engraving by Adolf van der Laan in N. Burman, *Flora Indica*, 1768.
[MEL 22.7 x 16.6]

Frans Stafleu[7] paraphrased Carl Linnaeus, saying: 'The illustration must be of natural size and show the parts in their proper places; it must also show even the most minute parts of the fructification.' Stafleu also observed that: 'The art of making herbaria developed from botanical illustration: pressed plants were originally used as material for the botanical artists.'

During the early- and mid-eighteenth century, two artists were instrumental in establishing the style of illustration that was ultimately adopted by descriptive botany.

In England, George Dionysius Ehret[8] proved to be an influential figure in establishing style and standard for English botanical artists. He was a self-taught, German-born, botanical artist who settled permanently in England in 1736. Ehret's work illustrated a number of significant botanical books, for example Carl Linnaeus' *Hortus Cliffortianus* (1737).

Towards the end of his career Ehret worked for Joseph Banks painting plants brought back from Banks' Newfoundland–Labrador expedition of 1766. Banks had also employed Sydney Parkinson to paint some of the zoological specimens from this voyage. The two men almost certainly interacted over the project. Parkinson went on to make a very significant contribution to Australian botanical illustration.

Nicholaus Joseph Jacquin (1727–1817) was also particularly influential in establishing tradition, style and standard of scientific botanical illustration. He was born in Holland and spent most of his working life as a botanist in Austria. Early in his career Jacquin executed his own illustrations. Later he employed and directed a number of highly competent artists to illustrate his books including the Bauer brothers, Ferdinand and Franz, who later made a very important contribution to Australian botanical illustration.

Seventy years after Dampier visited New Holland, James Cook[9] was commissioned by the British Government to carry out an expedition in the South Seas.

The voyage to the Pacific had three main aims. Cook was instructed to observe the transit of Venus across the Sun (in eclipse) in order to calculate the distance between the Earth and the Sun and facilitate the development of more accurate navigation tools. He was also requested to prove the existence of the great southland between South America and New Holland and make anthropological and natural history observations where possible. Third, he was to demonstrate that proper diet would eliminate scurvy, a classical disease of long voyages where there was little other than salt meat to eat. Cook's diet included sauerkraut, malt, citrus extract, raisins and mustard seeds, all of which were vegetable in origin and non-perishable.

Cook's commission took place when the art of botanical illustration was well established and the ship, the *Endeavour*, carried a team of scientists and artists: Charles Green, an astronomer; Joseph Banks[10] and Daniel Carlsson Solander,[11] natural historians; Sydney Parkinson[12] and Alexander Buchan, artists; and Herman Spöring, a trained surgeon, naturalist and draftsman.

Joseph Banks (1743–1820) was born in London of a wealthy family. He was a naturalist rather than a botanist and early in his career he accompanied several expeditions as the naturalist. From these voyages he accumulated a significant personal herbarium which he housed at his home in Soho Square. Banks soon became a philanthropist and played a 'political' and benevolent role in science. He was instrumental in setting up and directing the Royal Botanic Gardens at Kew; he was elected a Fellow of the Royal Horticultural Society in 1766; in 1788 he became President of the RHS and a Fellow of the newly formed Linnean Society; and he was knighted in 1795. His importance to Australia was that he appointed a high quality taxonomist (Solander) and a competent artist (Parkinson) to accompany him on the first scientific expedition to Australia; the British Government sought his knowledge of Botany Bay when it was planning a penal colony; and he later commissioned a number of competent collectors to carry out botanical exploration of Australia.

Daniel Solander (1733–1782) was born in Sweden. He studied under Carl Linnaeus in 1750 and moved to England in 1760 to work in the British Museum as Assistant Librarian.[13] Here he met and worked with Joseph Banks and through this connection Solander was sponsored to join the *Endeavour* in 1768.

Sydney Parkinson (1745–1771) was born in Edinburgh and became a woollen draper. Parkinson developed an interest in drawing plants and ultimately took his talents to London when his widowed mother moved there to live. When Parkinson exhibited some of his flower paintings he came to the attention of James Lee of the Vineyard Nursery at Hammersmith. Arising from this, among other commissions, he was employed to teach Lee's daughter. Through this connection Parkinson was recommended to Joseph Banks in 1767 as a natural history artist. Parkinson first worked for Banks on illustrations of fishes, birds and insects from Banks' Newfoundland expedition. He was therefore an obvious choice to accompany Banks on the *Endeavour*.

As botanical artist on the voyage, Sydney Parkinson was prolific in his output making a total of 1350 paintings and sketches. Of these 280 are finished plant paintings. Buchan, who was employed to illustrate landscapes and peoples, was less productive; he was not in good health, being prone to epilepsy, and a severe attack brought about his death in Tahiti in 1769. Buchan's death resulted in extra work for Parkinson during the rest of the voyage and restricted him to making only notes, sketches and drawings.

Cook sailed from Plymouth in August 1768. The transit of Venus was observed and measured on 3 June 1769 in Tahiti. During their stay there the men purchased or collected as much fresh fruit and vegetables as possible before sailing for New Zealand. Cook's considerable success in managing scurvy was marred by some of the men contracting venereal disease while in Tahiti.

Cook landed in New Zealand on 6 October 1769 to a hostile reception. However a bonus while there was the observation of the transit of Mercury across the Sun at Mercury Bay in November 1769. Before leaving New Zealand he circumnavigated both islands and disproved their connection to Staten Landt. And so the magnitude of the mythical great southland was shrinking.

Cook finally set sail for the east coast of New Holland. He had sailed too far north to land on Van Diemen's Land and first sighted the coast at Point Hicks on 17 April 1770. The explorers then sailed north, charting the coastline as they went for a further ten days before landing at Botany Bay. They named the bay Sting Rays Harbour but Banks named it Botany Bay in his Journal for the wealth and interest of the plants. There was an overwhelming amount of work for Parkinson. As in New Zealand, sketches had to suffice. His method was to sketch the whole plant and paint small significant parts as a colour record for completion back in England.

At Botany Bay the explorers received a mixed welcome. In a minor skirmish Parkinson was very nearly speared. Fortunately the only tragedy while there was the death of one of the crew from tuberculosis. In his Journal Banks observed:[14]

> The soil, whenever we saw it consisted either of swamps or light sandy soil, on which grew very few species of trees — one, which was large, yielding gum much like Sanguinis draconis — but every place was covered with vast quantities of grass ... The trees overhead abounded very much with loryquets and cockatoos, of which we shot several.
> ...[We] found also several trees which bore a fruit of the Jambosa kind, in colour and shape much resembling cherries ... which we ate with pleasure, though they had little to recommend them but a slight acid.

Cook sailed north a week later, charting more of the coast and landing occasionally as he went. On the way, Banks wrote:

> In another species of tree, *Xanthoxyloides* mite, a small sort of black ant had bored all the twigs, and lived in quantities in the hollow part where the pith should be; the tree, nevertheless flourishing and bearing leaves and flowers upon those very branches as freely and well as upon others that were sound ...

Further north:

> We went ashore and found several plants which we had not before seen; among them, however, were still more East Indian plants than in the last harbour; one kind of grass which we had also seen there was very troublesome to us. Its sharp seeds were bearded backwards, and whenever they stuck into our clothes were by these beards pushed forward till they got into our flesh. This grass was so plentiful that it was scarcely possible to avoid it, and, with the mosquitoes that were likewise innumerable, made walking almost intolerable.

In tropical waters Cook ran into trouble — bad weather and coral shoals. Early in June the ship struck a reef. After a massive struggle of discipline, effort and ingenuity Cook and the crew evaded disaster, cleared the reef and limped into the Endeavour River where Cook beached the *Endeavour* for repairs. They were a sorry lot. There was major damage not only from the grounding and holing but also from teredo wood-worm; some of the crew were suffering the effects of scurvy; and water had flooded the scientific collections. They spent two months mending the ship, regaining their health and restocking with wholesome food and water. Banks succeeded in saving most of the scientific collection by taking it ashore and drying it. He recorded:

> Since the ship has been hauled ashore, the water has, of course, all gone backwards; and my plants, which for safety had been stowed in the bread-room, were this day found under water. Nobody had warned me of this danger, which never once entered my head. The mischief, however, was done, so I set to work to remedy it to the best of my power. The day was scarcely long enough to get them all shifted, &c., many were saved, but some were entirely spoiled.

The scientists were able to perform their observations more thoroughly than on other landings. According to Banks:

> The wild plantain trees, though their fruit does not serve for food are to us a most material benefit. We made baskets of their stalks (a thing we had learned from the Islanders), in which our plants, which would not otherwise keep, have remained fresh for two or three days; indeed, in a hot climate it is hardly practi-

> cable to manage without such baskets, which we call by the Island name of *papa mija*. Our plants dry better in paper books than in sand, with the precaution that one person is entirely employed in attending them. He shifts them all once a day, exposes the quires in which they are to the greatest heat of the sun, and at night covers them most carefully from any damp, always being careful, also, not to bring them out too soon in the morning.

Parkinson described the vicinity of Endeavour River as having several useful species, 'many gum-trees, and a great number of other plants, among which was a beautiful *Nymphaea*, with blue and white petals.'[15]

Bad weather continued to frustrate Cook, but he finally made an escape on 4 August. Through the following weeks the sailors had to negotiate a vicious, treacherous maze of uncharted coral reefs and unreliable weather. An extraordinary effort of seamanship and discipline saw them through. With an obvious sense of relief Banks recorded:

> For the first time these three months we were out of sight of land, to our no small satisfaction. A reef such as we have just passed is a thing scarcely known in Europe, or indeed anywhere but in these seas.

Once successfully clear of the coral they set sail for New Guinea, which proved inhospitable. Cook abandoned a landing and sailed direct to Batavia, arriving in October 1770.

In Batavia, the ship underwent major repairs and the men enjoyed rest and recreation. During that time only Parkinson seems to have carried out much natural history work. Soon ill health overtook the men and many were stricken with either, or both, malaria and dysentery. Even before they left Batavia men began to die and others, including Banks and Solander, became seriously ill.

Cook restocked the ship, engaged some new crew members to replace those who had died, and managed to get away from Batavia at the end of December. Almost all those on board who were not ill before leaving Batavia, now became so. The water taken on board proved to be contaminated and infested with mosquitoes. Many died including Charles Green, Herman Spöring and Sydney Parkinson. What a bitter irony, that Cook had managed to more or less control scurvy but lost so many of his crew to tropical diseases picked up in port.[16]

Considering the length of the voyage, the sometimes treacherous sailing conditions, the ill health and death of his artist colleague and his own ill health, Parkinson's achievement was outstanding. In the words of Sir Rutherford Robertson:[17]

> It is one thing to be able to depict accurately and pleasingly an animal or a plant, or a person in costume, in the unhurried and quiet abstraction of the studio or museum, quite another to draw and paint with unusual distinction in the confines of a small vessel with limited resources, overcrowded with officers and crew, let alone the scientists on board. To cope with the sudden arrival of perishable objects whose colours must be recorded at once or fade beyond recollection,

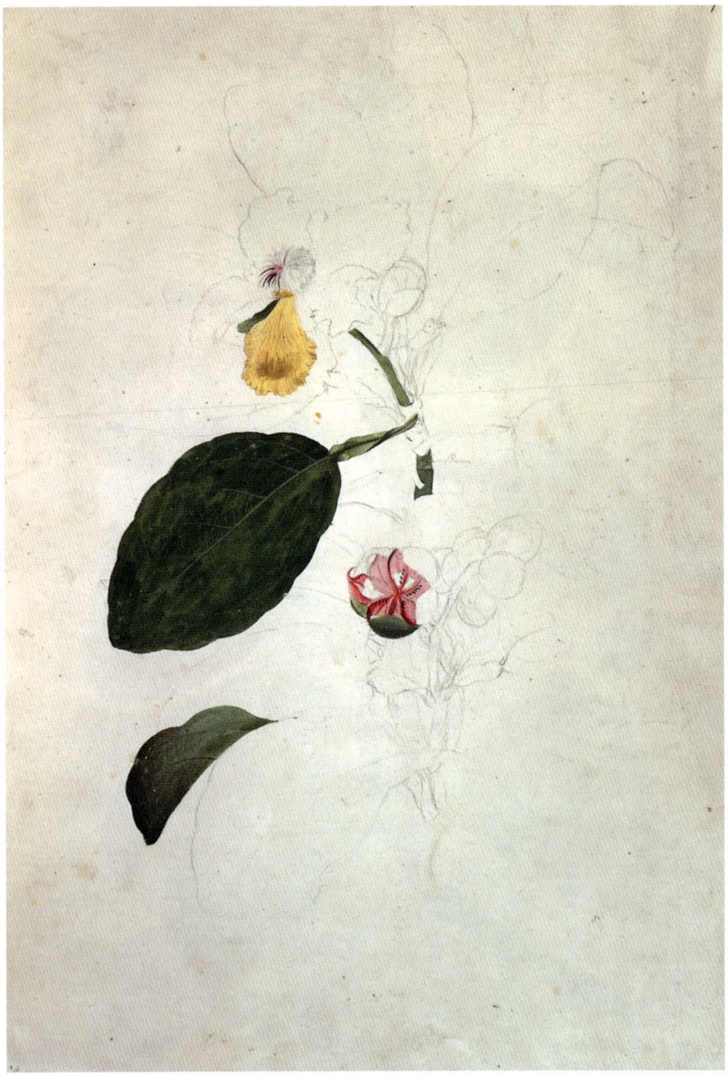

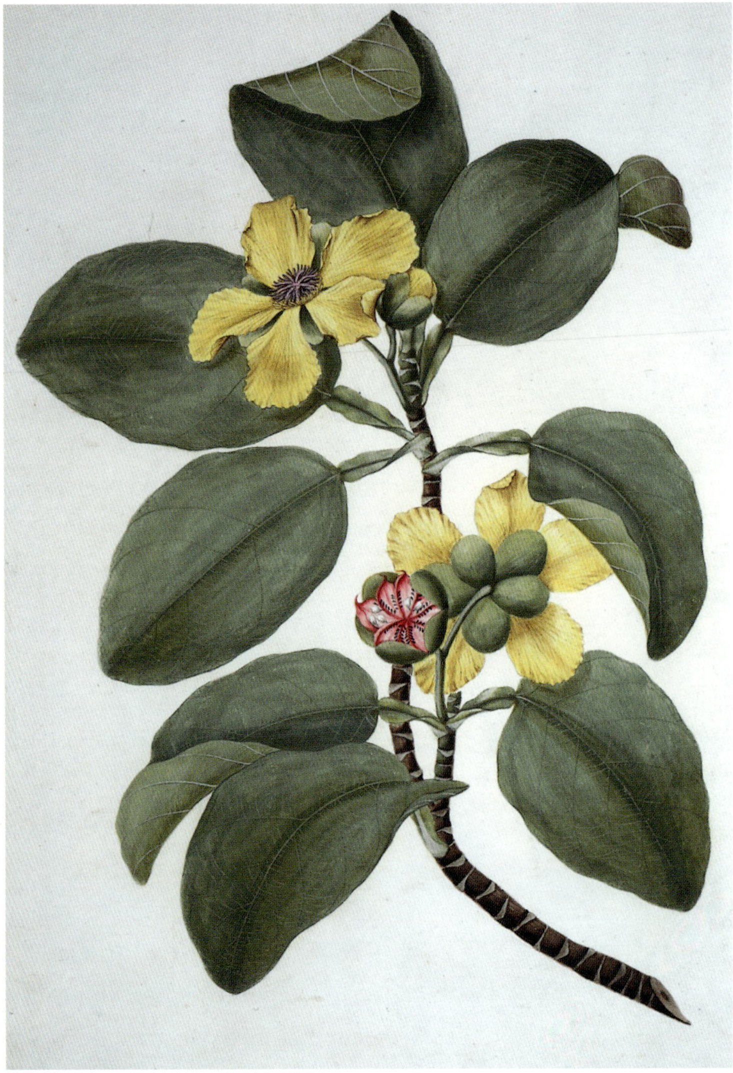

Dillenia alata (DC.) Martelli (= *Wormia alata* DC.)
Drawing (left) made by Sydney Parkinson while on James Cook's first voyage to the Pacific in the *Endeavour.* Parkinson colour-coded his drawing to aid the completion of the painting at a later time. The painting was completed by Frederick Nodder (right).
{BM 54.5 x 36.5}

> whose very forms might be as evanescent as their colours, to deal patiently with massive collections gathered at frequent intervals by exuberant hunters of plants and animals, anxious to have their new-found treasures recorded, to observe minutely and record accurately the important details of specimens, the full study of which might not be undertaken for decades and to do all this not only successfully but brilliantly requires genius.

Cook docked in Dover in July 1771. With the return of the collection to England horticulturists began to grow and botanists to study Australian plants in earnest. The first to be cultivated are reputed to have been *Casuarina torulosa, Casuarina stricta,*[18] *Eucalyptus obliqua* and *Leptospermum lanigerum.*

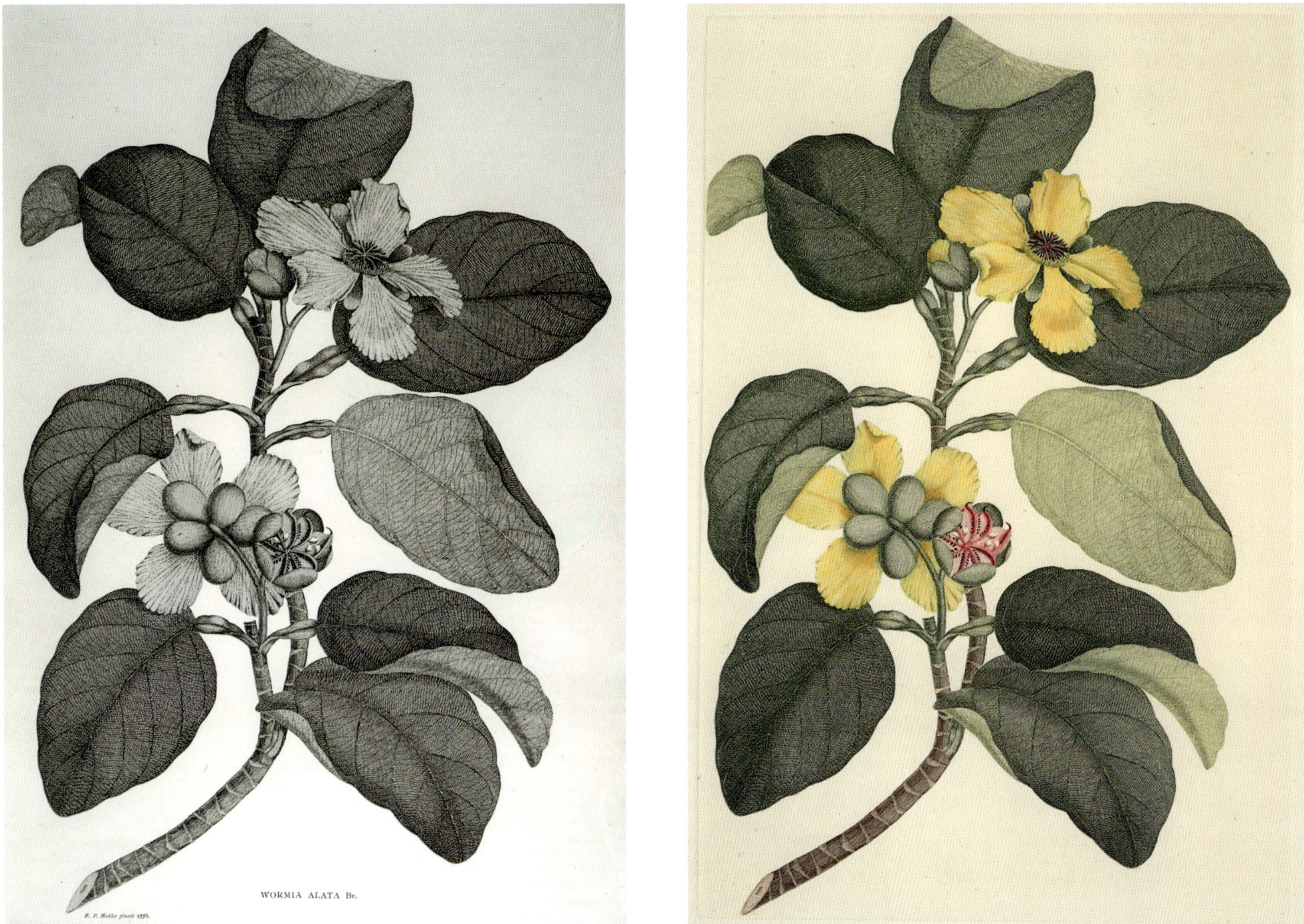

Dillenia alata (DC.) Martelli (= *Wormia alata* DC.)
Engraving executed by Frederick Nodder in 1778 for Joseph Banks' proposed Florilegium (left). The engraving was first published in J. Britten, *Illustrations of the Botany of Captain Cook's Voyage Round the World,* in 1900–1905. In the 1980s the original plates were cleaned and inked with colour to print the *Banks' Florilegium,* 1981–1988 (right). It is probable that Banks envisaged hand-coloured engravings two hundred years earlier.
[NLA] [MEL 46.4 x 30.0]

The loss of Parkinson at sea began a series of impediments to the publication of his paintings and drawings. Banks proposed the publication of a Florilegium and in 1773 he engaged several artists to complete the paintings. Some of the paintings were sent to Germany to be engraved at the Berlin Akademie der Wissenschaften but that proved too expensive. Banks then engaged local engravers, ten to begin with (eighteen in all), who were not especially experienced with botanical illustration. The work proceeded steadily and by 1787 at least 550 plates were engraved.

Meanwhile Solander prepared descriptions of the plants. He completed the manuscript before the engravings were ready. Banks, however, became increasingly preoccupied with other interests and responsibilities. It was gradually becoming evident that he had lost the will to organise the publication of his Florilegium. With the death of Solander in 1782 the project languished and Parkinson's name slid into obscurity. So too did the scientific effort of Solander. Thus, it was only the collection which made a significant contribution to science.

The artists who converted the 362 Australian plant drawings and sketches into paintings were Frederick Nodder, John Cleveley, John Frederick Miller and James Miller. From the Australian point of view, Nodder was the most significant artist and engraver to work on the Parkinson drawings. He converted 210 sketches into paintings, though not all of these were Australian. Little seems to be known about Nodder. He is reputed to have been botanical artist to the Queen and certainly contributed to several botanical works of the late 1700s.

The total engraving output was 742.[19] Proofs from the plates were pulled at the time of completion. The plates, however, remained stored and deteriorating at the British Museum. In 1900–1905 James Britten had 328 of the engravings reproduced by photolithography in his *Illustrations of the Botany of Captain Cook's Voyage Round the World*. The Natural History Museum published the complete set during the 1980s in Banks' Florilegium. These are of great historic and artistic interest but they were too late to fulfil their potential role in the science of taxonomic botany.

In the meantime James Cook received another commission — to prove or disprove the existence of the great southland. On 13 July 1772 he sailed from Plymouth for the southern seas again, this time on the *Resolution* and accompanied by the *Adventure*, under the command of Tobias Furneaux, and this time west to east. The natural historians, Johann Reinhold Forster and his son Johann George Adam Forster,[20] sailed on the *Resolution* with Johann senior reputedly behaving in a somewhat irascible manner. Fortunately Cook, being an excellent leader of men, proved able to manage him.

Early in 1773 the ships separated and the *Resolution* worked east and north to Dusky Bay, New Zealand. Meanwhile the *Adventure* made landfall in Van Diemen's Land. Furneaux did not disprove land connection with New Holland, being misled as to the extent of the Furneaux group of islands. The *Adventure* then sailed for Charlotte Sound, where the *Resolution* had been working. The ships met up in Queen Charlotte Sound late in May 1773.

It is unfortunate that all the natural historians were on the *Resolution* because it spent most of its time at sea. Even so, the discovery of New Zealand's natural history was advanced more than Australia's. The Forsters returned with thousands of plants and about two hundred and seventy animals in their collection. George drew 301 plants and 268 animals. Many of the collections and drawings were sold to Banks in 1776. George worked similarly to Parkinson by making patch-coloured pencil sketches, though he completed some of his animal paintings. Generally he was a capable, but not outstanding, draftsman.

The book, *Characteres generum plantarum* (1775) of Johann and George, included seventy-eight plates representing plant genera they encountered on the voyage. None of the genera is peculiarly Australian though many are represented in Australia, e.g. *Casuarina* (Sheoaks). To produce this work, they complemented their own collection with that of Banks and Solander, making the study more comprehensive. The engraver is not acknowledged.

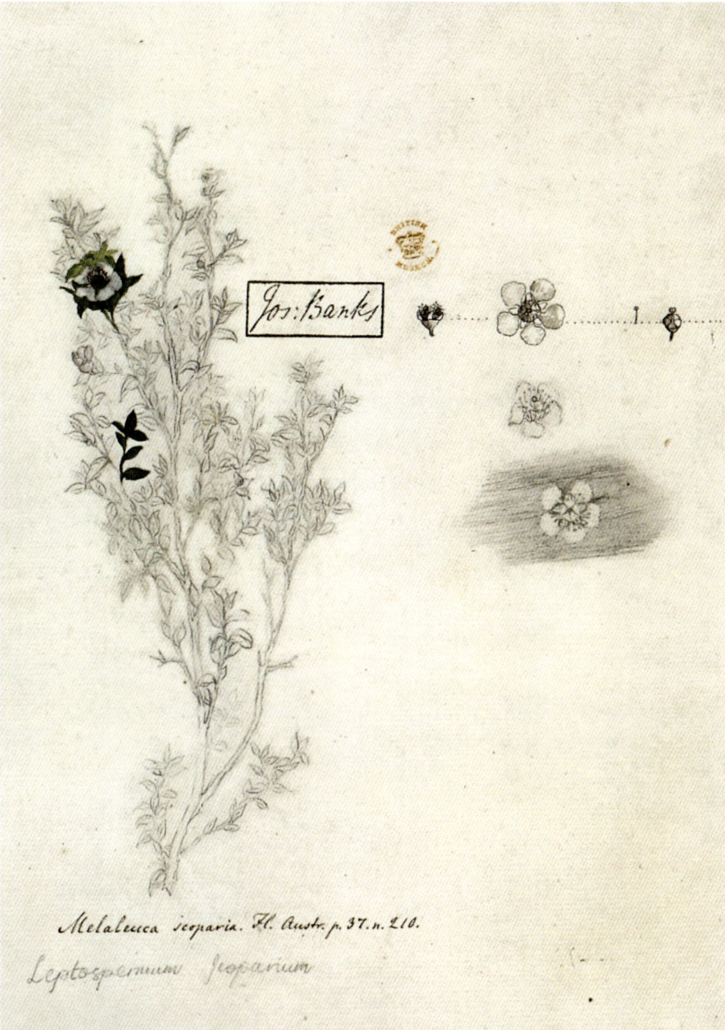

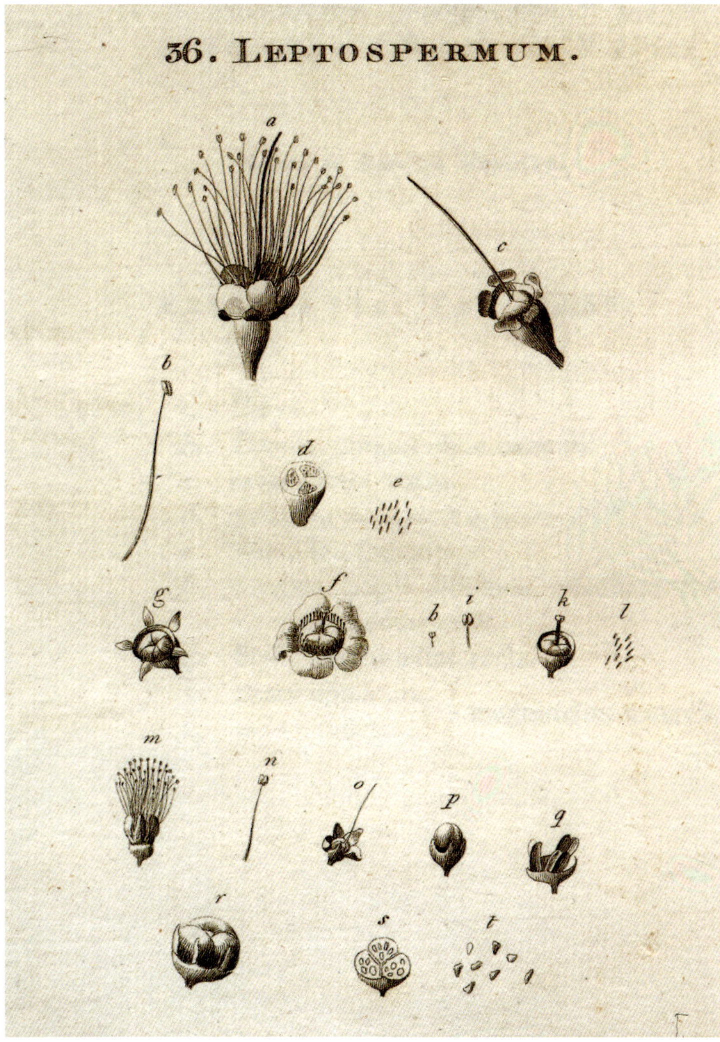

Leptospermum scoparium Forst. & G.Forst. (= *Leptospermum*)
Patch-coloured pencil sketch (left) made by George Forster while on Cook's second voyage to the Pacific. The copper engraving (right) drawn by him was published in J.R. Forster & J.G.A. Forster, *Characteres generum plantarum,* in 1775. The Forsters described the genera they encountered in the south-western Pacific. This species grows in south-eastern Australia and New Zealand.
[BM 33.5 x 48.0] [NLA 18.6 x 12.8]

In 1776 James Cook made a third commissioned voyage to the Pacific, this time to explore its northern regions. He sailed on the *Resolution* accompanied by the *Discovery* under the command of Charles Clerke. The *Discovery* had on board a surgeon-naturalist, William Anderson,[21] and a natural history draftsman, William Ellis. Cook spent some time in the southern Indian Ocean, especially on the Austral and the Kerguelen groups of islands. On the way to the Pacific *Discovery* anchored for four days in Adventure Bay, Van Diemen's Land, late in January 1777. The natural history legacy from this stopover does not seem to have been particularly significant, with one notable exception. William Anderson, together with a midshipman, David Nelson,[22] made a collection of *Eucalyptus obliqua* which was later to become the type of the genus *Eucalyptus*. The ill-fated voyage went on to carry out its mission, but not before James Cook was killed on Hawaii on 14 February 1779.

William Anderson studied medicine at Edinburgh University and accompanied Cook's second and third voyages in the capacity of surgeon. He died at sea. David Nelson, however, went on to become a gardener at Kew. The call of the sea was too alluring for him and he joined the *Bounty* under Captain William Bligh in 1787. After the mutiny in 1789, Nelson escaped with Bligh in the longboat, only to die a little later in Timor.

A sequel expedition to the third Cook voyage was carried out by George Vancouver in the *Discovery*, accompanied by the *Chatham* under the command of Captain Broughton. Vancouver was charged with completing the exploration of the western coast of North America. On board was Archibald Menzies,[23] a surgeon-naturalist. Vancouver sailed from Falmouth in April 1791 and returned in September 1795. On the voyage out they discovered King George('s) Sound on the south-west coast of New Holland. While there, Menzies collected plants which were ultimately incorporated in Joseph Banks' collection.

The scientific collection in London (including that of Banks and Solander, the Forsters, Anderson and Menzies) was freely available for other scientists to work on, as C. Koenig and J. Sims noted in 1806:[24]

> ... these treasures ... have by no means been lost to the public, having, by their liberal possessor [Joseph Banks], been rendered easily accessible to every one desirous of informing himself in the natural history of New Holland.

Carl Linnaeus jr was among the first and most prominent to do so. He included Australian plants from the Banks' collection in his *Supplementum Plantarum* (1781). This work is not illustrated.

Joseph Gaertner from Germany was another. Gaertner liaised with numerous collectors and botanists, including Joseph Banks in London, in the compilation of his major work *De Fructibus et Seminibus Plantarum* (1788–1792). Gaertner used fruit and seed characters (characters new to science) to reinforce and extend the Carl Linnaeus classification of flowering plants. His research included species which had been unavailable to Linnaeus. Some were Australian. Johann Georg Sturm is the major engraver for this work.

Two other engravers, Georg Paul Nusbiegel and Hermann Jakob Tyroff, prepared some plates including some depicting Australian material. Gaertner died in 1791 and his son, Carl Friedrich Gaertner, completed a *Supplement* which was published between 1805 and 1807. Johann's son, Jakob Sturm, took over the engraving for the *Supplement* when his father's health failed. Jakob went on to become an engraver of some note in Germany during the early-nineteenth century with major contributions to *Deutschlands Flora*.

Charles-Louis L'Héritier de Brutelle[25] (1746–1800), of Paris, also made use of the collection in his research. While employed in legal and administrative roles for the French Government, he also pursued an interest as an amateur botanist becoming highly competent and forward thinking.

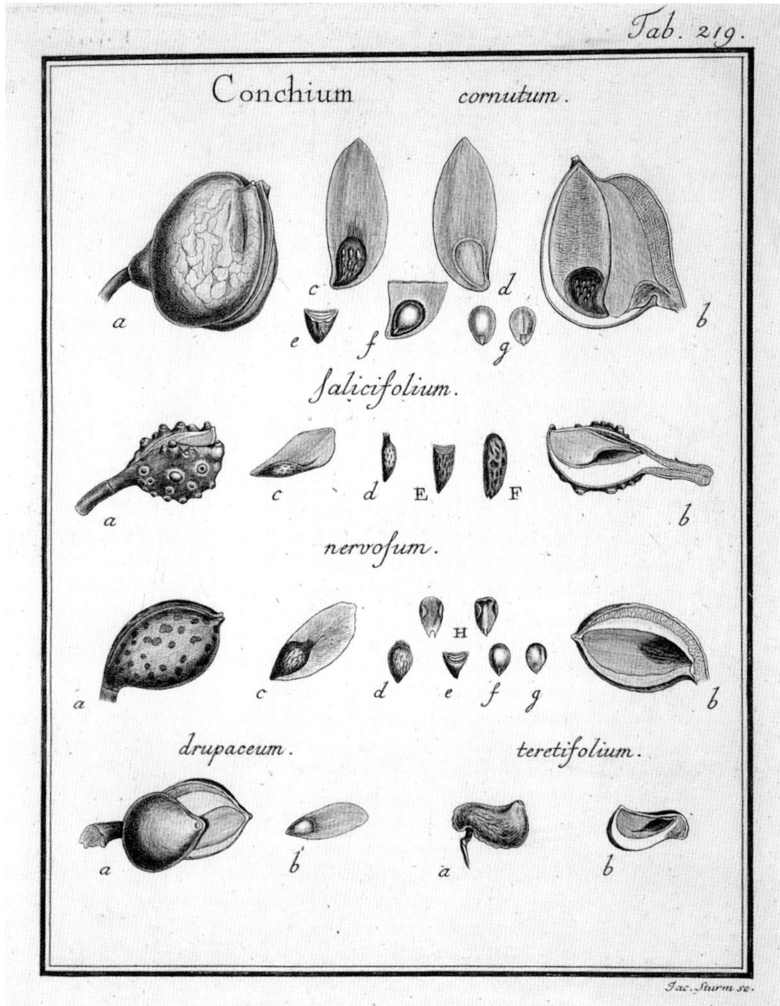

Hakea fruits and seeds
Copper engraving of fruits, engraved by Jakob Sturm and published in C.F. Gaertner, *De Fructibus et Seminibus Plantarum, Supplement*, 1805–1807. Joseph Gaertner and his son Carl were describing fruit and seed characters that previous botanists were overlooking. The species are: *Hakea gibbosa* (Smith) Cav. (= *Conchium cornutum* C.F.Gaertn.); *Hakea salicifolia* (Vent.) B.L.Burt (= *Conchium salicifolium* C.F.Gaertn.); *Hakea dactyloides* (Gaertn.) Cav. (= *Conchium nervosum* C.F.Gaertn.); *Hakea drupacea* (C.F.Gaertn.) Roem. & Schult. (= *Conchium drupaceum* C.F.Gaertn.); *Hakea epiglottis* Labill. (= *Conchium teretifolium* C.F.Gaertn.).
[CANB 21.4 x 16.6]

He was an active supporter of the Linnaean system of classification which was artificial, being based on the number, position and fusion of the sexual parts of the flower. As a result of work by a fellow countryman, Antoine Laurent de Jussieu, this system was being replaced by the natural system of classification, which emphasised similiarities of flower structure.

L'Héritier liaised with Joseph Banks and Jonas Carlsson Dryander,[26] Banks' librarian, in the publication of his *Stirpes Novae* (1785–1805) and *Sertum Anglicum* (1789–1792). He spent about a year in London working on the latter which aimed to document rare plants growing in the gardens of London. Some of these plants, of course, were Australian in origin.

One of L'Héritier's claims to fame is that he discovered Pierre-Joseph Redouté at the Jardin du Roi, recognised his ability, and trained him in the techniques of botanical illustration. Redouté received the highest accolades as a botanical artist/illustrator. L'Héritier is also thought to have been influential in the career of the English artist, James Sowerby.

The first published Redouté drawings of plants are in the L'Héritier publication, *Stirpes Novae*. The first Redouté drawing of an Australian plant, however, is of *Eucalyptus obliqua* in *Sertum Anglicum*, together with the original description of *Eucalyptus*. The plate is engraved by François Hubert and is based on the collection made in Van Diemen's Land by Anderson and Nelson while on Cook's third voyage.

It took almost a hundred years to get the scientific record of Australia's plants underway. Slow and frustrated as this achievement was, the process was irreversible and was about to gain a dramatic momentum.

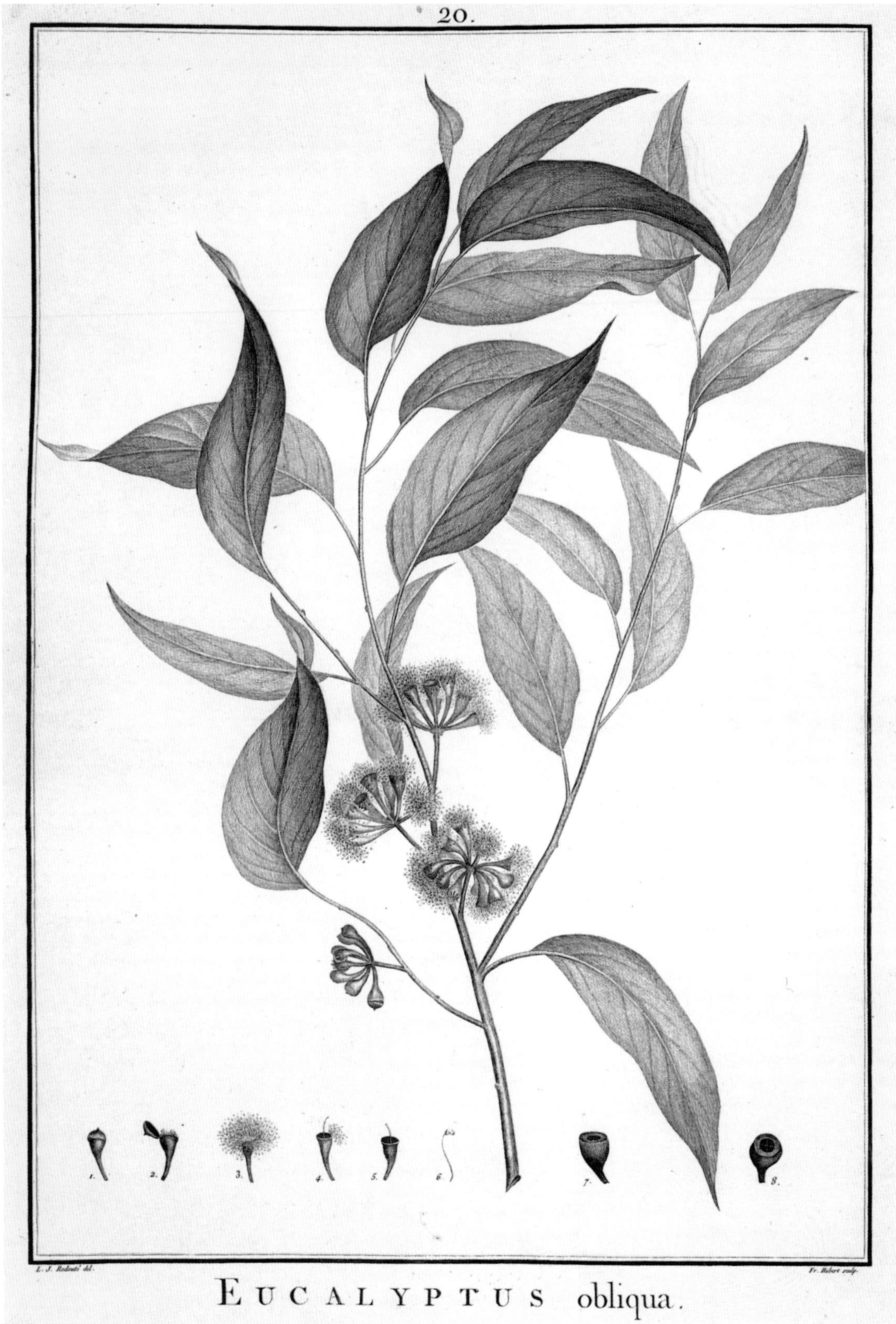

Eucalyptus obliqua L'Hér.
The first published drawing by Pierre-Joseph Redouté of an Australian plant, collected during Cook's ill-fated third voyage. The plate, engraved by François Hubert, was published in C.L. L'Héritier, *Sertum Anglicum,* in 1779–1792.
[CANB 38.5 x 27.5]

3. Colonisation

Of plants in general the country affords a far greater variety than its barren appearance seemed to promise; many of these no doubt possess properties which might be useful for physical and economic properties which we were not able to investigate.

Joseph Banks[1]

Towards the end of the eighteenth century, for a variety of reasons, England was looking for an offshore gaol for its undesirables. The possibility of a colony in the southern hemisphere had appeal and Joseph Banks, who had observed several lands in the South Seas, was consulted. In 1786 the *Heads of a Plan for effectually disposing of convicts, by Establishment of a Colony in New South Wales*, was approved by the British Government and planning for a colony at Botany Bay began.

The First Fleet under the command of Captain Arthur Phillip sailed from Plymouth in May 1787. Phillip was commissioned to found a British convict settlement at Botany Bay and he became the first Governor of the colony. The Fleet landed in Botany Bay on 18 January 1788 and Phillip decided it was unsuitable. It had little fresh water, an open anchorage and no suitable farming land. Phillip decided to try to find a better location for the colony and investigated overland. The next inlet to the north looked very promising so the Fleet moved north into Port Jackson anchoring in Sydney Cove on 26 January. As the Fleet was abandoning Botany Bay, by extraordinary coincidence, they were visited by Jean-François La Pérouse in the *Boussole* and the *Astrolabe*. In his Journal John White, First Fleet surgeon, wrote:[2]

> While these preparations were making, every person in the fleet were stupified to see, in this part of the world, two large ships plying hard in the offing to get into the bay.

In Port Jackson, conditions were somewhat more satisfactory. However, the history of the establishment of the colony reveals great hardship, illness, starvation and misfortune. By dogged determination the colony survived and, in due course, thrived. Its survival saw the return of plants and animals to England destined for the halls of herbaria and museums, the cages of zoos, and the beds of botanic gardens. It is this legacy which is of interest here.

The Fleet did not have the services of a botanist yet John White was an amateur naturalist and proved an active participant in recording the natural history and in collecting plants and animals for return to England. White's *Journal of a Voyage to New South Wales* (1790) led a burgeoning record of Australian plants and animals. Of the vegetation White wrote on 24 August 1788:

> During our return [from Broken Bay] we picked up, in the distance of about half a mile, twenty-five flowers of plants and shrubs of different genera and species, specimens of which I have transmitted to Mr Wilson, particularly Red Gum Tree.

White sent his plant collections to Thomas Wilson in London, who in turn passed them to James Edward Smith[3] for study. In his *Journal*, White not only recorded the experiences of the first few months, but also had plants and animals described and illustrated. Seven new plant species were described by Smith. Of the nine plant plates in the *Journal*, six are signed by Frederick Nodder (as *F.P.Nodder delin.*), one by E.Kennion, and the others are unsigned. Sarah Stone signed several of the bird illustrations. The engraver is not acknowledged.

It is thought by some that Thomas Watling, a convict artist of some ability, may have done the original artwork, but he did not arrive in the colony until late in 1792. Watling did, however, end up in the employ of White and did natural history paintings for him. John Calaby[4] believes that White had Watling copy the plates as published in the *Journal* for practice, because Watling was trained in landscape painting not natural history painting. This set of paintings is stored at the Natural History Museum, London, and has caused the confusion.

The anonymously produced *Naturalist's Pocket Magazine* (1798–1803) has several plates which bear a strong resemblance to Watling originals. Scrutiny of the Watling originals surviving in the Natural History Museum does not bear out the author's claims to be 'faithful, accurate copies', if indeed Watling's work served as the artwork for the engraver. In some cases Watling's work is superior, in others, the engraver's work is superior. The author refers to a draftsman working in the field in the colony but neither a draftsman, engraver, nor indeed the author, is acknowledged. There is speculation that George Shaw had some participation in the publication, but the quality does not compare with Shaw's other contemporary works, such as *The Naturalist's Miscellany* (1790–1813).

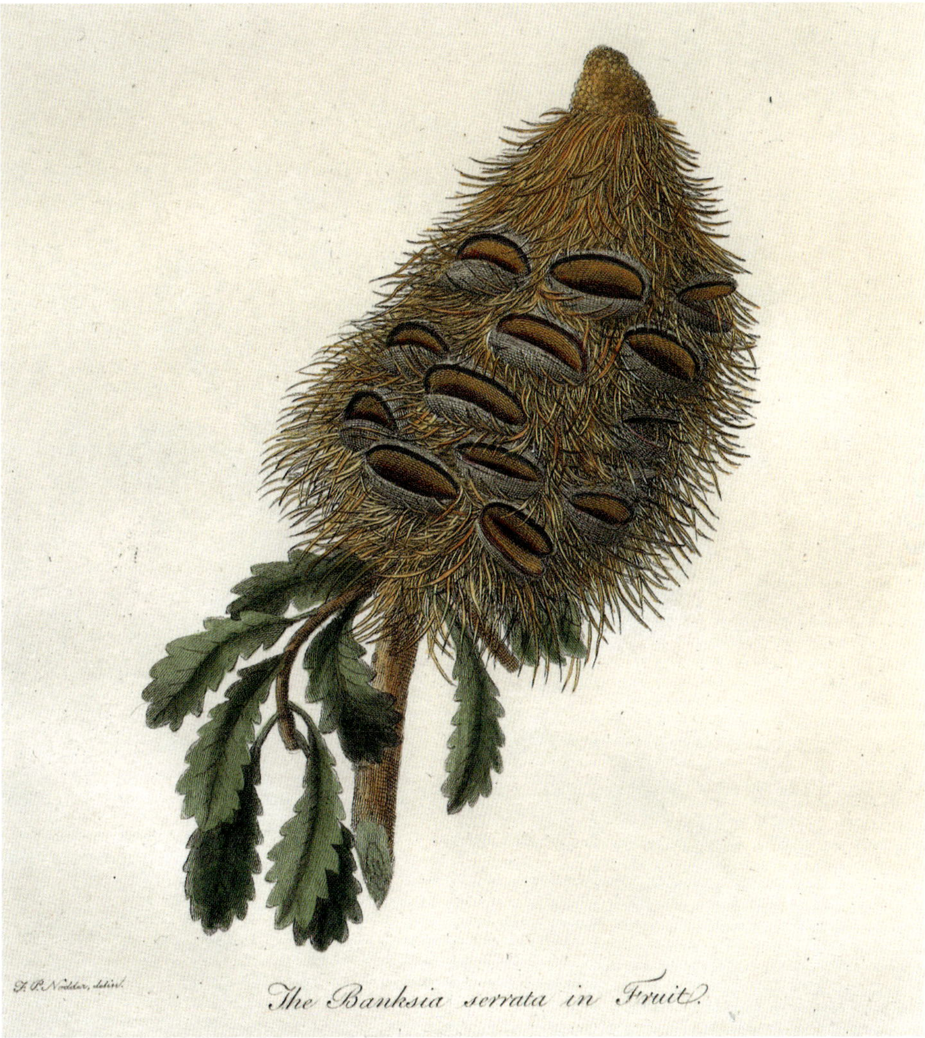

Banksia serrata L.f.
Hand-coloured copper engraving by Frederick Nodder drawn from plants collected during the first months of colonisation by John White, surgeon with the First Fleet. Published in J. White, *Journal of a Voyage to New South Wales*, 1790.
{NLA 22.8 x 17.8}

Back in England several scientists began to describe the wealth of Australian material arriving from the colony, notably James Edward Smith (1759-1828). Smith was an active and prolific botanist who had studied medicine in Edinburgh and Leiden. He must have been a man of means because, on the recommendation of Joseph Banks, he purchased the Linnaean Herbarium following the death of Carl Linnaeus jr in 1783. He then founded the Linnean Society of London in 1788 and remained its founding president until he died. With the Linnaean collection as its nucleus, Smith accumulated a significant personal herbarium which included some Australian material. It was ultimately purchased by the Linnean Society.

For Australian botany Smith's book, *A Specimen of the Botany of New Holland* (1793–95), is significant. It is Australia's first Flora. It was published in four parts, the first two being

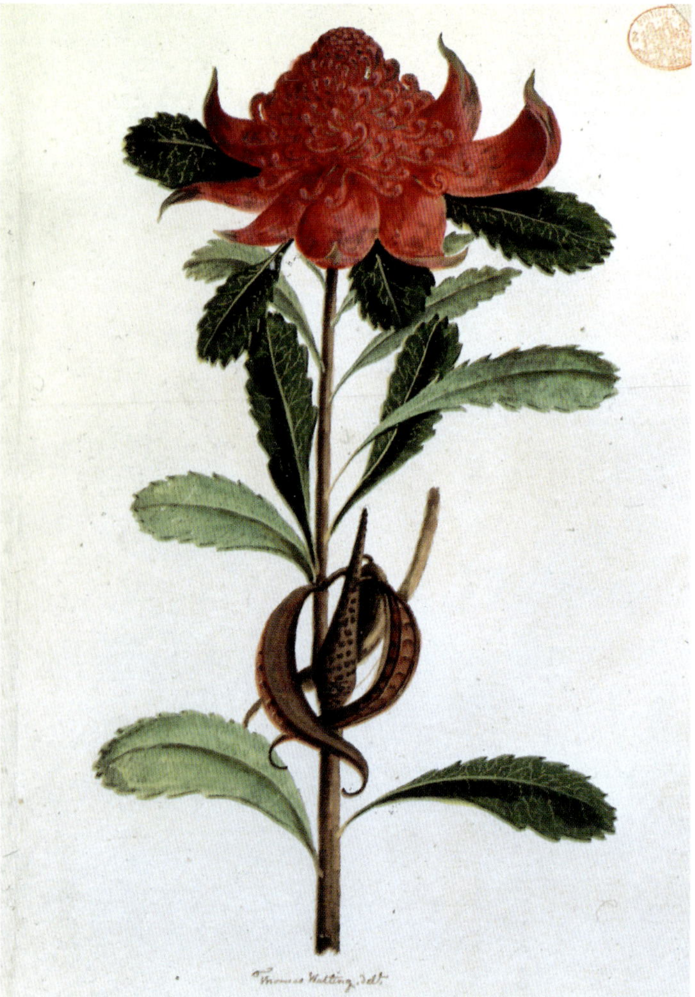

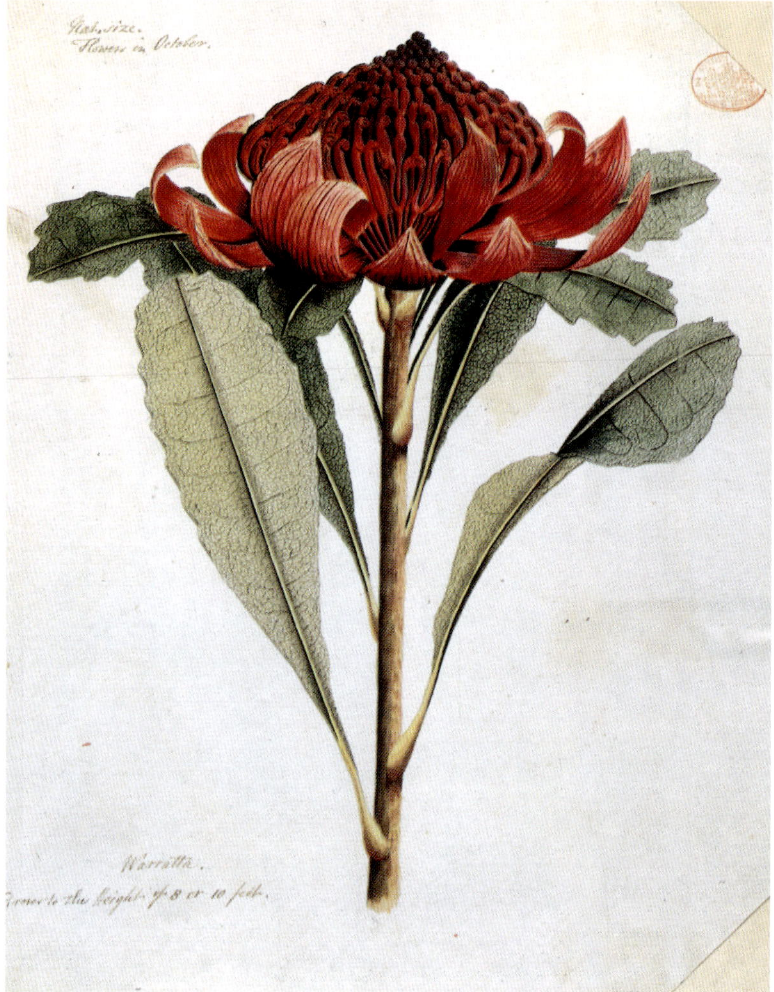

Telopea speciosissima (Smith) R.Br. (= *Embothrium speciosissima* Smith)
Two watercolour paintings (above) of Waratah by Thomas Watling for John White. Watling's paintings were used, together with dried specimens, by James Sowerby and James Smith in the preparation of the engraving (opposite) and descriptions of the plants published in Smith's *A Specimen of the Botany of New Holland*. The engraving by James Sowerby shows that elements from each Watling painting were used to create the published image.
[BM]

originally published as a component of *Zoology and Botany of New Holland* (1793). Smith prepared the descriptions of the plants. James Sowerby prepared the engravings (sixteen plates), based on plant specimens and drawings done in Australia and sent to Thomas Wilson by John White. Wilson, in turn, passed the material to Smith. Some drawings by Thomas Watling at the Natural History Museum can be identified as probably assisting Sowerby in his interpretation of the plants' colour and form.

Thomas Watling (1762–c.1814) appears to be the only convict artist to have some of his work form the basis of botanical scientific publication. Watling was born in Dumfries, Scotland, and trained to become a professional artist. Among other outlets, Watling applied his ability to forgery for which he was transported to New Holland for fourteen years.

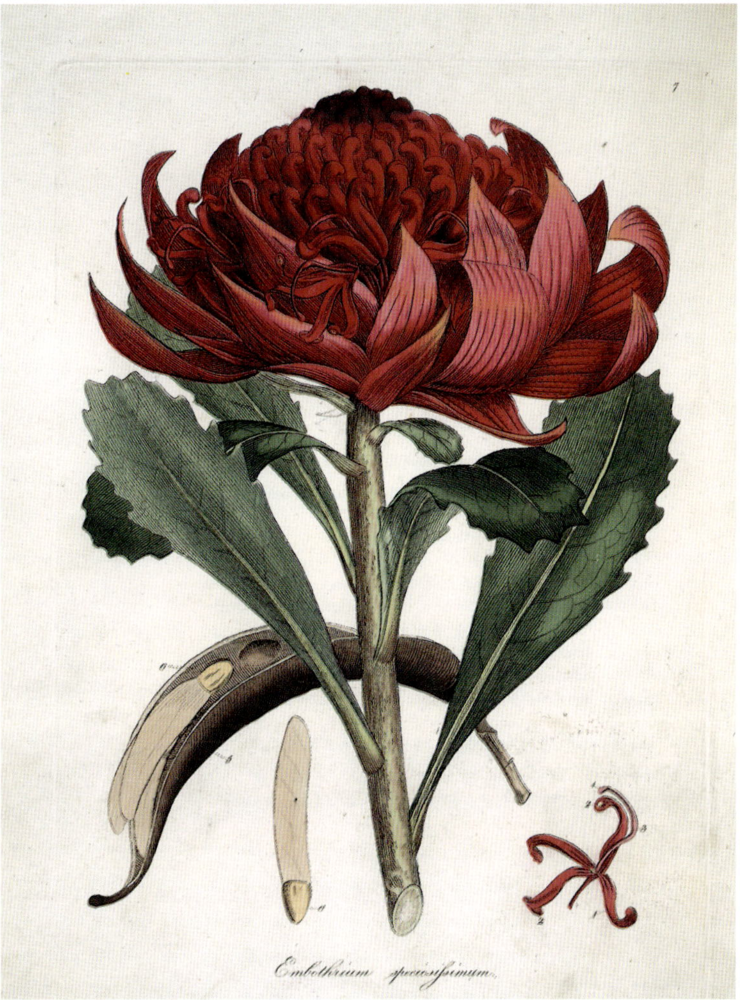

[NLA 23.5 x 15.5]

Watling arrived in October 1792 and was appointed to the employ of John White to paint the natural history and landscape of the colony.

When White returned to England in 1794, Watling was assigned to David Collins, Judge Advocate. In 1796 Watling received a Conditional Pardon. When this was made Absolute, in 1797, he left the colony. During Watling's service as a convict he described himself as 'the principal limner in New South Wales'. In letters to an aunt back in Scotland, Watling complained about White as an employer and about the stimulus provided by Port Jackson:[5]

> ... the landscape painter, may in vain seek here or there for beauty which arises from happy opposed off-scapes ... I however confess, that were I to select and combine, I might avoid that sameness, and find engaging employment.

By virtue of his employment, Watling may have been the principal limner (painter) in the colony. Watling's work certainly is professional. However, John Lewin and Joseph Lycett seem to have been the more competent natural history artists. Unfortunately none of their work has been identified as having contributed to the botanical scientific record of Australian plants. Lewin did, however, make a name for himself in publishing zoological work.

James Sowerby[6] made a huge contribution to English botanical illustration. The range of Sowerby's talent was significant — he was a competent natural historian, artist and engraver — and the volume of his work enormous. James trained at the Royal Academy School and ultimately found his niche as a botanical illustrator. In this he teamed with James Smith in the publication of several illustrated works. *A Specimen of the Botany of New Holland* was one.

In addition, members of Sowerby's family also became noteworthy botanical and natural history artists.

Meanwhile at Port Jackson the colony was becoming established. Governor Arthur Phillip was succeeded by Governor John Hunter in 1795. Hunter was an enthusiastic amateur naturalist. He had made friends with George Bass, the ship's surgeon, and Matthew Flinders, a midshipman, on the *Reliance* which had brought them all to Australia in 1795. At Port Jackson, both explored and both accompanied Hunter on minor expeditions around Sydney. Hunter painted some of the curiosities that he observed.

Indeed, drawing and painting were the best way of communicating images and many people were quite competent with the brush. Sometimes in the course of communication, paintings were copied. It is interesting to study some of the collections of paintings from the first ten or so years of the colony. Particular images appear more than once and deciding the original artist becomes problematic. Hunter made and collected natural history paintings and had them bound into a booklet entitled *The Hunter Sketchbook*.[7]

Independently, George Bass set out from Port Jackson in 1797 to prove the existence of a channel between the mainland and Van Diemen's Land. Bass failed due to insufficient supplies. The following year he tried again with Flinders in the *Norfolk*. They proved the existence of the Strait and went on to circumnavigate Van Diemen's Land.

Britain was becoming increasingly un-nerved by French activity in her waters, believing that French colonisation of Van Diemen's Land would threaten British control of the southern sea passages of New Holland.

Whatever the politics of the issue, Flinders returned to England in 1800 and lobbied Joseph Banks with a proposal to have the coastline of New Holland explored and completely charted. Banks continued to hold a proprietary interest in New Holland. He recognised the significance of the proposal and took it to the Lords of the Admiralty. Aware of the French expedition set up earlier in 1800 under the command of Nicolas Baudin, the Lords decided that it would be prudent to act.

In July 1801 the *Investigator* sailed for New Holland under the command of Matthew Flinders,[8] and a prediction of Watling was about to come true:[9]

> Should ... the prying Botanist emigrate here [he] could not fail of deriving ample gratification in [his] favorite pursuits in this luxuriant museum. Birds, flowers,

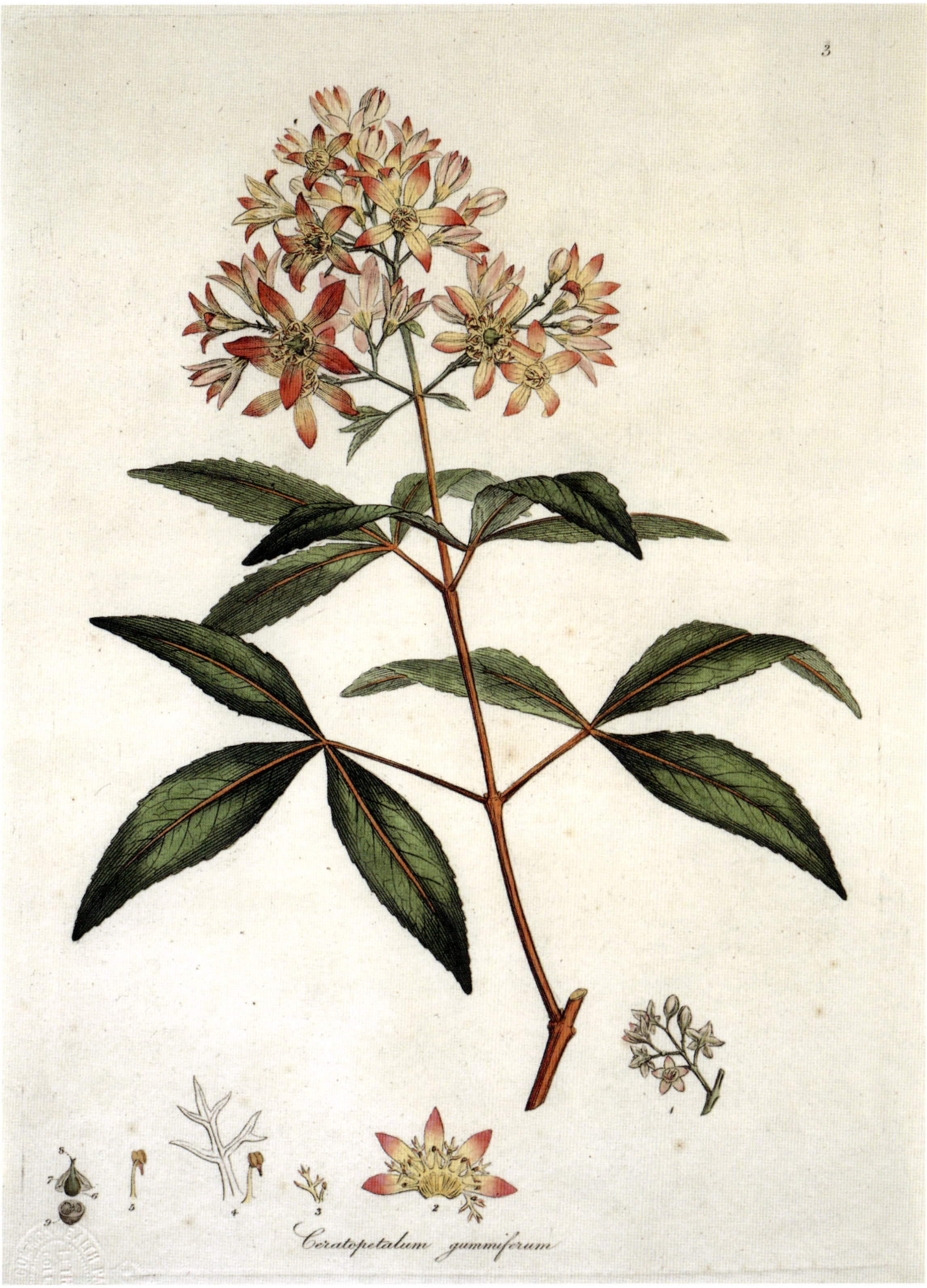

Ceratopetalum gummiferum Smith
Hand-coloured engraving by James Sowerby in J.E. Smith, *A Specimen of the Botany of New Holland*, 1793–1795. This book is the first Australian Flora.
[NLA 22.3 x 15.0]

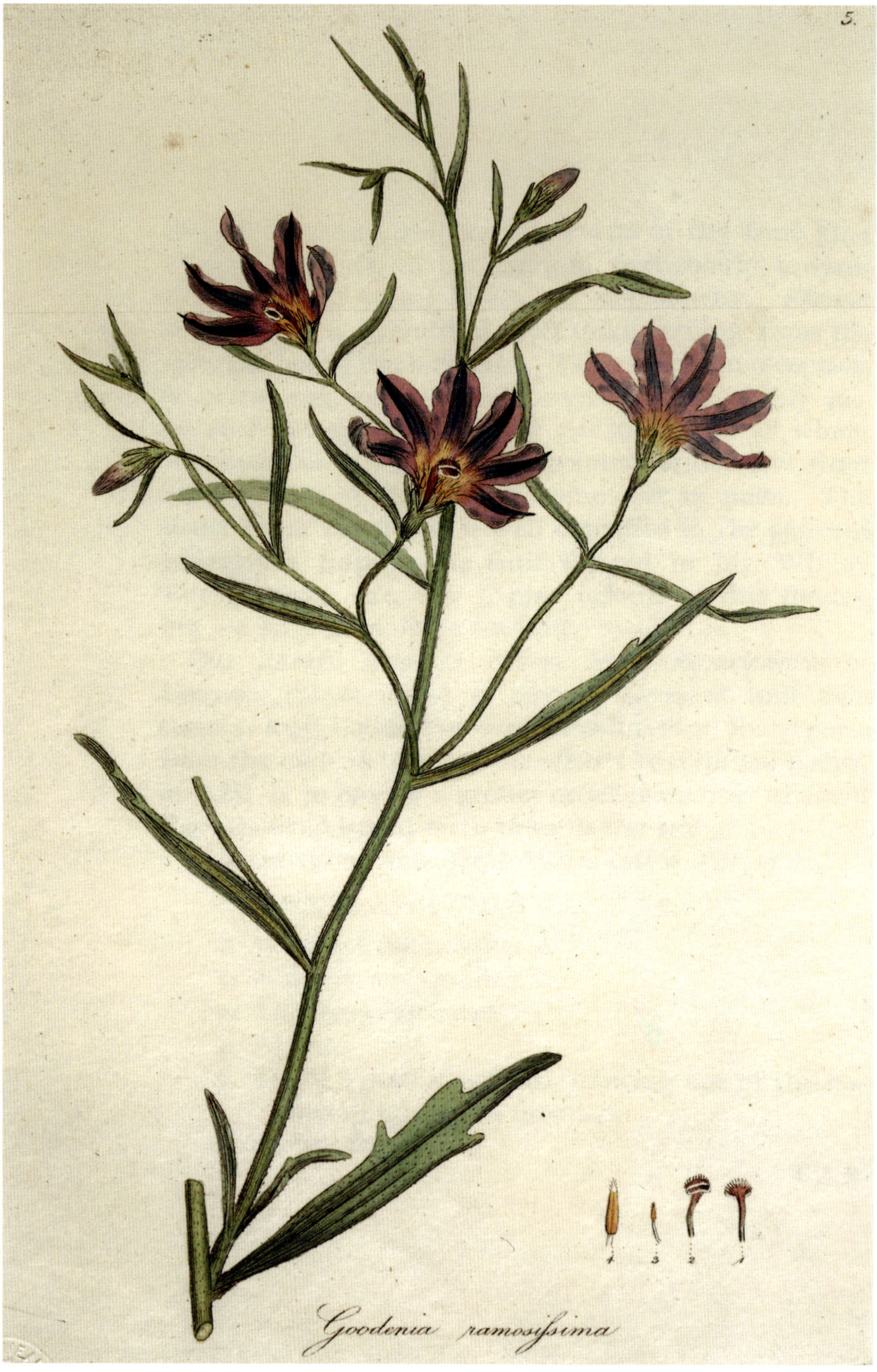

Goodenia ramosissima Smith
Hand-coloured engraving by James Sowerby in J.E. Smith, *A Specimen of the Botany of New Holland*, 1793–1795.
[NLA 22.3 x 15.0]

> shrubs and plants; of these many are tinged with hues that must baffle the happiest effects of the pencil.

On board were Robert Brown,[10] naturalist, Ferdinand Lucas Bauer[11] and William Westall, artists, John Crosley, astronomer, John Allen, mineralogist, and Peter Good,[12] gardener and assistant botanist. Flinders reached Cape Leeuwin early in December and sailed on to spend three weeks in King George Sound. They sailed east, charting the coastline and collecting and recording enthusiastically as they went. Tragedy befell them in February. A long boat with John Thistle and five other sailors on board was lost when they attempted to obtain wood off what is now known as Thistle Island.

The expedition worked on, with Flinders being supportive of the naturalists in their collecting endeavours. While they were at Kangaroo Island they encountered Baudin in the *Géographe*.

The *Investigator* landed at Port Jackson in May 1802, having had twenty anchorages between (and including) King George Sound and Port Jackson. The *Naturaliste*, of the French Baudin expedition, was there at the time.

Governor Philip Gidley King supplied Flinders with the *Lady Nelson*, under the command of Captain James Grant, to accompany him in further exploration. Flinders sailed north on 22 July and reached the tip of Cape York Peninsula on 29 October having anchored six times. The *Lady Nelson* accompanied them only as far as the Cumberland Islands.

Flinders then spent four months in the Gulf of Carpentaria. While there the crew effected repairs to the ship on Sweers Island. By January 1803 they were opposite Groote Eylandt. Flinders further explored Arnhem Land before sailing to Timor, arriving there at the end of March. As previous expeditions had experienced, landing there caused ill health and death. The *Investigator* was in a very bad state of repair so Flinders set sail, in haste, for Port Jackson with a crew sick from dysentery. Three died at sea and a further four died after they reached Port Jackson early in June. Peter Good, the gardener, was one of those who died.

The loss of Good was keenly felt because he was such a competent and supportive collector for Robert Brown. There are no Good herbarium specimens and it is generally believed that this is due to Brown incorporating them into his collection. However, the strength of Good's contribution was his seed collection which resulted in important plant establishments in the Royal botanic garden at Kew.

At Sydney the *Investigator* was beyond repair and had to be abandoned for another vessel. In August Flinders sailed for England, as a passenger on the *Porpoise*, to obtain another ship. The naturalists stayed and were to be taken on board the replacement ship when she returned. On this circumstance, Robert Brown reported to Joseph Banks:[13]

> I have in conjunction with Mr Bauer, stated to him [Flinders] the advantages of our remaining here during his absence, and after consulting with Governor King, he had acquiesced in the sufficiency of our reasons, and we accordingly remain, either till his return in the prosecution of the voyage, or if the idea of completing it should for the present be given up till instruction relative to us

> arrive from England, limiting however, the period of such arrival to eighteen months from the date of his departure from this port in the Porpoise ...

Two other ships, the *Cato* and the *Bridgewater* from the British East India Company, accompanied the *Porpoise*. A week later the *Porpoise* and the *Cato* came to grief on what is now known as Wreck Reef in the Great Barrier Reef. The *Bridgewater* sailed off without stopping to help. She, in turn, was lost at sea after sailing from Bombay for England.

Before she sank the crew rescued sufficient supplies and equipment from the *Porpoise* to sustain them for three months. To escape from this desperate situation some of the crew decided to sail the ship's cutter, named *Hope*, back to Port Jackson for help. Flinders was in command and they arrived safely early in September.

Collections destined for the British Museum were lost in the grounding of the *Porpoise*. If it had not been for the careful practice of Robert Brown, the effort and industry of the past couple of years would have been wasted. The best set was lost but fortunately Brown had carefully made a duplicate set of his collections and had retained one of these at Port Jackson.

Flinders returned to Wreck Reef in the *Cumberland* accompanied by the *Rolla* and the *Francis*. Once he had rescued the remaining crew, Flinders sailed on for Timor arriving in November 1803. Four days later he set sail for England, but was obliged to put in at Ile de France (now Mauritius) because the *Cumberland* had been damaged and was hard to handle. At Ile de France Flinders was caught up in the political situation of the day (England and France were at war again) and he was kept prisoner there for over six years.

Back in New Holland, Brown and Bauer worked on.

The British had decided to found a new penal colony in the south of New Holland, thereby thwarting any aspirations the French might have to set up a colony in the same region. In 1803 John Bowen was sent by Governor King to establish a colony in Van Diemen's Land. The party was shipwrecked, rescued, and finally settled at Risdon Cove on the Derwent. David Collins sailed from England in April 1803 in the *Calcutta*, accompanied by the *Ocean*, to establish a colony at Port Phillip. They landed near Sorrento. The site proved to be unsatisfactory and Collins sought permission to move the settlement to Van Diemen's Land. This was granted and early in 1804 they moved to Sullivan Cove on the Derwent River.

On 28 November 1803, aboard the *Lady Nelson*, Robert Brown went south with Lieutenant-Colonel William Paterson[14] to found another British colony in Van Diemen's Land on the Tamar River. Paterson called in at Port Phillip Bay and took on board a small

Passiflora aurantia G.Forst. (= *Murucuia baueri* Lindl.)
Hand-coloured copper engraving of a painting by Ferdinand Bauer, originally drawn in 1804–1805 while he was on Norfolk Island. The engraving by Weddell was published in J. Lindley, *Collectio Botanica*, 1821–1826.
[NLA 47.4 x 35.0]

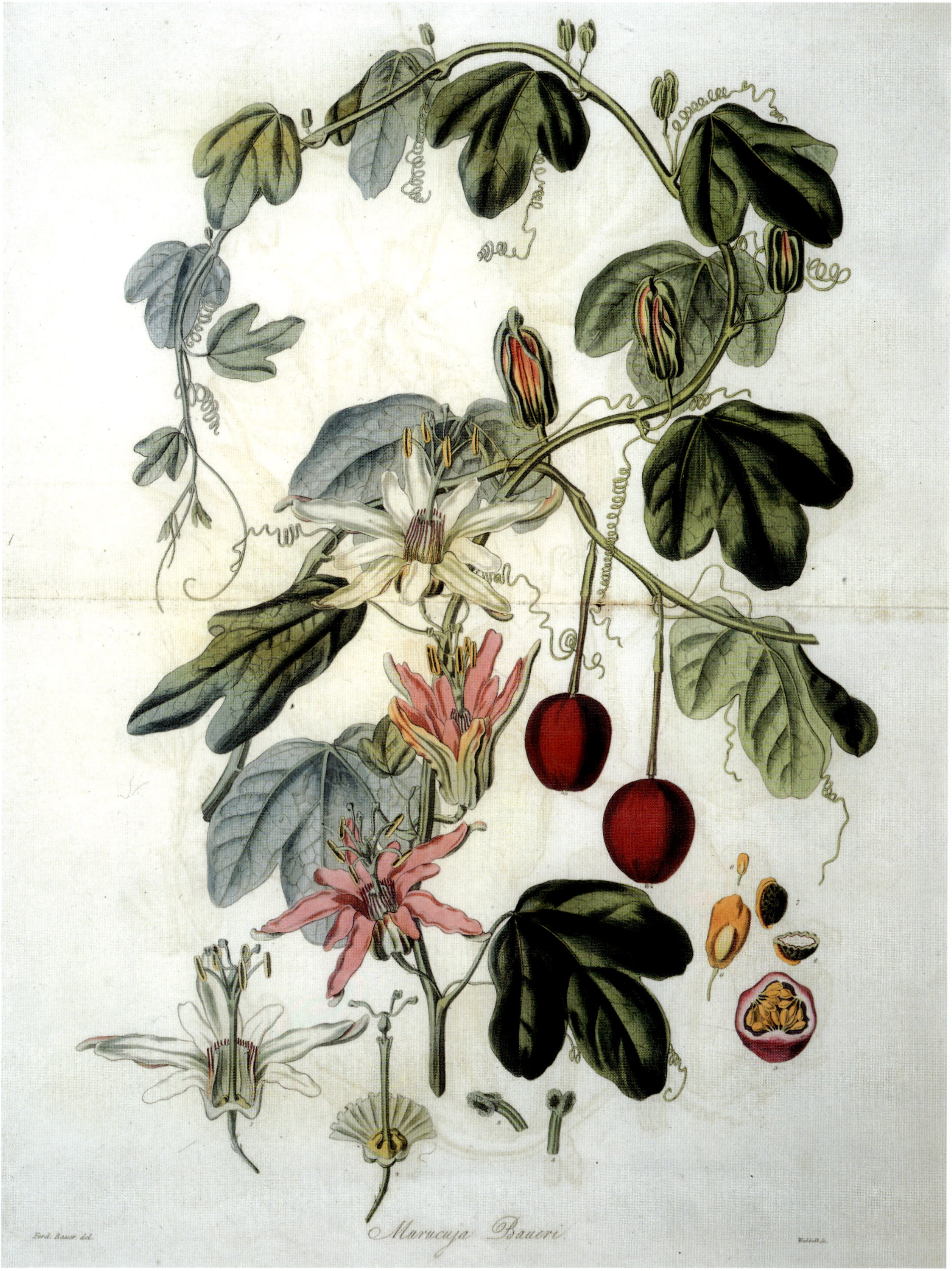
Ferd. Bauer del.
Murucuja Baueri

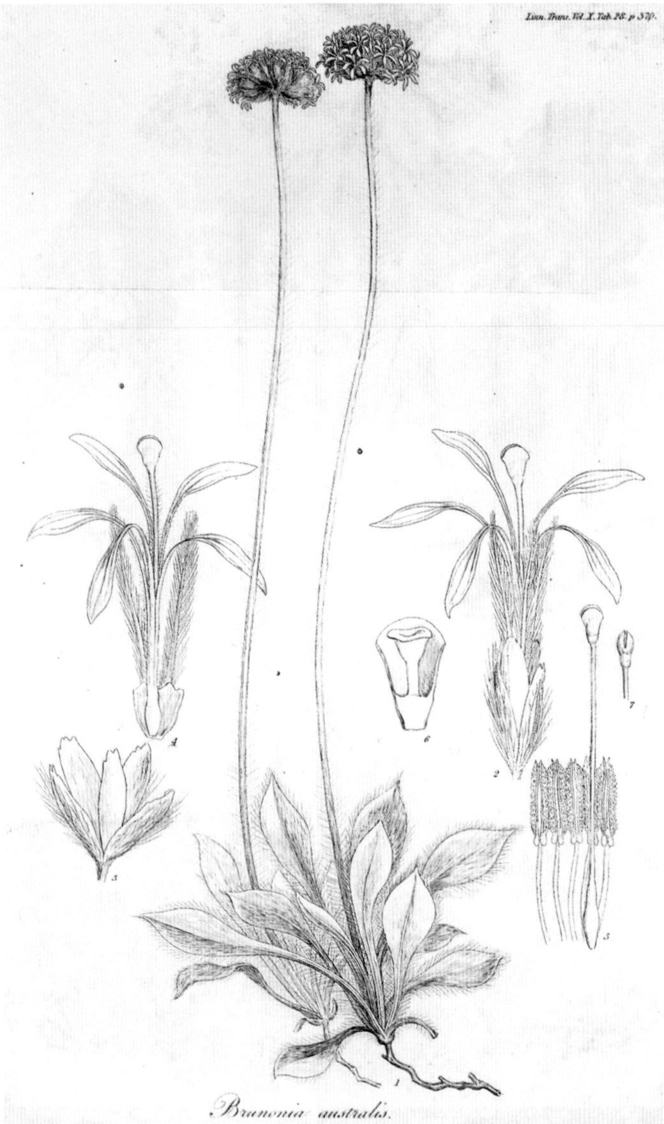

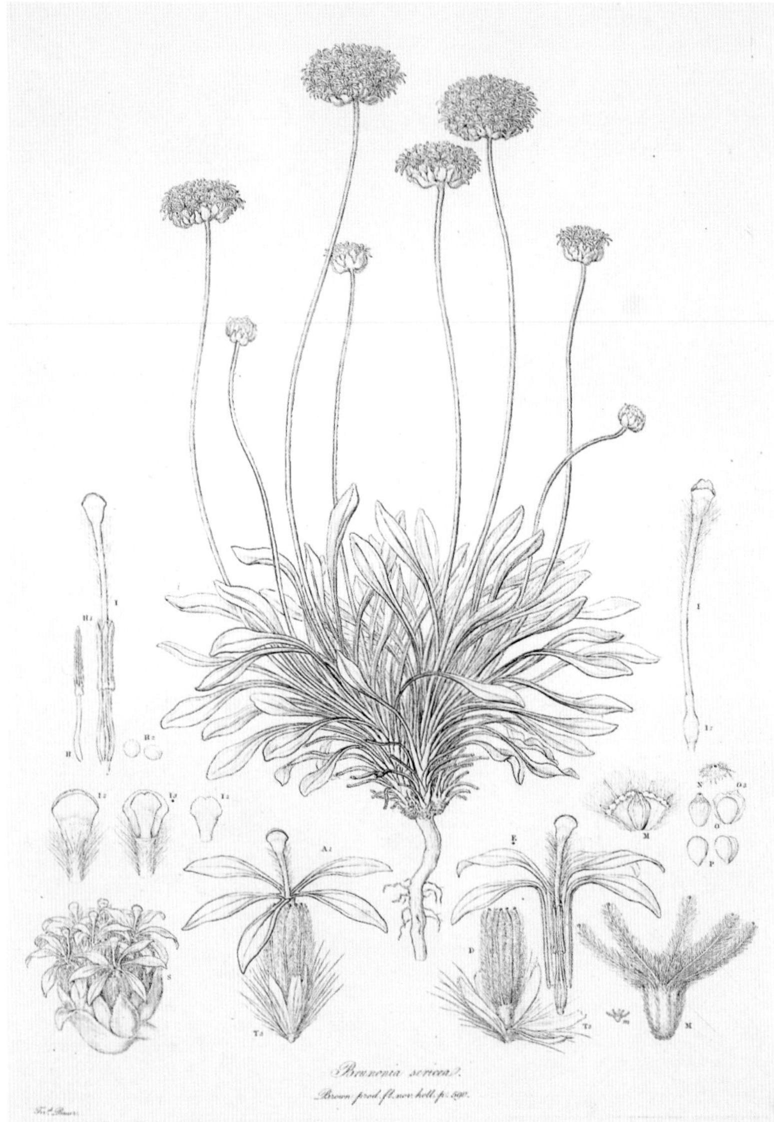

Brunonia australis Smith ex R.Br. (= *Brunonia sericea* Smith)
Copper engraving (left) drawn by Ferdinand Bauer for the *Transactions of the Linnean Society of London*, Volume 10, 1811. Engravings such as this are ideal for hand colouring, even though this was not done in the *Transactions*. Bauer executed a more detailed engraving (right) for his *Illustrationes Florae Novae Hollandiae*, 1813, some copies of which were not coloured.
[CANB 31.5 x 22.5] [NLA 39.0 x 25.0]

group of convicts left behind by Collins to relocate them to Risdon Cove. Brown remained in Van Diemen's Land for approximately nine months. No doubt conditions were raw during that first year of establishment.

Ferdinand Bauer stayed on at Port Jackson. In the following year, August 1804, he sailed east in the *Albion* to Norfolk Island where there was another British penal colony. He spent the following eight months making collections and drawings.[15] He returned to Port Jackson in March 1805.

The same year Brown joined Bauer again at Port Jackson. As there was no sign of Flinders, they decided to return to England. Ironically, the *Investigator* had been made seaworthy in the interim and they returned on her, leaving in May. Brown had collected over three thousand five hundred plants and Bauer had drawn over fifteen hundred plants.

Flinders remained imprisoned on Ile de France for six years, an ordeal which broke his health. When he finally returned to England in 1810, he prepared a journal of the voyage, *A Voyage to Terra Australis*, which was published a day or so before he died in 1814. Banks wrote a preface, Brown provided the botanical entry and Bauer prepared ten plates. The latter were engraved by Elizabeth Byrne, Francis Sansom, Pye, and an unacknowledged engraver. These plates are uncoloured metal engravings and do not measure up to Bauer's own standard.

In London, Brown prepared and published volume 1 of his *Prodromus Florae Novae Hollandiae et Insulae van-Diemen* in 1810. This was to be illustrated with paintings by Ferdinand Bauer in a complementary work. In 1813 Bauer published three parts (fifteen plates) of his *Illustrationes Florae Novae Hollandiae*. A sixteenth plate (*Lambertia*) was bound with some copies. The plates were metal engravings with the drawing and engraving being done by Bauer. The engraving technique was a combination of line and stipple with an apparent touch of mezzotint work on the stipple. Some sets were hand-coloured, the original sets being coloured by Bauer.

Neither of these complementary works was a commercial success and, in spite of scientific acclaim, the whole project was abandoned. Thus the third Australian Flora remained incomplete and very sparingly illustrated. Yet another outstanding opportunity for science was lost. John Lindley later wrote:[16]

> With Flinders's voyage, a Botanist and a Botanical draughtsman were associated, at the public charge; and both of these officers discharged their duty of collecting materials most zealously. But if we inquire for the public Botanical advantage which has been derived from that expedition, either to ourselves or to the New South Wales colonists, we find it to consist of little... That full account of the Botanical discoveries made during Flinders's expedition, which the public had a right to expect, has never appeared. The dried plants, or what remains of them, are to be found in the cases of the British Museum; and the beautiful plates made by Mr Ferdinand Bauer with such care and admirable skill, after slumbering for years at the Admiralty, are now (we understand) also at the Museum.

A supplement to Brown's *Prodromus* was published in 1830, but it too was unillustrated.

Bauer completed about 236 paintings from his over two thousand drawings for the British Admiralty (1542 of Australian plants).[17] Not many of his drawings were published in his lifetime. In addition to his own *Illustrationes Florae Novae Hollandiae* (1813), Bauer had approximately one hundred Australian plant plates published.

Ferdinand Bauer's published contribution to botanical illustration (during the early-nineteenth century) also included a similar number of plates of non-Australian material.[18] It appears that much more of Bauer's work has been published in various other books, journals and periodicals, frequently unsigned. Unfortunately he was very erratic about

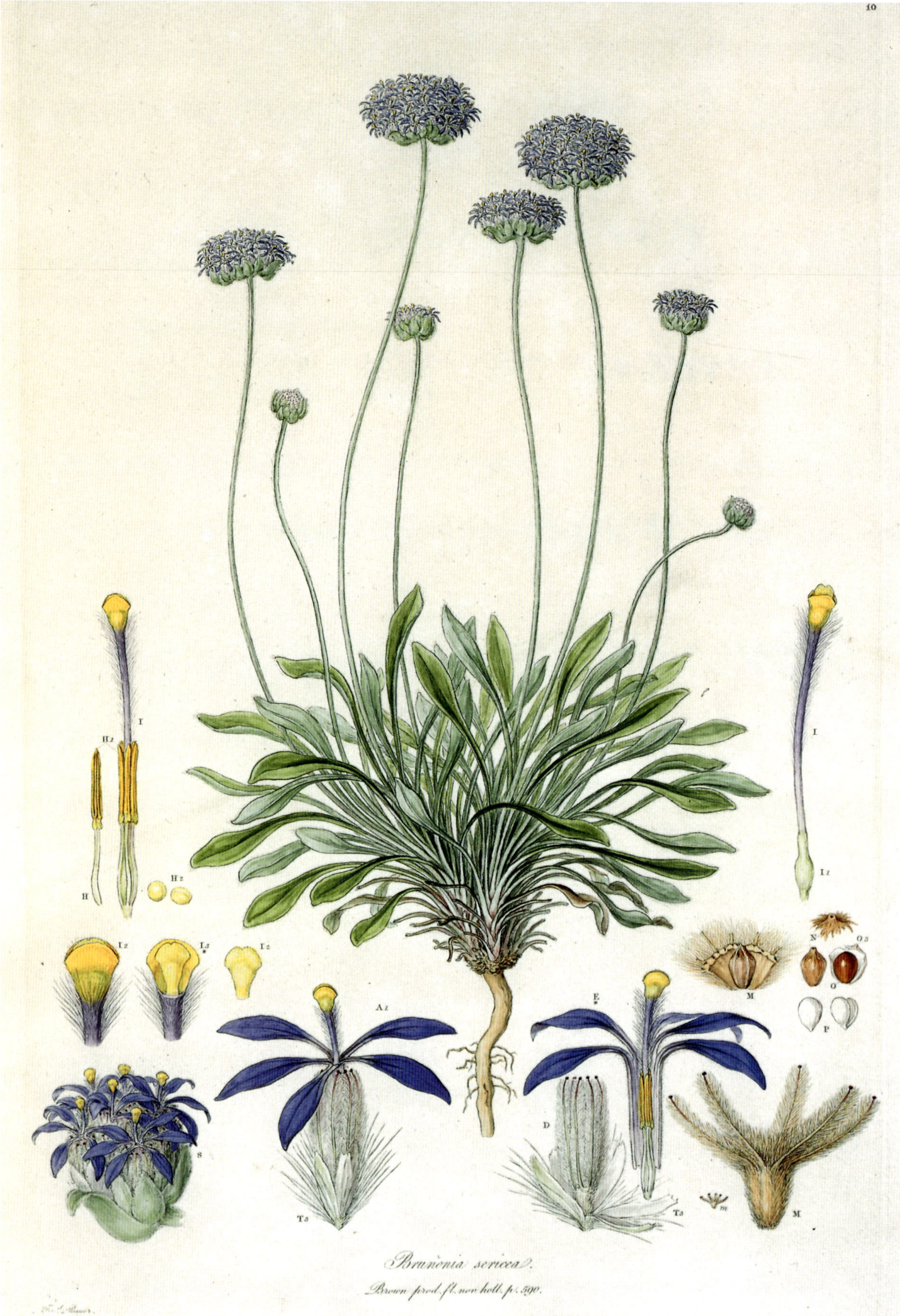
10
Brunonia sericea.
Brown prod. fl. nov. holl. p. 590.

signing his work. For example, in T. Horsfield's *Plantae Javanicae Rariores* (1838), Robert Brown provided the botanical descriptions. Under *Loxotis obliqua* he wrote:

> With respect to the generic name *Loxotis* here adopted, it is that which I first gave in my manuscripts to the plant now described. This, however, I many years ago changed to *Antonia*, in compliance with the request of my lamented friend and fellow-traveller Mr. Ferdinand Bauer, to whom I was indebted for the figure here published. But as that name, by which it was introduced into a celebrated flower piece, painted in the honour of the late Baron Jacquin at Vienna, and well known to botanists of that capital, was never otherwise made public ... I am obliged to recur to my original name.

The plate is unsigned, as indeed are the first twenty-five plates in the book; the remaining twenty-five are all signed by other illustrators.

In 1976 W.T. Stearn and W. Blunt published *The Australian Flower Paintings of Ferdinand Bauer* in which twenty-five plates of his original paintings are reproduced. The publication of this book brought an awareness of this great and almost forgotten artist. However, this twentieth-century publication of Bauer's work, like the *Banks' Florilegium* for Sydney Parkinson, makes a greater contribution to art and history than to science.

Robert Brown (1773–1858) was born in Scotland. He studied medicine and became a surgeon in the British army. He received a posting to Ireland and indulged in botanical studies in his spare time. Brown was also a bibliophile and it was fortuitous that he obtained a copy of Antoine de Jussieu's *Genera Plantarum* (1789), the work which presents an alternative classification to that of Carl Linnaeus.

In the 1790s Brown collected and studied plants which attracted his attention wherever he happened to be working or visiting. William Withering[19] became something of a mentor. Through Withering, Brown learned much and developed a network of botanical mentors. Ultimately he found his way to the Banks library and collection in London and soon came to the attention of Joseph Banks. In 1801 Brown was chosen by Banks to be the naturalist on Matthew Flinders' voyage to New Holland.

Brown gave every indication that he was disappointed with the extent of the Australian flora and with the size of his own collections. Nevertheless, the experience proved to be invaluable. Brown struggled with the classification of Australian plants. He was most comfortable with the natural classification system of Antoine de Jussieu but, when confronted with so much new material while working in Australia, he was obliged to use the artificial Linnaean system for practical purposes. All the while he was working towards a better under-

Brunonia australis Smith ex R.Br. (= *Brunonia sericea* Smith)
Ferdinand Bauer drew, painted, engraved and hand-coloured his own work in his *Illustrationes Florae Novae Hollandiae*, 1813.
[SLV 39.0 x 25.0]

standing of the variation observed in nature and developing his own philosophies. In England he rearranged his collection into the natural system.

In England Brown and Bauer worked together in the one laboratory at the British Museum and the mental and observational stimuli of the one enhanced the output of the other. In keeping with his developing beliefs, Brown undoubtedly influenced the extent and level of observation needing to be recorded in the drawings by Ferdinand Bauer. This, therefore, proved to be one of the greatest teams ever sent into the field in hitherto unknown, or little-known, parts of the world.

Robert Brown went on to become Joseph Banks' librarian following the death of Dryander in 1810. He also became one of the most prolific and greatest botanists of all time. Frans Stafleu wrote of him:[20]

> British romantic empiricism, embodied for instance in Robert Brown, ultimately reached the same conclusions as the French philosopher-naturalists from Buffon and Adanson, to Lamarck. This outcome was a vindication of the theoretical basis of biological systematics laid by the French taxonomic *philosophies*, replacing the scholastic classification of the past. Adanson and Buffon recognised diversity on the basis of relatively small practical experience, but with superior intuitive reasoning; a man like Robert Brown built his diversitarian view of nature on the basis of superior experience both from exploration and experiment, and discovered the new theory to be a fitting framework for his discoveries. The victories of empirical science and romanticism went hand in hand.

Ferdinand Lucas Bauer (1760–1826) was born in Austria into an artistic family in Feldsberg. He was the youngest of three brothers. All three, Josef Anton, Franz Andreas and Ferdinand Lucas, became renowned artists in their own right. An apparently excellent grounding was provided for the boys by Father Norbert Boccius, a local prior.

In 1780 Franz and Ferdinand moved to Vienna to work with Nicholaus Jacquin at the University of Vienna. The influence of Jacquin set them on the road to being among the great botanical illustrators. Jacquin was a leading botanist of the time and also an accomplished botanical illustrator. He taught the Bauers to observe botanical idiosyncrasies and to record these observations in accurate and minute detail. To achieve the latter, Jacquin encouraged the use of the microscope.

Ferdinand soon joined John Sibthorp, from Oxford, in the field in the Mediterranean to study the plants for Sibthorp's *Flora Graeca* (1806). Ferdinand did over fifteen hundred drawings in the field, in one and a half years, for this project. In England he busied himself with the preparation of these for publication.

The well-networked botanical community was soon aware of him and his talent. Through the contacts of his employers, Jacquin and Sibthorp, he became known to Joseph Banks. When Banks agreed to support Flinders' expedition to circumnavigate New Holland, Bauer was a well-qualified and experienced artist and an appropriate choice for natural history illustrator.

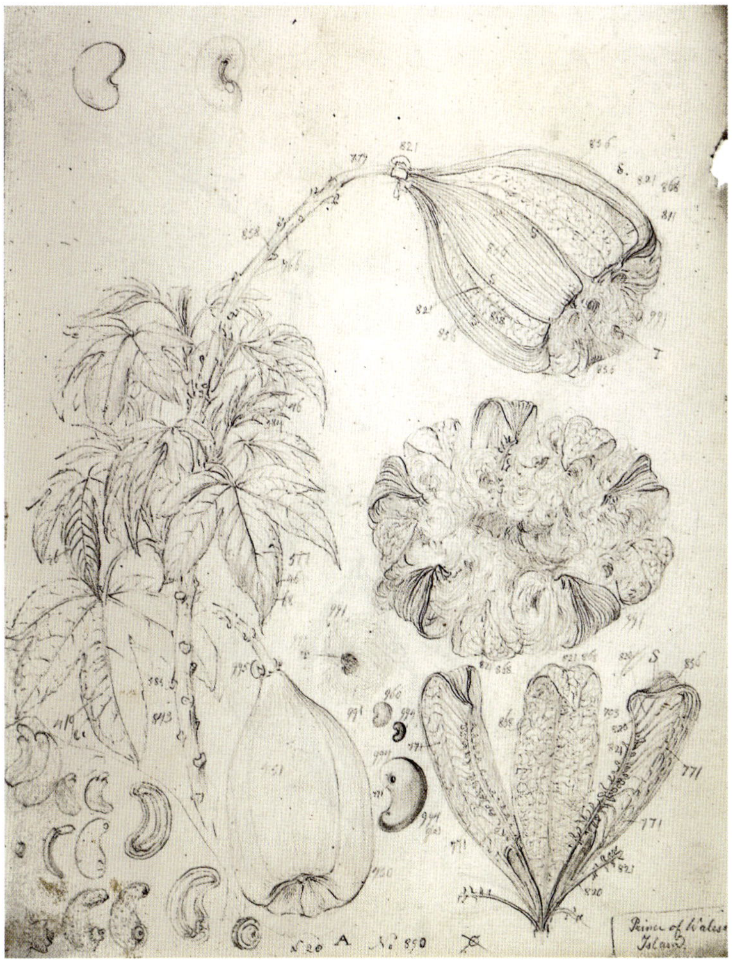

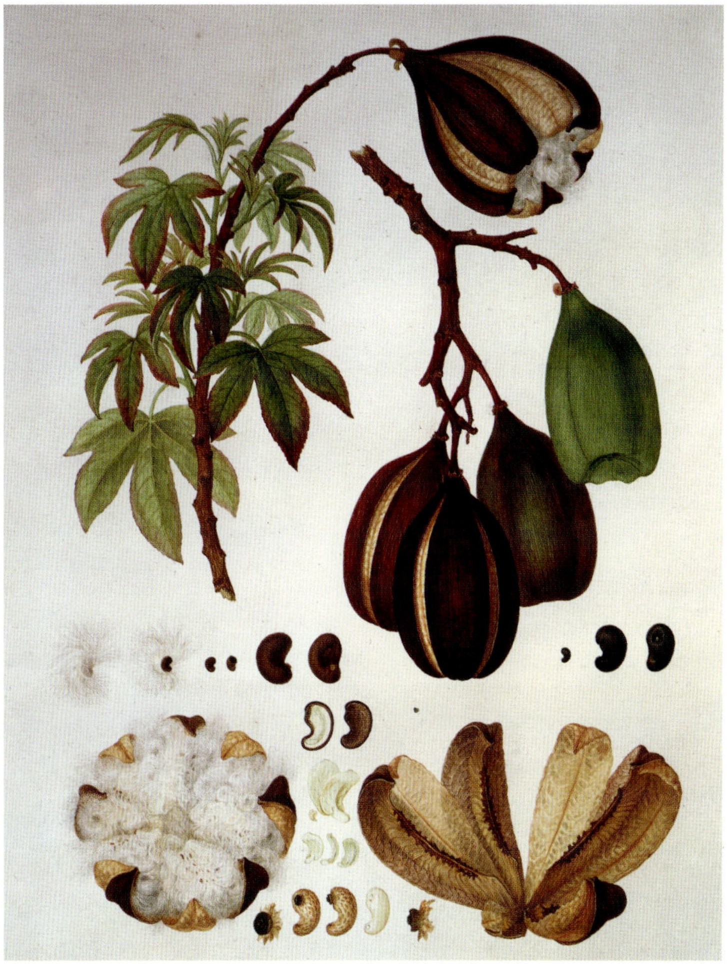

Cochlospermum gillivraei Benth.
Pencil sketch made by Ferdinand Bauer while on the voyage of circumnavigation of Australia with Matthew Flinders aboard the *Investigator*. Bauer's method was to draw in pencil and code the colours according to a colour grid of some 1 000 or so colours. In this way he was able to create an accurate painting at a later time. The resultant watercolour painting, executed by Bauer back in London, was created using dried specimens and the colour-coded sketch.
[WIEN] [BM 52.5 x 36.5]

Bauer's method was to sketch the plant, life size, in pencil in the field. He then colour-coded the sketch so that it could be accurately represented when painted. His code consisted of a set of rows of numbered colours. A prototype for earlier work is still extant, but the Australian code has not been located. The latter had in excess of a thousand individual codes (ascertained by studying the original sketches). A 'signature' of his work was the inclusion of at least one structure at natural size to give scale. In engravings his signature was *Ferd. Bauer* in contrast to *F.* or *Fr. Bauer* of his brother Franz.

Ferdinand's versatility was remarkable. He is renowned for his draftsmanship, the level of detail, and the degree of accuracy in the execution of his drawings, paintings and engravings, including that of colour and of form, both combining to give a true three-dimensional image. In 1998 Walter Lack postulated that Bauer made use of a camera lucida.[21] Such an aid

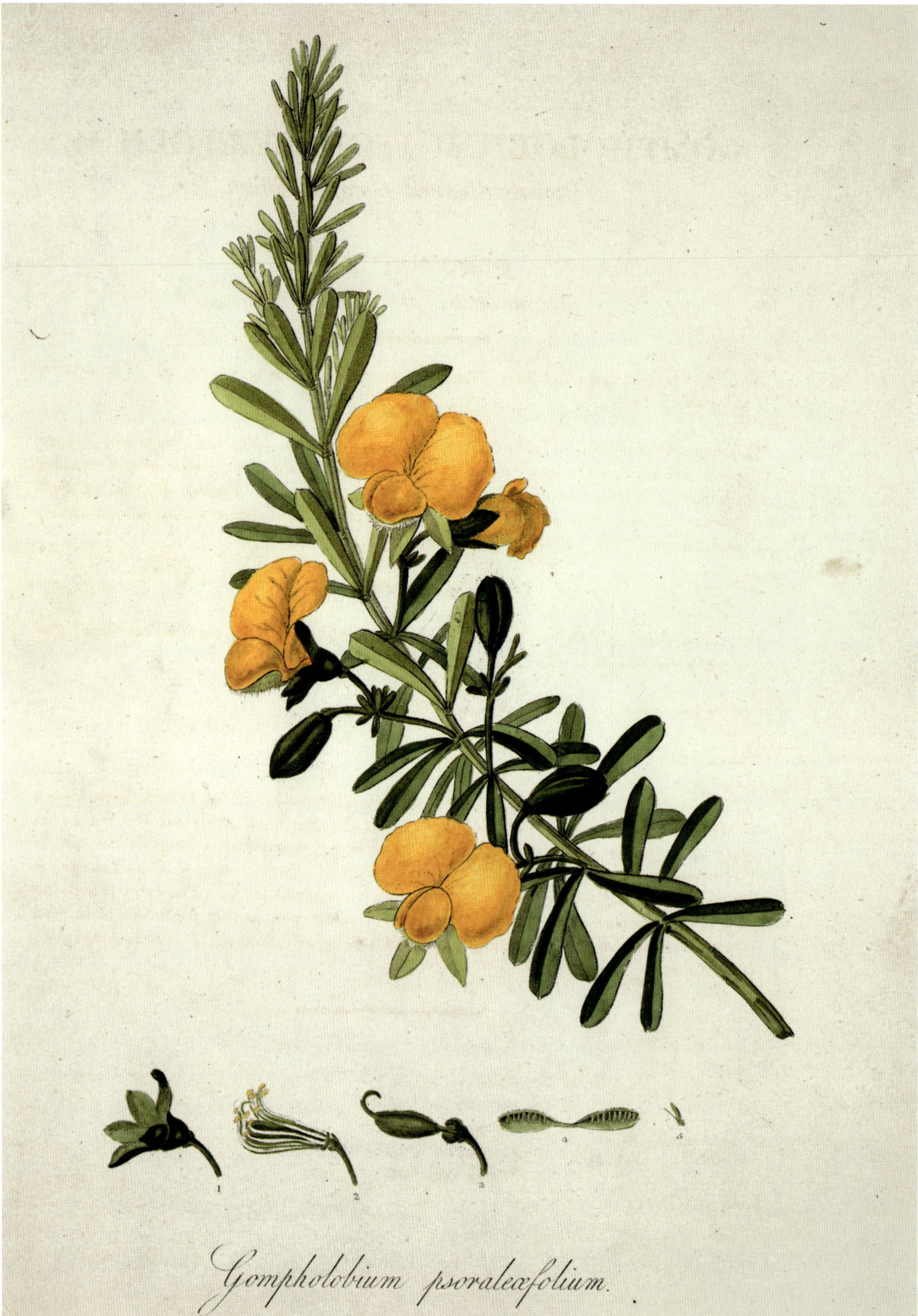
Gompholobium psoraleæfolium.
Drawn & Pub. by W. Hooker, July 1.1805.

in the right hands would surely have facilitated the achievement of accuracy and the feeling of reality for which Ferdinand is so renowned.

The detail in Bauer's paintings is analytical in the scientific sense and certainly reflects the influence of Robert Brown and, in turn, of Antoine de Jussieu. Brown encouraged Bauer to record the observations necessary for depicting the classification of the subject within the natural system. This frequently required a greater range of observations than were being depicted for the Linnaean sexual-system classification. Many of these extra observations are necessarily minute, even microscopic. Brown was a master of the microscope and encouraged Bauer to make good use of it too.

The number of significant observations to be recorded afforded Bauer the opportunity to create a more pleasing and balanced composition than some of his predecessors and contemporaries had achieved — even better than he created for Jacquin and Sibthorp.

In a letter to Joseph Banks, Brown described Bauer as 'indefatigable', working with 'minute accuracy', and 'the number of drawings he made, astonishing'.

Following the commercial failure of his *Illustrationes Florae Novae Hollandiae*, Bauer returned to Austria taking his drawings and some of his paintings with him. Once there, he does not seem to have been particularly active, but he cooperated with the contemporary botanists, especially Stephan Endlicher, before he died in 1826.

Other horticultural and scientific outcomes from Matthew Flinders' voyage of circumnavigation were manifold. Peter Good, the gardener, had returned volumes of seed to William Townsend Aiton at the Royal botanic garden at Kew. From this valuable contribution emerged many benefits. Much of the seed was successfully germinated and established horticulturally and Aiton had many species, arising from the successful germinations, described.

Aiton's father, William, was the Royal Gardener at Kew until his death in 1793. William sr had 'a catalogue of the plants cultivated in the Royal botanic garden at Kew' published in 1789. William jr succeeded his father as Royal Gardener at Kew. He edited a revised edition of the *Hortus Kewensis* published in five volumes between 1810 and 1813. Robert Brown and others contributed descriptions to both editions. The original had some plates, none Australian, and the second edition, despite plenty of opportunity and an emerging entrepreneurialism, was unillustrated. However, the Royal botanic garden provided many of the vouchers for illustration in other works and some of these were from Good collections.

Ferdinand Bauer's brother, Franz, had William Hooker as a student of botanical art. William went on to become an outstanding illustrator of fruits for the Royal Horticultural

Gompholobium latifolium Smith (= *Gompholobium psoraleæfolium* Salisb.)
Hand-coloured copper engraving of a painting by William Hooker, published in R.A. Salisbury & W. Hooker, *Paradisus Londonensis*, 1805–1808. This William Hooker became a famous painter of fruit and should not be confused with William Jackson Hooker.
[CANB 26.0 x 18.0]

Society. In the meantime, one of his earlier enterprises was to paint 'plants cultivated in vicinity of The Metropolis' of London. Hooker sought the aid of Richard Anthony Salisbury to provide the descriptions. These were published in *Paradisus Londonensis* (1805–1808). Some Australian plants were published in this book and included *Goodia lotifolia*, named in honour of, and collected by, Peter Good. Salisbury stated that:

> I know nothing more of Peter Good than that he left a very lucrative situation to collect seeds for his Majesty in New South Wales where he died, and that all the best and rarest plants from that country now at Kew, have his name attached to them.

The plates, hand-coloured copper engravings (a mixture of line and stipple), do not measure up to the standards of either Franz Bauer or Hooker's own later fruit paintings. It seems that Hooker did both the original drawing and the engraving for the *Paradisus*.

Australian plants were being described in a number of places, with and without illustrations, *Transactions of the Linnean Society of London* being a frequent outlet. An illustrated example is provided by Edward Rudge, who published descriptions of Australian plants in the *Transactions* in 1811. These were illustrated by his wife, Anne Rudge.

The bounty of Flinders' and concurrent French voyages for horticulture and science seems endless. Certainly it was much more sought after than the spoils from Cook's voyage.

4. The French Impression

With all the exertions that were to be expected from his {Joseph Banks} known love for the sciences, I soon had the satisfaction of finding myself in possession of the requisite materials, for making known to the world the natural productions which I had discovered in the different countries we had visited during the course of our expedition.

Jacques J.H. de Labillardière[1]

While the British were struggling to establish the penal colony at Port Jackson, European exploration was continuing in Southern Hemisphere waters. French interest, in particular, paralleled that of the British.

In 1785 the French had sent Jean-François Galaup de La Pérouse on an expedition to explore the Pacific Ocean as potential for French exploitation. He sailed in the *Boussole* together with the *Astrolabe* under the command of Fleuriot de Langle. The expedition was very well equipped to study physics, geology, climate, natural history and botany and had the services of three artists.[2] La Pérouse criss-crossed the Pacific for three years. In mid-expedition, while at Kamchatka, La Pérouse received instructions to sail to New Holland to investigate the proposed new British colony. On the way La Pérouse stopped at Samoa to renew water supplies. The natives had other ideas and attacked. In the skirmish the natives killed twelve expeditioners and severely injured twenty others. At the same time the French began to suffer from the effects of scurvy.

La Pérouse's intelligence must have been remarkably accurate because he crossed paths with the First Fleet at Botany Bay a few days after the Fleet's arrival. At Botany Bay La Pérouse and his men spent a few weeks recovering. The expeditioners also seized the

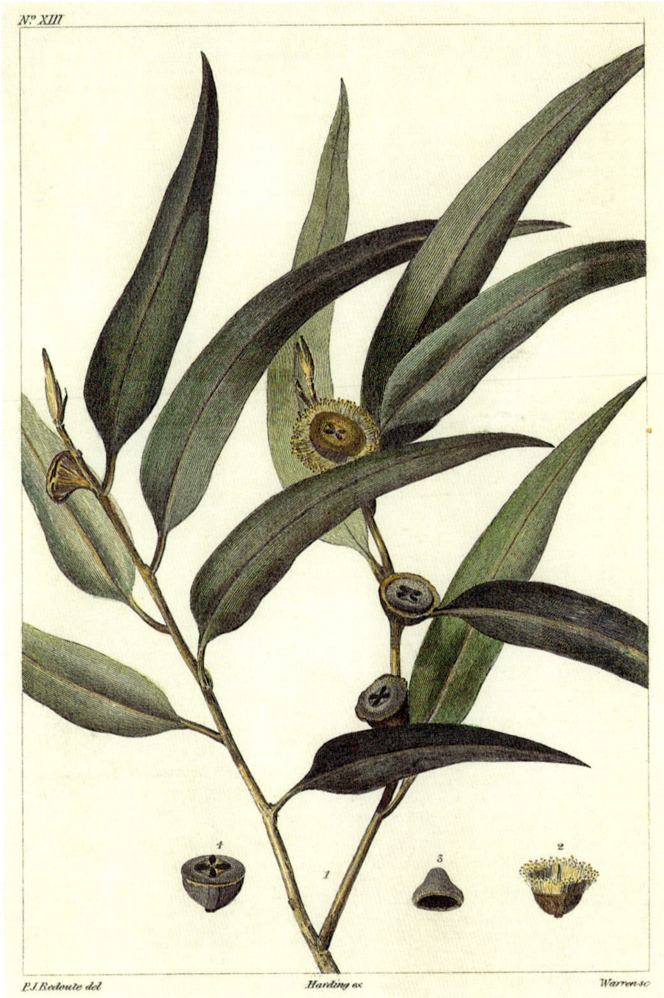

Eucalyptus globulus Labill.
The Tasmanian Blue Gum was named *Eucalyptus globulus* by J.J.H. de Labillardière in 1792. This coloured copper engraving by Warren, based on a drawing by Pierre-Joseph Redouté, was published in Labillardière's *Voyage in search of La Pérouse*, 1800, by Debrett.
[PC 28.0 x 22.0]

opportunity to forward some of their recent journals, collections and artworks to Europe on British ships returning from delivering the First Fleet consignment. Early in March, they sailed for the Friendly Islands.

No more was heard of the La Pérouse expedition. It was lost off the coast of the Santa Cruz Islands, destined not to be found for decades.

After nearly two years the French Government decided to search for La Pérouse and mounted an expedition in 1791. They commissioned A.R.J. de Bruni Chevalier d'Entrecasteaux to lead the search in the *Recherche* accompanied by the *Espérance* under the command of J.M. Huon de Kermadec. On board were the naturalists Jacques Julien Houtou de Labillardière,[3] Louis Ventenat (brother of Étienne), Louis-Auguste Deschamps[4] and Claude-Antoine-Gaspard Riche.[5] A gardener, Felix de Lahaie, and an artist, Nicolas Piron, also accompanied the naturalists. The voyage inherited problems before it began. The ships did not handle well and the men were divided along the lines of the politics of the prevailing French Revolution. d'Entrecasteaux departed in September 1791 and, after difficult sailing and changes of plan, reached Van Diemen's Land in April 1792. He spent a month in the region of Bruny Island and Recherche Bay.

Labillardière made important collections in the area. He was particularly impressed by the eucalypts, writing:[6]

> The finest trees in this country are the different species of eucalyptus. Their ordinary thickness is about eighteen feet: I have measured some that were twenty-five in circumference. The spongy bark of the eucalyptus resinifera, becoming slippery in consequence of the moisture that constantly prevails in the heart of these thick forests, renders it still more difficult to penetrate into them. This bark very readily peels off into pieces that have a great degree of flexibility, and are used by the natives for covering their huts. They often find long stripes of it about a foot in breadth, which spontaneously shell themselves off from the lower part of the trunk. They might easily peel it off in pieces of twenty-five or thirty feet in length.

On another occasion Labillardière wrote:

> I had not as yet been able to procure any flowers of a new species of *eucalyptus*, remarkable by its fruit, which very much resembles a coat-button in shape. This tree, which is one of the tallest in nature, as it grows sometimes to the height of 150 feet, blossoms only near its summit ...We were obliged to cut down one of these trees in order to obtain its blossoms.

He named it *Eucalyptus globulus* (Tasmanian Blue Gum and floral emblem of Tasmania).

Later, Labillardière wrote:

> Piron, the painter to the expedition, who was of our party took several drawings of the landscape. The round hills, covered with tall trees which bounded the horizon, added greatly to the beauty of the prospect.
>
> We were obliged to return back by the road we had come, in order to arrive at the opposite side of the lake. Piron returned on board.
>
> I discovered an evergreen tree, which has its nut situated, like that of acajou, upon a fleshy receptacle much larger than itself. I therefore named this new genus *exocarpos*. [Wild Cherry or Cherry Ballart]
>
> It has hermaphrodite flowers upon the same peduncle with others which are distinctly male and female.

In May they sailed on to New Caledonia and other islands of the western Pacific in a futile search for La Pérouse. d'Entrecasteaux arrived in Amboina in September. Here he replenished food and water and encountered tropical disease with Louis Ventenat nearly losing his life to an apparent attack of malaria. The expedition left in October and sailed south to New Holland. Following bad weather d'Entrecasteaux made anchor in Esperance Bay on the south-west coast of New Holland. Again Labillardière made valuable collections. The stay was short and would have been shorter if Riche, one of the crew, had not got lost in the scrub.

While searching for Riche, Labillardière wrote of the experience:

> We traversed a track wholly covered with sand, where we found extensive spaces absolutely destitute of vegetables. But I saw with surprise, on those distant shores, the grass known by the name of *Spinifex squarrosus*; and thus had a new and admirable instance of the facility with which plants, which grow on the sea shore, diffuse themselves to prodigious distances.
>
> In those arid wastes, grows a fine plant which nearly resembles the iris, and which naturally classes itself with the genera *dilatris* and *argolasia*. It forms, however, a new and very distinct genus, principally by its irregular corolla.
>
> I have delineated it under the name *anigozanthos*. [Kangaroo Paw and Western Australia's floral emblem]
>
> Its flowers have no calix. The corolla has the form of a tube ...

d'Entrecasteaux set sail for Van Diemen's Land once again, leaving in December 1792. Had they followed the tides and winds they experienced, they would certainly have discovered Bass Strait. Labillardière speculated upon it.

This time they stayed at Recherche Bay for a month. In February 1793 Labillardière wrote:

> Several inscriptions, engraved on the trunks of trees, acquainted us, that Captain Bligh had anchored in this bay in the month of February, 1792; when he was on his voyage to the Society Islands for bread-fruit trees, which he was to carry to the English colonies in the West Indies lying between the tropics.
>
> Bligh had with him two botanists, who sowed, at a little distance from the shore, cresses, a few acorns, celery etc. We saw three young fig trees, two pomegranate trees, and a quince tree they had planted, which had thriven well ...
>
> Near the north shore of this bay we landed a young he-goat, and a she-goat big with kid, putting up our prayers that the savages might allow them to propagate their breed in this island.

In March they set sail again to search the islands of the western Pacific for signs of La Pérouse. In July d'Entrecasteaux abandoned the search and headed for Java but he was ill with scurvy and dysentery and died at sea. Others of the crew were ill too and six died before they landed at Surabaja in October. There the expeditioners learned that France and Holland were at war. The political situation led to the disintegration of the expedition. Eventually, in February 1794, seven officers and naturalists (including Labillardière) were placed under arrest and the natural history collections confiscated. The fate of each of the men and the collections is complex. Suffice to say that the gardener, Lahaie, succeeded in getting living and dried specimens back to Europe, with the herbarium collection going to the British Museum, London.

Labillardière eventually returned to France in 1796 and immediately sought to have the herbarium collection restored to his homeland. With the support of Joseph Banks this was achieved, and Labillardière set to work on publishing his record of the voyage in *Relation du voyage à la recherche de La Pérouse* (1799). This included some plant descriptions and twelve

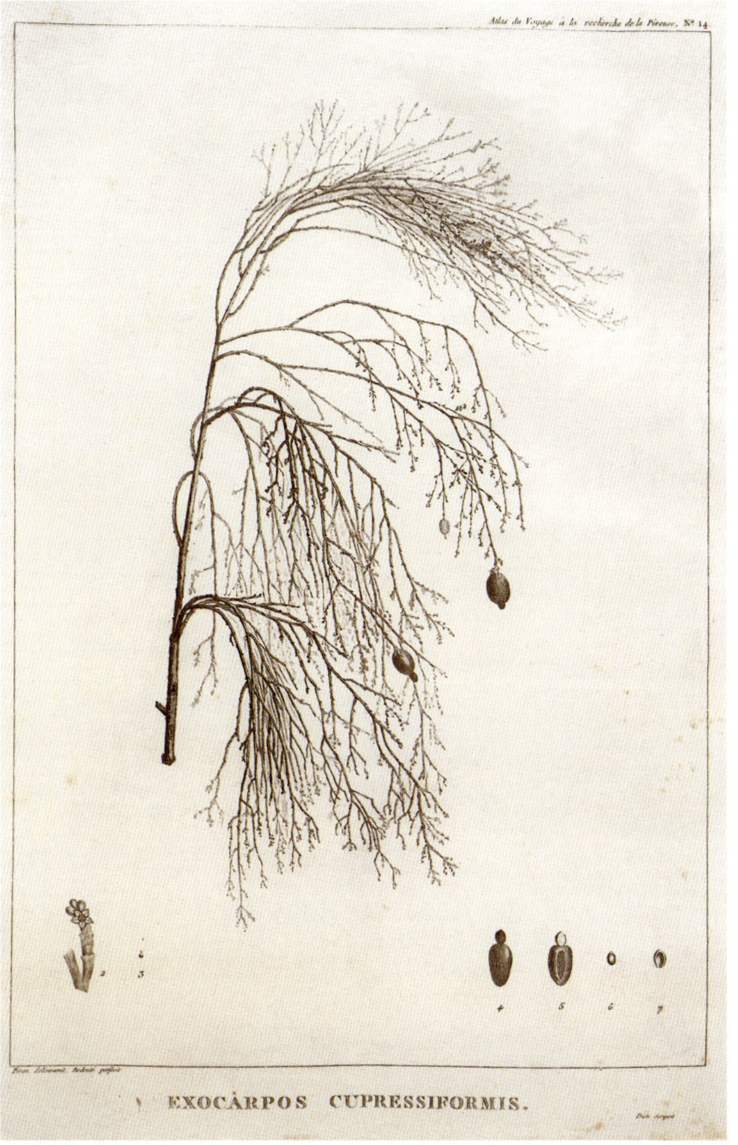

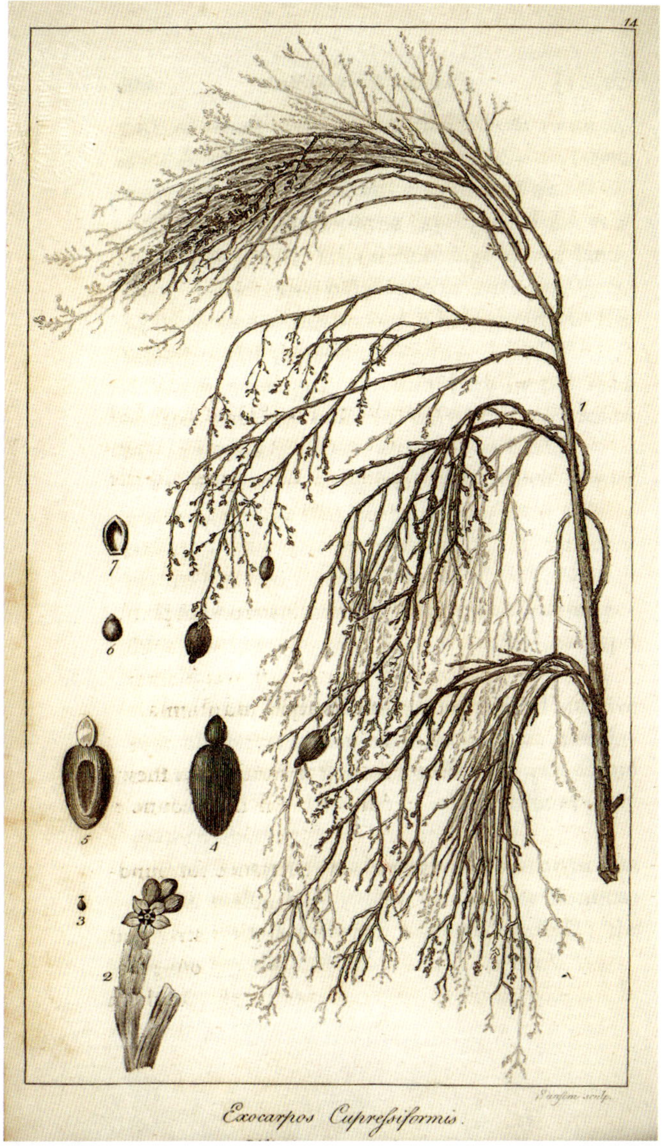

Exocarpos cupressiformis Labill.
Copper engraving of a drawing by Pierre-Joseph Redouté which formed part of the original description of the genus and species — the Wild Cherry. The engraving was first published by J.J.H. Labillardière in his *Relation du voyage à la recherche de La Pérouse*, 1799. The journal was translated into English and republished as *Voyage in search of La Pérouse* by John Stockdale in 1800. This book is much smaller than the original French edition. The original engraving was by J.B. Dien and the re-engraving by Francis Sansom resulted in a 'mirror image' print.
{NLA 44.5 x 31.0 NLA 22.5 x 13.6}

illustrations, the latter bound in an elephant-sized *Atlas*. The illustrations are uncoloured copper engravings of drawings by Nicolas Piron, Pierre-Joseph Redouté and Jean Baptiste Audebert. *The Relation* was translated into English and published in London in 1800 by John Stockdale. The botanical plates in this edition were re-engraved by Francis Sansom because the English version was much smaller and the plates were bound within the text. No credit was given to the original artists.

Of greater significance is Labillardière's *Novae Hollandiae Plantarum Specimen* (1804–1806), the second Australian Flora. It is the most exciting publication of illustration of Australian plants since the first publication of Dampier's collection, a hundred years earlier. Containing 265 plates, it is extensive and exquisite. The uncoloured copper engravings, mostly by Auguste Plée and his son Victoire Plée, J.B. Dien and Sohn Louis Sellier, are based on drawings by Nicolas Piron, Pierre-Antoine Poiteau, C.-G. Sauvage, Pierre-Jean-François Turpin, Pierre-Joseph Redouté and Labillardière himself. The combination of the engraver Plée with the artists Poiteau, Turpin and Redouté exemplifies great craftsmanship in botanical illustration. Nevertheless some of the work is flat indicating that the drawings were taken from herbarium specimens; others have great depth and surely were taken from living material.

The publication of the *Novae Hollandiae Plantarum Specimen* resulted in a total of 370 Australian species described by that time according to an estimate by Jonas Dryander (Banks' librarian). Approximately three-quarters of these plants were illustrated by published plates, with many species being illustrated more than once. The impressive total, though, is largely due to the efforts of Labillardière.

The associates of Labillardière and the artists contributing to the *Novae Hollandiae Plantarum Specimen* represent a 'who's who' of botany and botanical illustration of contemporary France. Frans Stafleu[7] described the years 1790 to 1840 as the 'golden half-century of French botanical illustration'. Certainly, the French artists associated with Australian plant material helped raise the lustre.

Jacques Julien Houtou de Labillardière (1755-1834) was born in Alençon of a wealthy family. He began studies in medicine and botany in Montpellier and completed them in Paris where one of his peers was René Louiche Desfontaines, later a prominent French botanist. In 1780 Labillardière travelled to England. There he met Joseph Banks and almost certainly became acquainted with James Smith. Back in Paris he met Louis-Guillaume Lemonier, another prominent botanist at the Jardin du Roi. Lemonier was influential in making academic contacts for Labillardière and for assisting the careers of the young botanists of the time, including Desfontaines and Antoine de Jussieu. Lemonier arranged for Labillardière to study the plants of Syria, the Lebanon and surrounds in the late 1780s. This led to Labillardière publishing his *Icones plantarum Syriae rariorum* (1791-1812).

Through his academic network, Labillardière also knew Charles-Louis L'Héritier. L'Héritier, in turn, introduced Labillardière to Pierre-Joseph Redouté. Redouté was one of the early artists commissioned to illustrate the *Icones*. Labillardière was then commissioned to join the search for La Pérouse. He was a Linnaean and a republican and moved in an elite and intimate circle of botanists and artists in Paris. By all reports he was an observer rather than a philosopher, being independent and perhaps aloof.

Nicolas Piron has been almost anonymous with his Christian name only known recently. It has also recently been learnt that he survived the privations of Batavia.[8] Unfortunately, given the significance and volume of illustration emanating from the d'Entrecasteaux voyage, the methods he used in the field are not known. Some engravings, based on Piron's drawing, are published in Labillardière's two books on the voyage. Whether this is work which survived the journey, or work that Piron did back in France, is not clear.

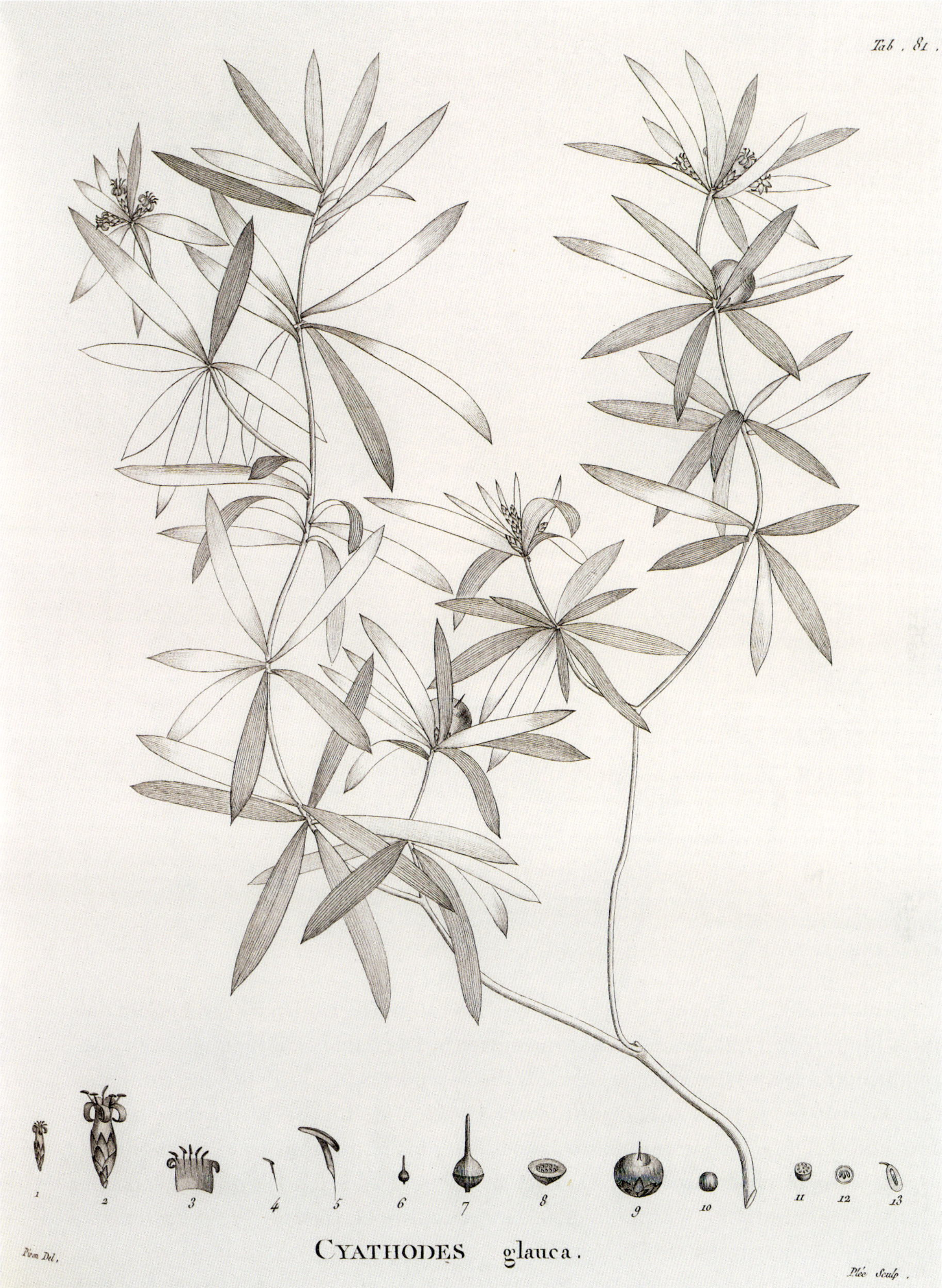

Cyathodes glauca Labill.
Engraving by Auguste Plée of a drawing by Nicolas Piron. The engraving was published in J.J.H. de Labillardière, *Novae Hollandiae Plantarum Specimen*, 1804–1806. This book represents the second Australian Flora and made an enormous contribution to the knowledge of the flora.
{NLA 28.3 x 21.7}

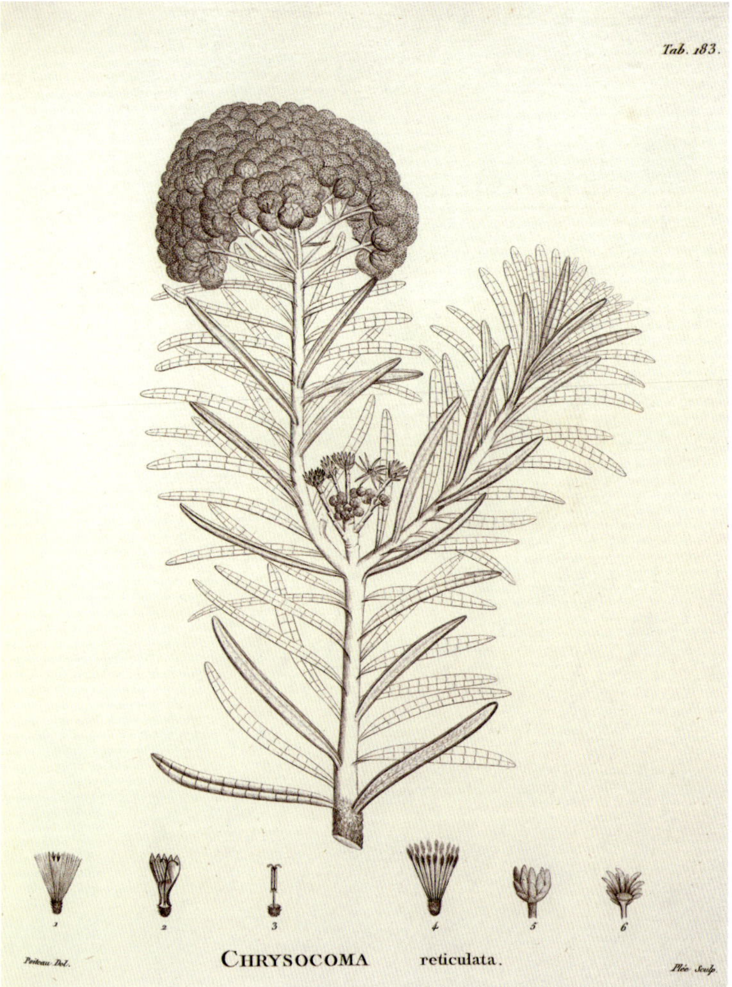

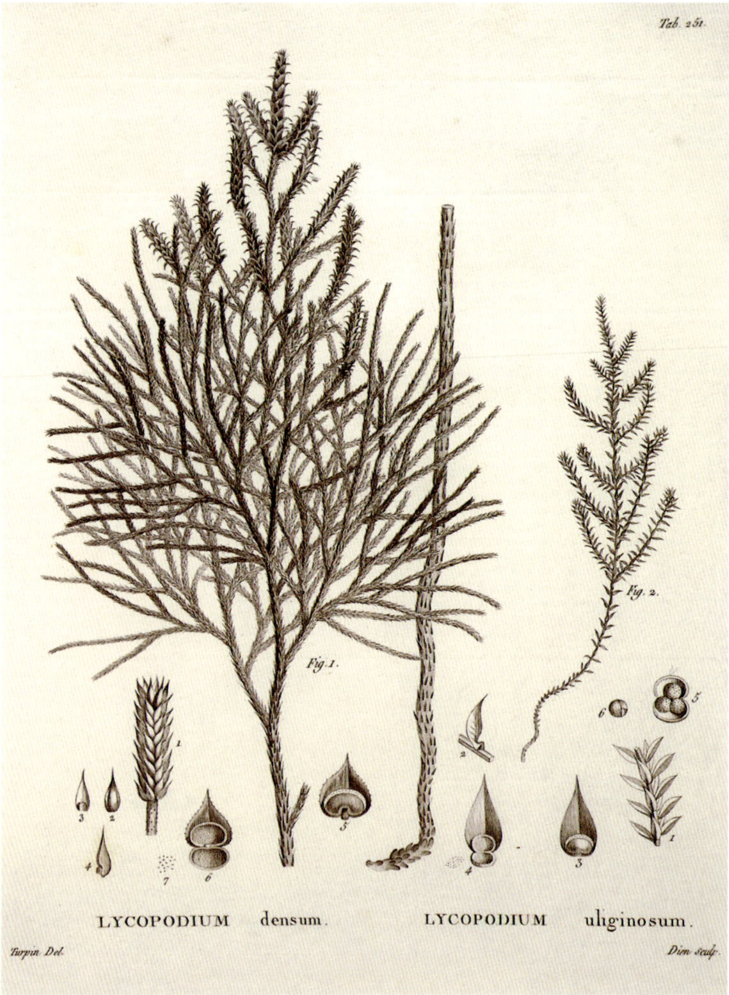

Helichrysum reticulatum (Labill.) Less. (= *Chrysocoma reticulata* Labill.) engraving by Auguste Plée of a drawing by Pierre-Antoine Poiteau (left). *Lycopodium deuterodensum* Herter (= *Lycopodium densum* Labill.) and *Selaginella uliginosa* Spring (= *Lycopodium uliginosum* Labill.) engraving by J.B. Dien of a drawing by Pierre Turpin (right). These engravings were published in Labillardière's *Novae Hollandiae Plantarum Specimen*, 1804–1806. [NLA 28.3 x 21.7] [NLA 28.3 x 21.7]

Pierre-Antoine Poiteau (1766–1854) began his career as a gardener at the Jardin des Plantes. Early in the 1790s he attended lectures given by Desfontaines. He is thought to have been influenced at the same time by Gerardus van Spaëndonck. In 1796 he was sent to the French West Indies to work as a botanist for the French Government. While there he met Turpin and they began a fortuitous association. After his return to France in 1801, he is thought to have become a pupil of Redouté and he certainly worked in the team of Redouté, Turpin and Plée illustrating for Labillardière. From an Australian perspective he became the major contributing artist for the *Specimen*.

Pierre-Joseph Redouté (1759–1840) was born into a family of artists in Saint Hubert, Belgium. He worked as an itinerant decorator and painter in his early years. He is reputed to have been influenced by the work of the Dutch flower painters Jan van Huysum and Rachel Ruysch. Redouté joined his older brother, Henri, in Paris in 1874. While there he discovered the Jardin du Roi where there were plants by the legion to paint and where several botanists

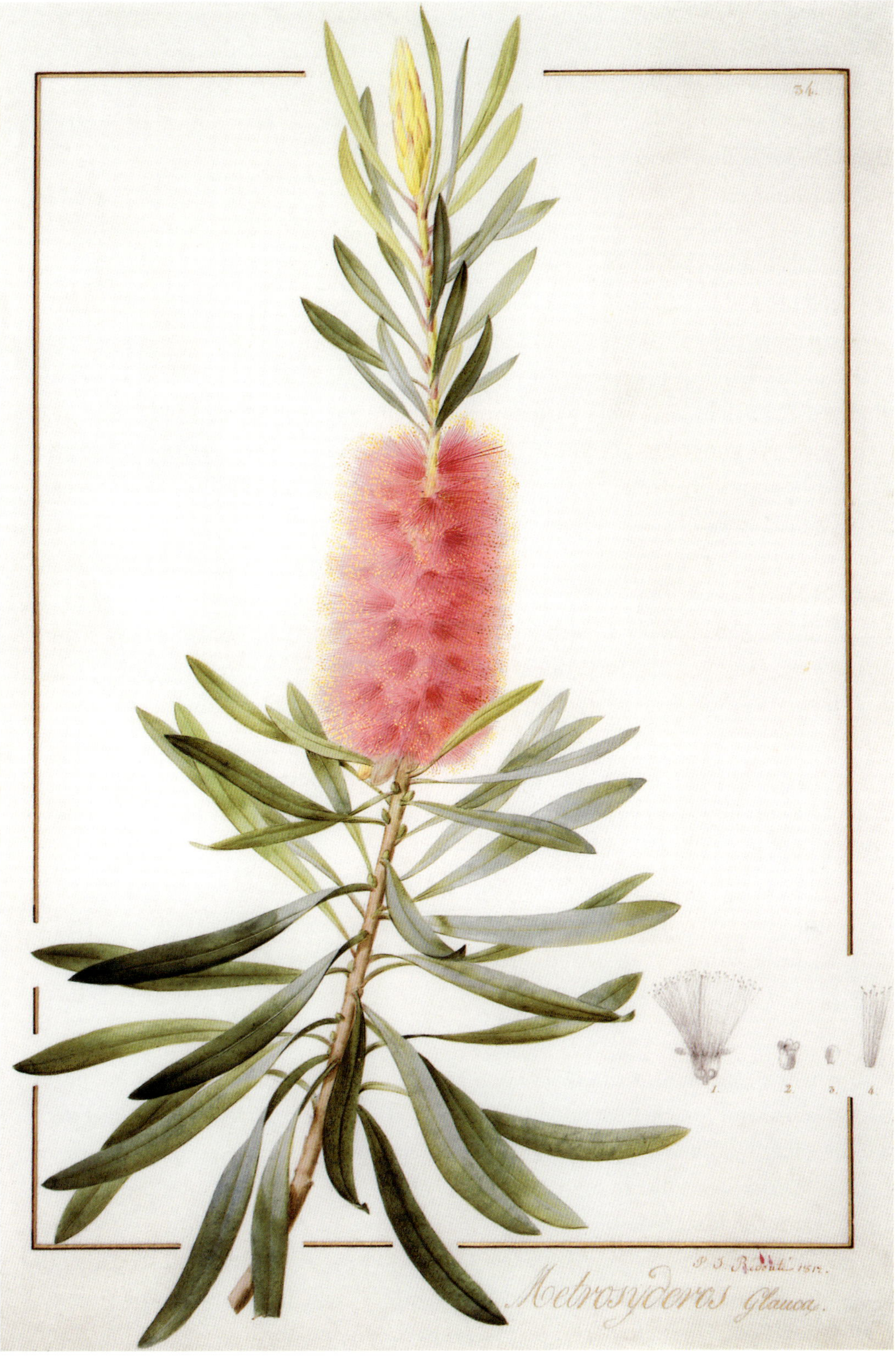

Callistemon glaucus (Bonpl.) Sweet (= *Metrosyderos glauca* Bonpl.)
Original watercolour painting by Pierre-Joseph Redouté with margins ruled in red and gold ink. The painting was later published in Aimé Bonpland's *Description des plantes rares cultivées à Malmaison et à Navarre* in 1813.
[FM 46.3 x 30.6]

and artists offered courses. One such was Gerardus van Spaëndonck. Fortuitously Redouté also met Charles-Louis L'Héritier. In 1783 L'Héritier recognised Redouté's ability and employed him to illustrate his forthcoming book, *Stirpes Novae aut minus cognitae* (1785–1805). In supervising him, L'Héritier taught Redouté the art of botanical illustration in contrast to the art of flower painting.

Pierre-Jean-François Turpin (1775–1840) was born in Vire. He had some education in drawing at school. However, he chose the army as a career, joined the French Revolutionary Army and was posted to Haiti. While there he met Poiteau and the two appear to have stimulated each other's capabilities. Certainly Poiteau inspired Turpin with an interest in botany and natural history. Turpin had a natural aptitude as an artist and probably encouraged Poiteau. Later in his career, Turpin worked as a draftsman for the army. As his interest changed, he took leave to work with Poiteau in Haiti and was subsequently employed as an artist with an American diplomat. This in turn led to his meeting Alexander Humboldt. By the time Turpin returned to France in 1802 he was already highly competent.

In Paris Turpin came under the influence of van Spaëndonck and Redouté and frequently collaborated with Poiteau. As a result he became an outstanding botanical illustrator — among the best. His finest work is considered to be in the *Nova genera et species plantarum* of A. Humboldt, A. Bonpland and C.S. Kunth (1815–1825). Another fine example is in *Sertum Austro-Caledonicum*[9] (1824–1825) by Labillardière for which Turpin expertly executed pencil drawings with pen to emphasise the darkest tones, placing great emphasis on the use of tone to convey form.

Jean Baptiste Audebert and C.-G. Sauvage are names leaving a legacy of a few illustrations. Little more is known of the engravers Auguste Plée, Victoire Plée, J.B. Dien and Sohn Louis Sellier.

Auguste Plée was a botanist and illustrator. He was employed by the Paris Muséum d'Histoire Naturelle and conducted fieldwork for the Muséum between 1820 and 1825. He produced a family of illustrators including Victoire, who was his oldest son, and François.

In 1793 a Spanish expedition in the *Descubierta* and the *Atrevida*, under the command of Alejandro Malaspina and José de Bustamente y Guerra, called at Port Jackson. On board were botanists Luis Née, who was born in France, and Thaddaeus Haenke who was born in Czechoslovakia. They made collections in the vicinity of several ports including Port Jackson. On return to Spain the material was passed to Antonio José Cavanilles and was described mostly in his *Icones et Descriptiones Plantarum* (1791–1801). All of the illustrations of Australian plants were metal engravings based on Cavanilles' own drawings. The plates were engraved by Vinz. Lopez Enguidanos and José de Fonseca.

Antonio José Cavanilles, born in Valencia, was a botanist and clergyman, both in Paris and Madrid. In 1801 he became Director of the Madrid Botanical Garden.

In 1800 the political situation in Europe motivated the French to send Nicolas Baudin with an expedition to New Holland and the Pacific to establish a French presence in those parts of the world. Baudin sailed in the *Géographe* accompanied by the *Naturaliste* under the command of Emmanuel Hamelin. The crews included Jean Baptiste Leschenault de la Tour,[10] André Michaux and Jacques Delisse as botanists and Charles Alexandre Lesueur, Nicholas Martin Petit, Jacques Milbert, Louis Lebrun and Michel Garnier as painters and draftsmen. The scientific team also had five zoologists, including François Péron and Louis de Freycinet,[11] five gardeners, two mineralogists and two astronomers. It appears to have been more of a natural history exploration than a political mission. Whatever its purpose, it was a tragic voyage, dogged by bad judgement and ill fate.

Baudin sailed in October 1800. When the ships reached Ile de France in March 1801, forty men abandoned the expedition, some being sick and others not prepared to endure Baudin's authoritarian leadership. Those deserting included one astronomer, two botanists, two zoologists and two artists — approximately a third of the scientific team. The plan to go

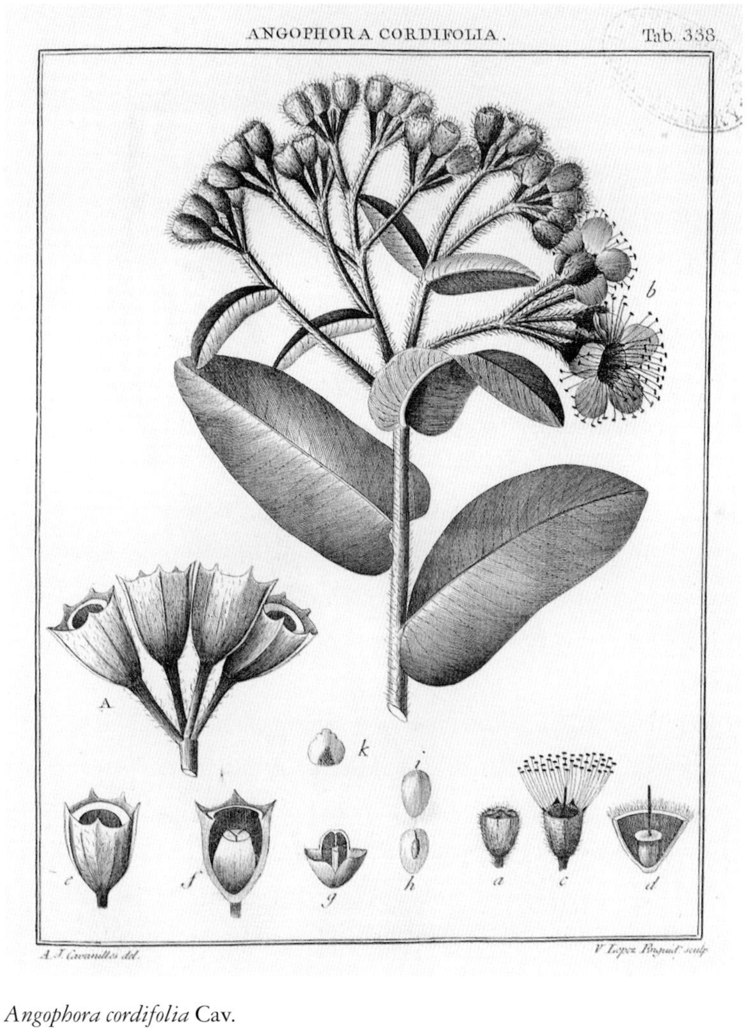

Angophora cordifolia Cav.
Metal engraving by Vinz. Lopez Enguidanos of a drawing by Antonio Cavanilles and published in his book, *Icones et Descriptiones Plantarum*, 1791–1801. This engraving is part of the original, or type, description of the genus and species.
[SLV 20.2 x 15.1]

to Van Diemen's Land was dropped in favour of exploring the western and north-western coast of New Holland. This, of course, proved to be largely inhospitable and barren.

In August 1801 the ships put in at Timor. There Baudin took on fresh supplies and the men recuperated from scurvy which had emerged during the long stint at sea. Many, however, succumbed to dysentery and several died, including some of the scientists.

Finally, Baudin did what he should have done in the first place. He sailed for Van Diemen's Land in November 1801, arriving two months later. On the way more men died of scurvy, thirst and hunger. In spite of the illness of many of the crew, Baudin explored up and down the east coast of Van Diemen's Land before sailing north into Bass Strait.

In his journal Baudin made very little botanical comment, however on Partridge Island, south-east Van Diemen's Land, on 14 January 1802 he observed:[12]

> Upon landing, we decided to cross over the island, but were soon forced back to the shore by finding the way extremely difficult and obstructed by grass and scrub. The trees on the island are tall and well-grown, most of them belonging to the *Eucalyptus* family, but we saw only a few plants in flower.

The two ships became separated early in March, largely due to the loss at sea of the topographer, Charles Pierre Boullanger, and six crewmen off Cape Tourville (to the north of Freycinet Peninsula). The *Naturaliste* stayed behind to search for them. Remarkably, Boullanger and his men had been picked up at sea by a British brig. Baudin continued on across Bass Strait to the mainland and named it Terre Napoléon.

Baudin decided to explore to the west. By yet another extraordinary coincidence, Baudin met Matthew Flinders in the *Investigator* at Kangaroo Island on 8 April 1802. By all reports it seems to have been a fairly cool encounter.

Baudin pressed on into the Great Australian Bight before returning to Van Diemen's Land in May. Still ill from scurvy and suffering from insufficient food and water, the men were basically too ill to man the ship properly. Baudin relented and sailed for Port Jackson. They arrived in June and the crew were in such poor health that they required assistance to land. Here they recuperated.

Baudin learned that the *Naturaliste* had been there earlier and had set off in search of him, advising that it would return to France if it failed. Actually the *Naturaliste* returned to Port Jackson in August. The crews were slow to recover despite encouraging support from the colony. The scientists and artists made the most of the opportunity and collected and recorded many specimens. In October 1802, Baudin decided to have the *Naturaliste* return to France as something of a hospital ship. She carried the sick members of the crews together with a large volume of live and preserved collections and arrived in France in June 1803.

Baudin left Port Jackson in November and the horror story went on. Of the total of twenty-three scientists who had joined this expedition, only three completed the entire journey and lived to tell of their experiences. Baudin himself succumbed on Ile de France in September 1803. The *Géographe* returned to France in March 1804. The extraordinary thing is that the expedition was extremely rewarding in the volume and quality of the collections

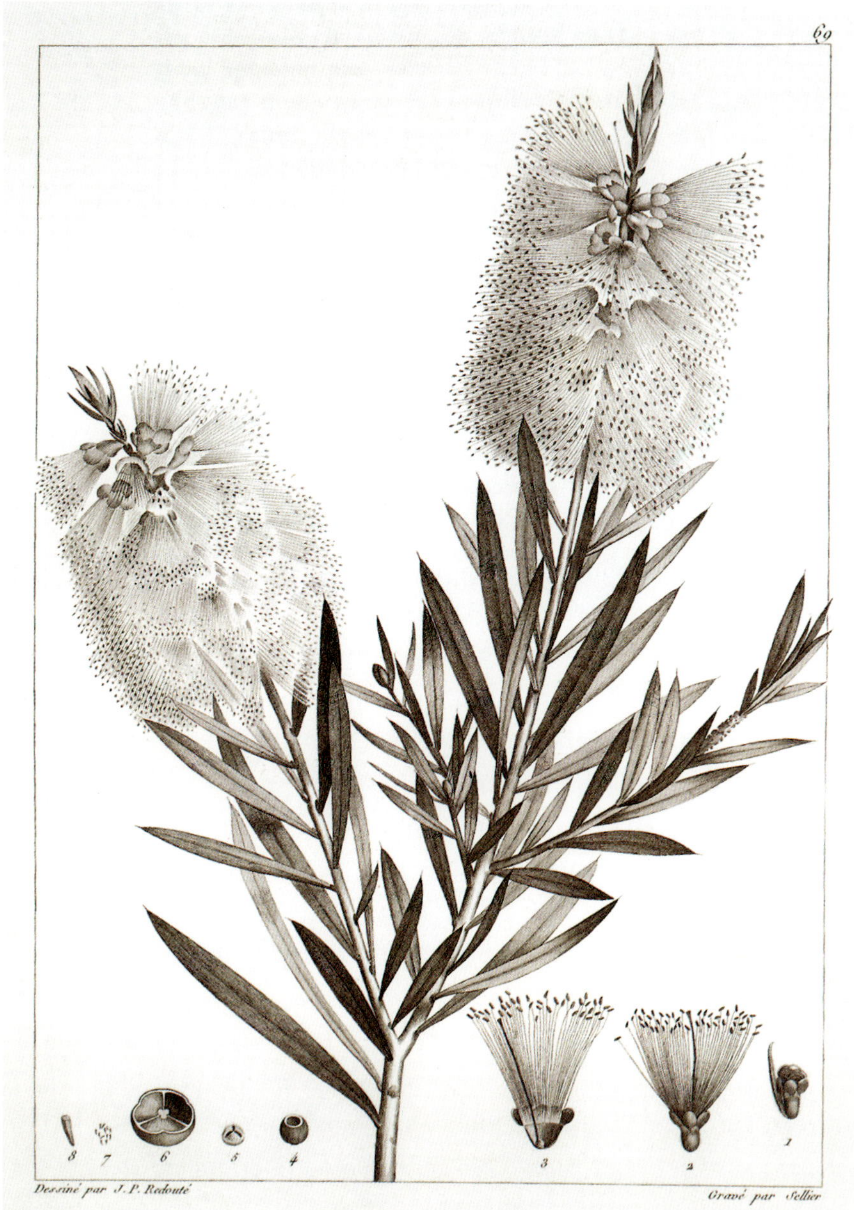

Melaleuca citrina Dum.-Cours. (= *Metrosideros lophantha* Vent.)
Copper engraving of a drawing by Pierre-Joseph Redouté, engraved by Sohn Sellier and published in É.P. Ventenat, *Description des plantes nouvelles et peu connues cultivées dans le jardin de J.M.Cels*, 1800–1803.
[CANB 28.0 x 19.5]

(especially zoological) which returned to France for science. The energy and determination of the zoologist, François Péron, together with his friend, the artist Charles Lesueur, ensured this. Even though no botanist completed the voyage, the botanical collections, especially the live plants, were significant too. The gardener, Antoine Guichenot, received some acclaim for caring for live plants on the *Géographe*'s return voyage. As a result, the botanic gardens, Jardin des Plantes, Malmaison and Chateau de Navarre, all became well stocked with Australian plants.

The herbarium specimens were passed to Labillardière which resulted in them making only a minor contribution to science. Labillardière, however, is believed to have used some of the material in the compilation of his *Novae Hollandiae Plantarum Specimen*.

Of greater import is the fact that material from the French expeditions succeeded in surviving in French gardens. Étienne Pierre Ventenat, a botanist in France at the time, began publishing grand, illustrated works portraying the exotic plants included in the great gardens.[13]

In 1800–1803 Ventenat published his *Description des plantes nouvelles et peu connues cultivées dans le jardin de J.M.Cels*. There are ten plates of Australian plants. Seven are drawn by Pierre-Joseph Redouté, two by Henri Joseph Redouté and one by Nicholas Maréchal. Seven are engraved by Sohn Louis Sellier[14] and three by Auguste Plée. Jacques Martin Cels was an horticulturalist-botanist, friend of L'Héritier, and adviser to the Empress Joséphine on the acquisition of plants for her gardens.

Empress Joséphine,[15] wife of Napoléon Bonaparte, took an interest in gardens and in rare and exotic plants. She employed a team to establish and record the plants in her gardens at Malmaison and Navarre in particular. The team included Étienne Ventenat and Pierre-Joseph Redouté. Étienne Ventenat was a man of many qualities, being a clergyman, amateur botanist and librarian in the Bibliothèque du Panthéon, Paris. Ventenat was a follower of Antoine de Jussieu and the natural system of classification.

The first product of their labours was the *Jardin de la Malmaison* (1804–1805). The plates are coloured stipple engravings and include a number of Australian plants. Each plate was stipple-engraved, then hand-coloured and, after the plate was printed, was finished by hand-touch. The entire process was executed by Redouté himself.

Josephine's book *Choix de plantes* (1803–1808) included four Australian plant plates. These are uncoloured copper engravings. Pancrace Bessa drew two plates under the supervision of Redouté and Redouté drew the other two. All were engraved by Sohn Louis Sellier.

When Ventenat died in 1808, Joséphine employed Aimé Jacques Alexandre Bonpland[16] as the botanist in her team, to provide the descriptions for the third book portraying the plants in her gardens. Bonpland was a French explorer and botanist. Most of his field work was done in central and southern America.

Bonpland published *Description des plantes rares cultivées à Malmaison et à Navarre* (1812–1817). There are sixty-four plates which are stipple engravings, colour-printed and finished by hand. Of these, Pierre-Joseph Redouté drew fifty-four and P. Bessa drew nine; one is not signed. Twenty-one of the plates are of Australian plants. These are engraved by six, or more, engravers.[17]

The political activities in Europe often disrupted the obvious ebb and flow of scientific process. L'Héritier was caught up in one such glitch over material collected by Joseph

Crowea saligna Andrews
Coloured stipple engraving by J.B. Dien of a painting by Pierre-Joseph Redouté published in É.P. Ventenat's *Jardin de la Malmaison*, 1804–1805. This book contains illustrations of many Australian plants growing in Empress Joséphine's garden.
[SLV 54.0 x 35.0]

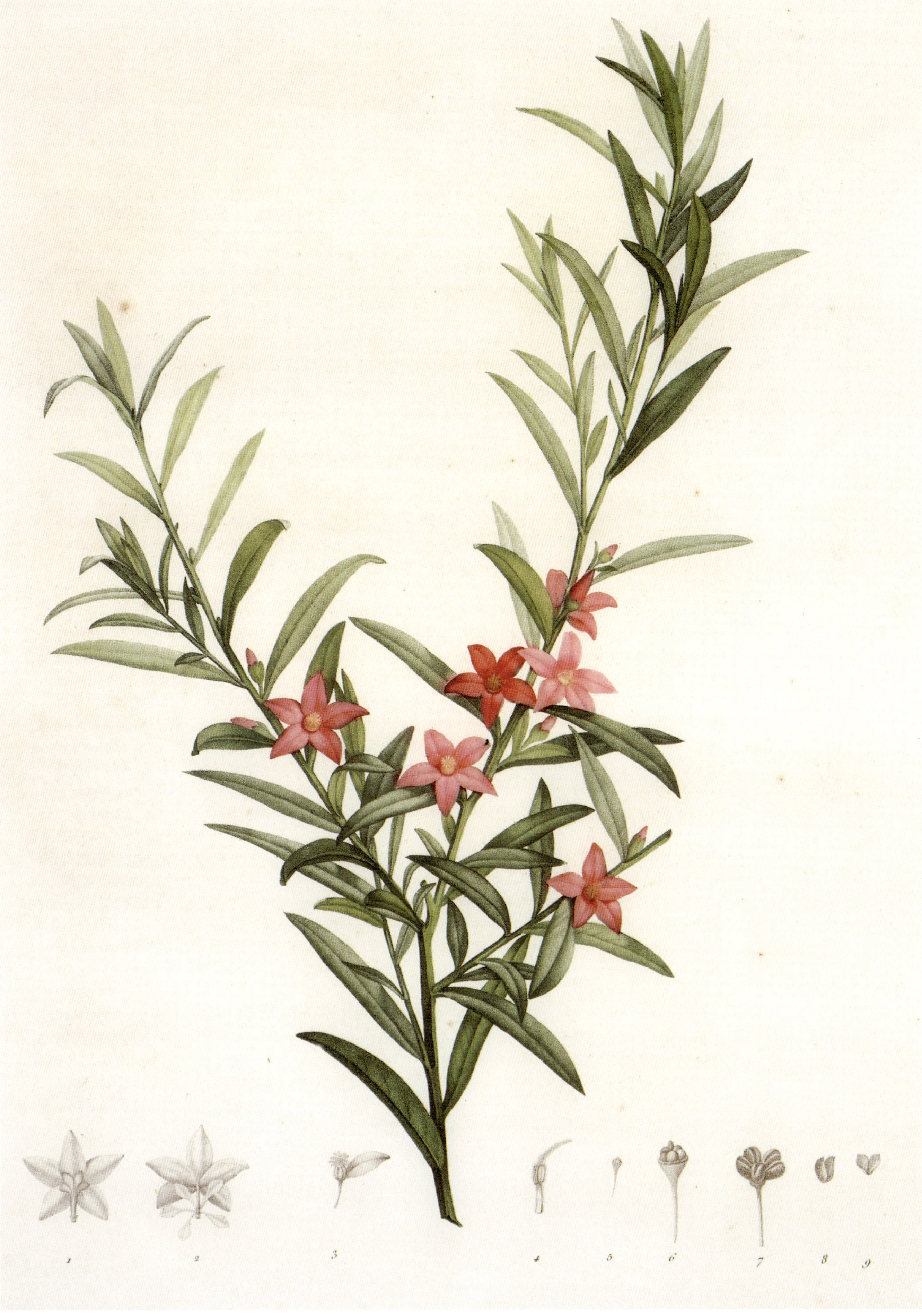
1
2
3
4
5
6
7
8
9

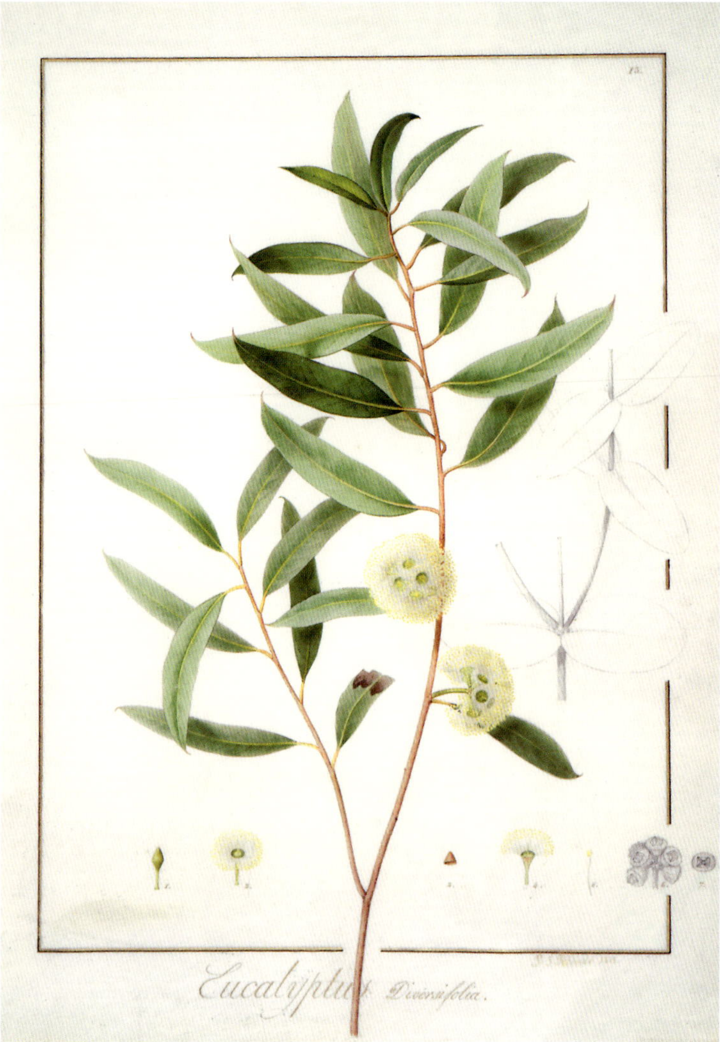

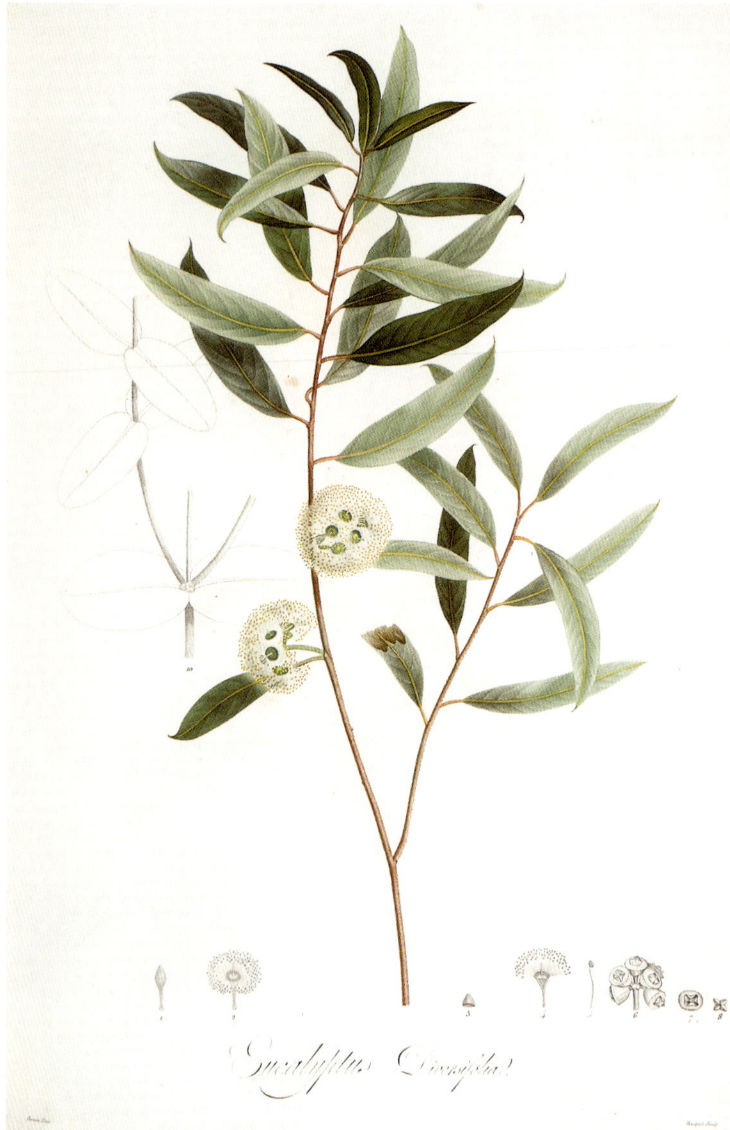

Eucalyptus diversifolia Bonpl.
Original watercolour painting (left) signed by Pierre-Joseph Redouté in 1811. This was published in Aimé Bonpland's *Description des plantes rares cultivées à Malmaison et à Navarre* in 1813 (right). Here, however, the painting is attributed to Bessa and engraved by Bouquet.
{FW 47.6 x 34.9} {SLV 51.0 x 33.0}

Dombey in South American Spanish countries. To avoid the French Government's impounding specimens to which the Spanish Government had some rightful claim, L'Héritier escaped to England carrying the Dombey collection with him.

Joseph Banks offered hospitality and in next to no time L'Héritier was engaged in preparing his *Sertum Anglicum* (1789–1792). To carry out the illustrations for this work, L'Héritier invited Pierre-Joseph Redouté to join him in England. While there Redouté worked with both James Sowerby and Francesco Bartolozzi. Bartolozzi was not a botanical illustrator but he was a master of stipple engraving and colour printing using stipple engraving. From Bartolozzi Redouté first learned the technique which he was later to adopt and execute with masterful skill.

The association between L'Héritier and Redouté grew and lead to the opening of other doors; doors leading to a network of botanists and benefactors.

The influence of van Spaëndonck on Redouté's technical expertise cannot be emphasised enough. Van Spaëndonck was Professeur de peinture de fleurs at Jardin du Roi. He taught the technique of painting on vellum using gouache and, later, watercolour. He also adopted and developed the technique of stipple engraving for botanical illustration. Through these influences and his own inherent ability, Redouté became a master.

Redouté belonged to the circle of academics while remaining aloof from politics. He had entry to both sides of the political scene. He was appointed as draftsman and painter to Queen Marie Antoinette and after she was executed Redouté fitted just as comfortably into the patronage of Empress Joséphine. Indeed the books commissioned by Joséphine really established him as one of the greats.

Redouté clearly established a school of engravers around him who were skilled in the art of stipple engraving and of colouring the plates for colour printing. While he is associated with some very significant scientific works, he is better known for his grand works, *Les Liliacées* (1802–1816) and *Les Roses* (1817–1824). Half a dozen Australian species are illustrated in *Les Liliacées*. These were engraved by de Gouy, R. Bessin, Vincent Marie Langlois and Jean Baptiste Chapuy.

Pancrace Bessa, also a student of van Spaëndonck, became a master of stipple engraving. He moved in elite circles, was a teacher of painting to the rich and is perhaps better known for his flower paintings than illustration of scientific works.

Henri-Joseph Redouté, younger brother of Pierre-Joseph, was appointed to the staff of the Museum National d'Histoire Naturelle to carry out zoological painting under the supervision of van Spaëndonck. His contribution to botanical illustration was minor. Instead, he became a great animal painter.

Nicholas Maréchal was a painter and engraver.

The French engravers of the works that included Australian plants (Sohn Louis Sellier, Auguste Plée, François Plée, Antoine Maxime Monsaldi, R. Bessin, Louis Bouquet, Jean Louis Denis Coutant, de Gouy, Vincent Langlois and Jean Baptiste Chapuy) all are less renowned, but most were quite prolific.

Plant material from French and British expeditions was making its way into private and public botanic gardens all over Europe. Johann Christoph Wendland was a horticulturalist and worked as a gardener at the Herrenhausen garden in Hannover for several years. He took a particular interest in rare and exotic plants, especially the representatives of the family Ericaceae. Wendland was also a competent artist. He published several books in which his paintings were engraved or etched. Australian plants appear in his *Collectio plantarum* (1805–1819). In this work the plates are etchings, printed in blue-green ink and hand-coloured.[18] Wendland was just one of many horticulturalists embracing the rare and exotic and working towards a permanent portrayal of them.

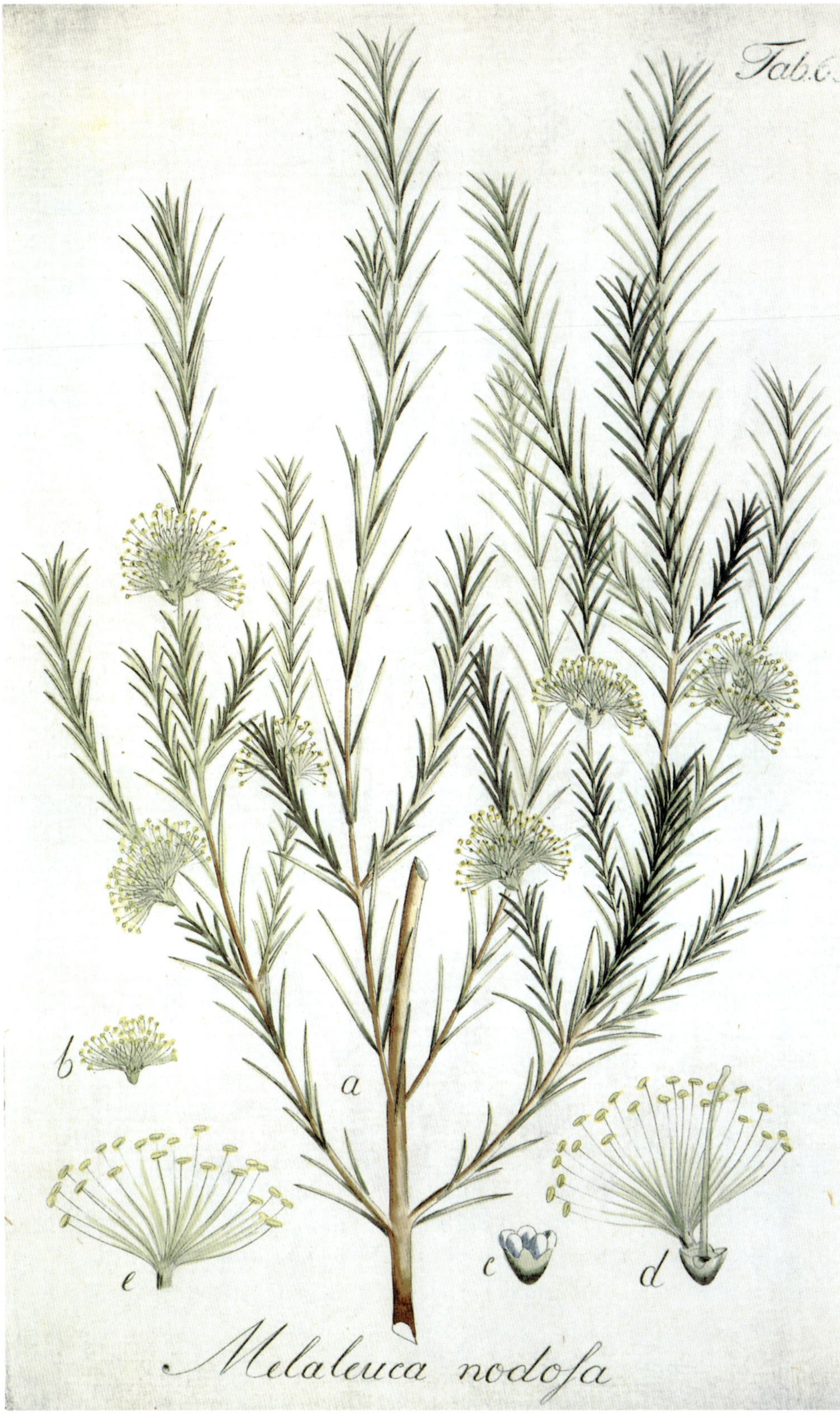

Melaleuca nodosa (Gaertner) Smith
Hand-coloured etching by Johann Wendland, published in his book, *Collectio plantarum*, 1805–1819. Engraving by acid etching seems to be rare, or unidentified, in botanical illustration.
[KEW 24.8 x 19.0]

5. The Banksian Collectors

The Banksian collections and the diverse array of flora brought back to England and Europe ... ushered in a new era in botanical science. Not only were there vast new materials to be delineated and compared, but the old taxonomic system of classification was being reshaped.

Ann Moyal[1]

Joseph Banks and Daniel Solander were the original Banksian collectors to take material from Australia and other exotic places back to England for its gardens and halls of science. In England, Joseph Banks settled into a position of authority on biological issues relating to countries like Australia and New Zealand, and established the focus for taxonomic research through his personal collection and library in Soho Square. The end of the eighteenth century saw the beginning of an explosion of scientific-periodical publications — both for scientific societies and for scientific entrepreneurs — as outlets for the research. Banks' librarian, Jonas Dryander, was kept busy overseeing much of this activity.

New scientific societies were being formed and they began to publish the transactions or proceedings of their meetings. Prior to this, the Royal Society of London published its *Philosophical Transactions of The Royal Society of London*, beginning in 1665. Joseph Banks was active in the Royal Society, being President from 1778 until he died in 1820.

The foundation of the Linnean Society of London in 1788, by James Edward Smith, resulted in the beginning of the *Transactions of the Linnean Society of London* in 1791. Journals produced by these and similar societies (some being splinters from the original parent society, the Royal Society) are still published today, though most have had at least one name

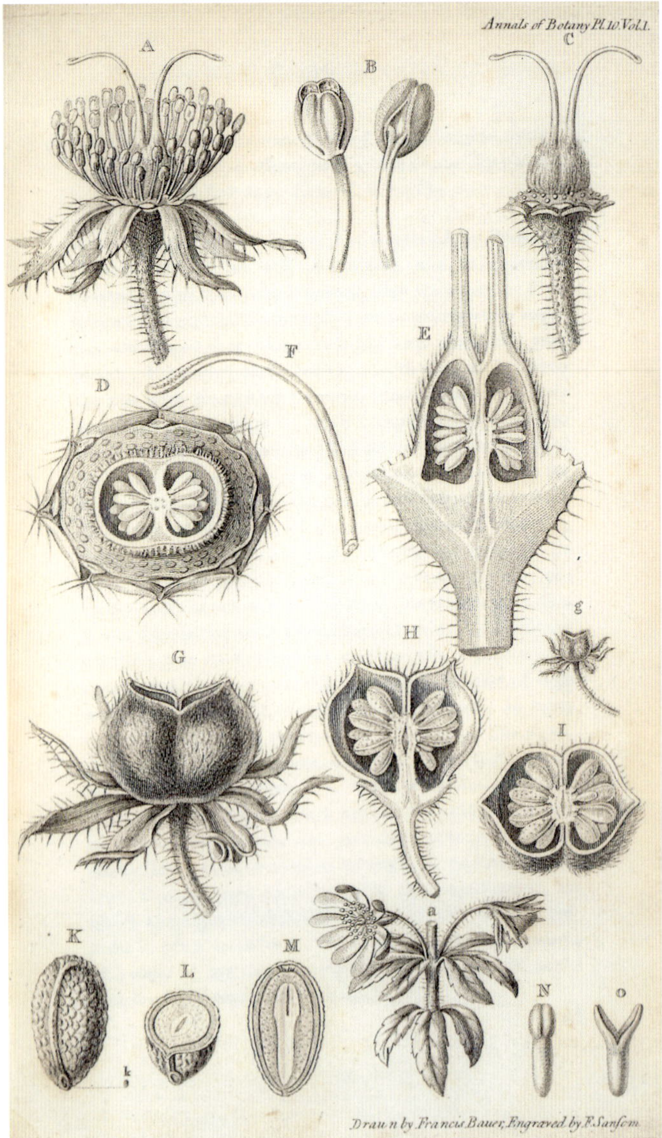

Bauera rubioides Andrews
Copper engraving of a drawing by Franz Bauer, engraved by Francis Sansom and published in C. Koenig & J. Sims, *Annals of Botany*, I, 1805–1806.
This plate demonstrates a detailed analysis for a classical botanical illustration of a species and its features.
[MEL 20.0 x 12.0]

change, usually coupled with a change of emphasis in content.

The *Transactions of the Linnean Society of London* had a strong, though not exclusive, botanical content. This periodical included much early description of Australian plants. Several artists, such as Franz and Ferdinand Bauer, contributed illustrations to the descriptions.

Franz Andreas Bauer[2] (1758–1840) was an older brother of Ferdinand. The influence of a prior at the local monastery at Feldsberg, Father Norbert Boccius, was substantial as Boccius recognised the brothers' artistic abilities, trained them, and had them paint a number of works for him. They moved to Vienna in about 1780. Both came under the influence of Nicholas Jacquin and received further training in botany and observation of detail in order to achieve excellence in botanical draftsmanship. While with Jacquin they both prepared illustrations for Jacquin's *Icones Plantarum Rariorum* (1781–1793). Ultimately they both ended up working for English scientists.

Ferdinand proved to be venturesome and indulged in voyages of discovery. Franz was more conservative, undertaking a European tour in 1788 together with Jacquin's son, Joseph Jacquin. Franz and Joseph travelled to England in 1789 and were discovered by Joseph Banks. Franz was persuaded to take a position as botanical artist at the Royal botanic gardens, Kew. He worked there until he died in 1840. Both brothers were outstanding illustrators and became esteemed for the detail embraced in their work, their styles being virtually inseparable. In due course, Ferdinand developed an analytical interpretation of the subject at a macroscopic level; Franz developed the skill of portraying the microscopic detail of the subject.

Having secure employment Francis, as Franz became known in England, was able to indulge his passion for seeking and portraying anatomical detail. The inscription in his memorial in St Anne's Church, Kew Green, states, 'in microscopical drawing he was altogether unrivalled and science will be ever indebted for his elaborate illustrations of animal

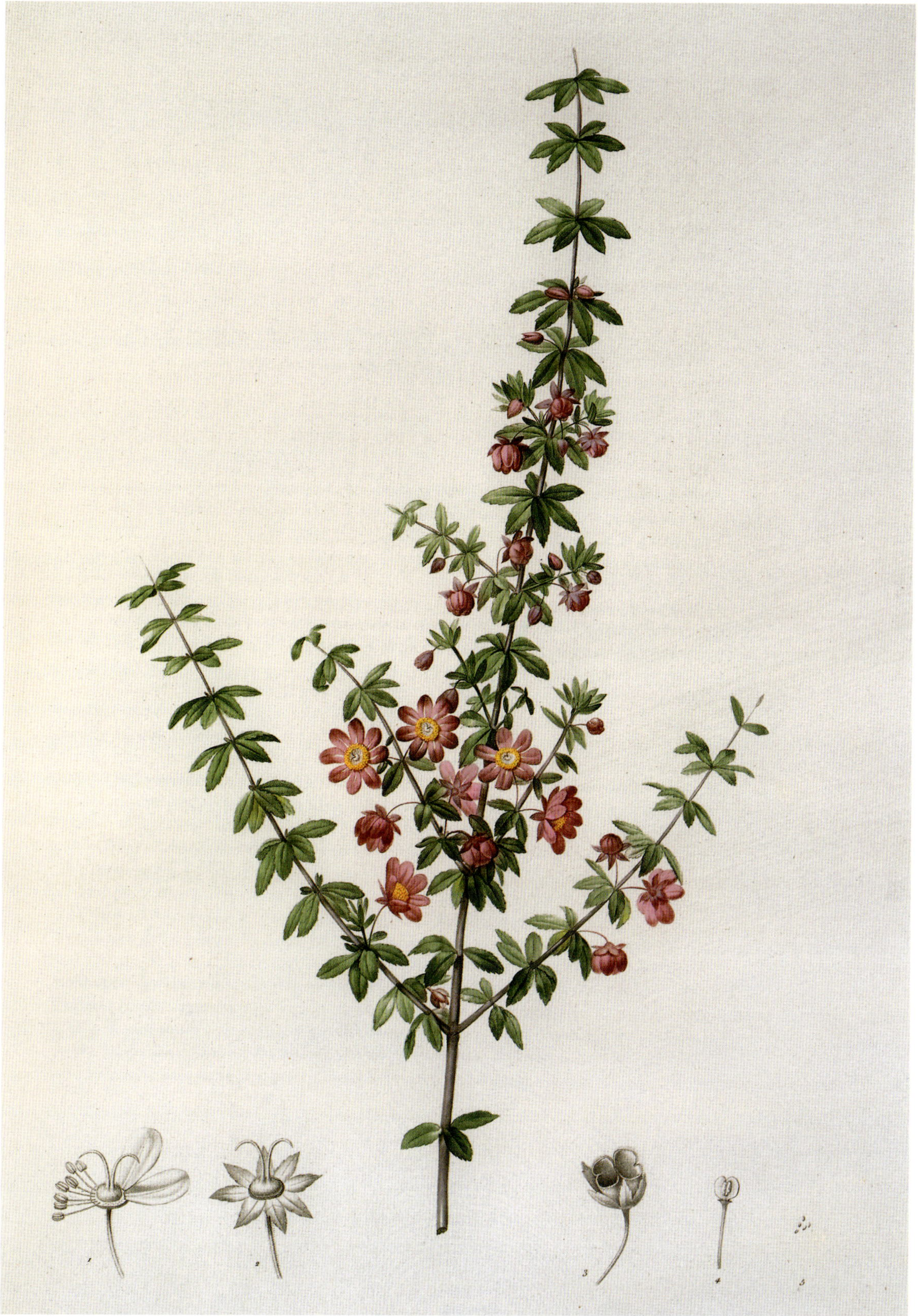

Bauera rubioides Andrews
Coloured stipple engraving by Pierre-Joseph Redouté from É.P. Ventenat, *Jardin de la Malmaison*, 1804–1805. The plant was named in honour of the Bauer brothers, Ferdinand and Franz.
[SLV 54.0 x 35.0]

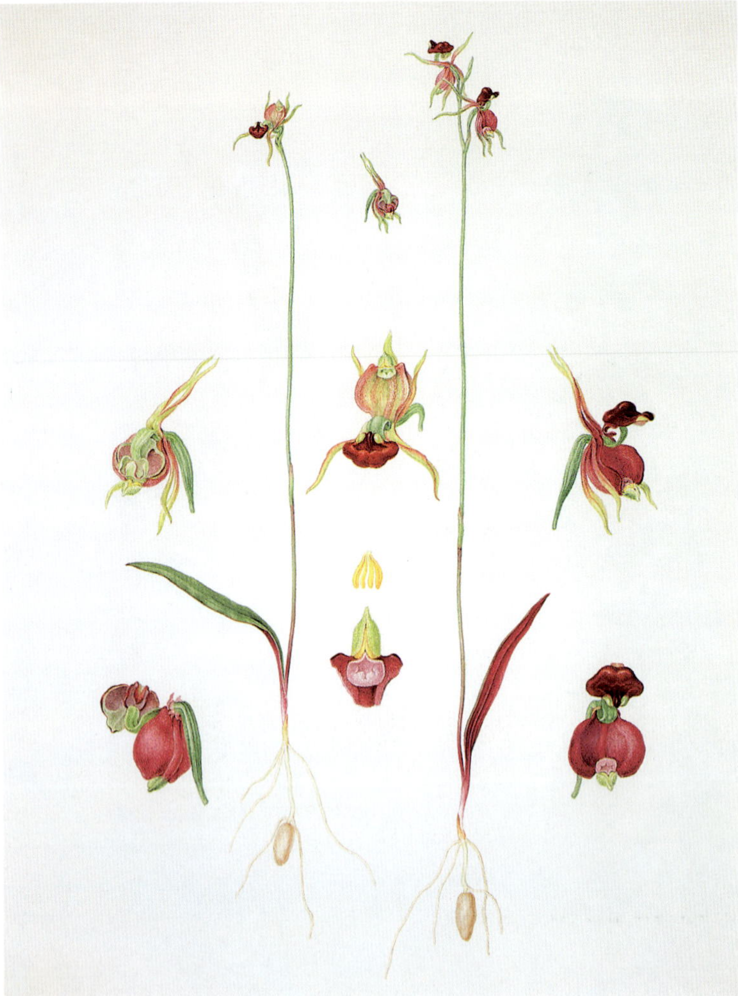

Caleana major R.Br.
Original painting attributed to Franz Bauer. The similarity of some of the original paintings of the Bauer brothers is disarming. It is unfortunate that both seemed casual about signing their work.
[BM]

and vegetable structures'. From an Australian point of view, Ferdinand is the more significant because he went to Australia, however over the fifty years he worked at Kew, Franz illustrated numerous of the Australian plants incorporated into the Kew collection. From a botanical illustration view, it is difficult to rate one above the other, both being outstanding. From the view of botanical art, Ferdinand perhaps had the edge, his subjects and composition frequently being the more appealing.

Entrepreneurs, too, indulged in regular publication of grand, highly illustrated series. William Curtis, the first such entrepreneur, initiated the *Botanical Magazine* in 1787. He owned and managed the publication of this magazine until he died in 1799. Sydenham Teast Edwards, William Kilburn and James Sowerby were the major early illustrators for this periodical and Francis Sansom was the engraver.

Curtis's Botanical Magazine, as it became known, was owned from 1801 till 1845 by Samuel Curtis, a cousin of William. It was then transferred to the publisher, Lovell Reeve, until 1920. It was then taken over by the Royal Horticultural Society[3] through timely support from Henry John Elwes.

In 1971 the magazine was transferred to the Bentham Moxon Trust of the Royal Botanic Gardens which continued to publish it, initially in conjunction with the Stanley Smith Horticultural Trust, until 1984. In that year Kew expanded the magazine to embrace a wider scientific coverage and retitled it, *The Kew Magazine incorporating Curtis's Botanical Magazine*. In 1995, it was titled *Curtis's Botanical Magazine incorporating The Kew Magazine. Curtis's Botanical Magazine* remains the longest running periodical of this type.

William Curtis (1746–1799) was trained as an apothecary with a greater interest and enthusiasm for botany, entomology and gardening. This led to his becoming a teacher for apothecaries at the Chelsea Physic Garden. His particular interest in British plants led to his famed publication, the *Flora Londonensis* (1775–1798). The *Botanical Magazine* was initially a

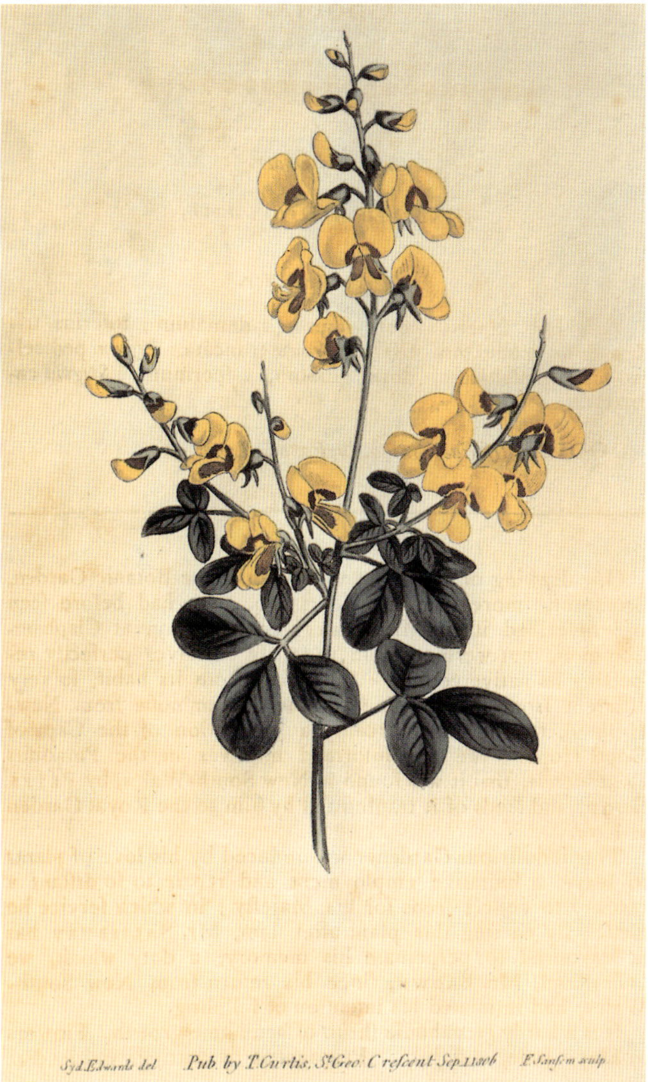

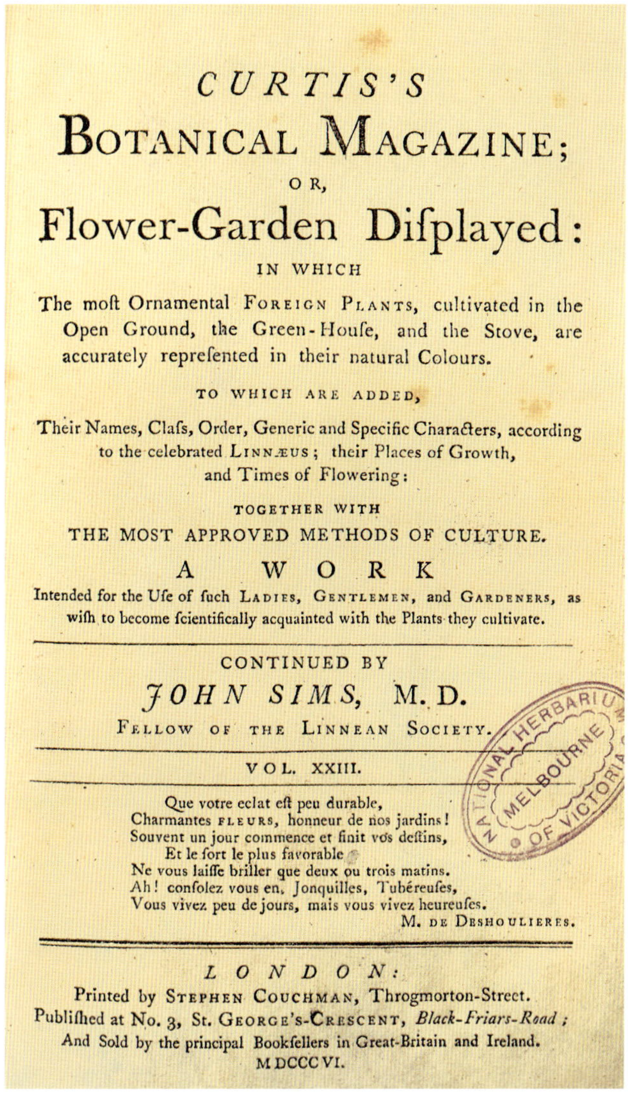

CURTIS'S

BOTANICAL MAGAZINE;

OR,

Flower-Garden Difplayed:

IN WHICH

The moft Ornamental FOREIGN PLANTS, cultivated in the Open Ground, the Green-Houfe, and the Stove, are accurately reprefented in their natural Colours.

TO WHICH ARE ADDED,

Their Names, Clafs, Order, Generic and Specific Characters, according to the celebrated LINNÆUS; their Places of Growth, and Times of Flowering:

TOGETHER WITH

THE MOST APPROVED METHODS OF CULTURE.

A WORK

Intended for the Ufe of fuch LADIES, GENTLEMEN, and GARDENERS, as wifh to become fcientifically acquainted with the Plants they cultivate.

CONTINUED BY

JOHN SIMS, M.D.

FELLOW OF THE LINNEAN SOCIETY.

VOL. XXIII.

Que votre eclat eft peu durable,
Charmantes FLEURS, honneur de nos jardins!
Souvent un jour commence et finit vos deftins,
Et le fort le plus favorable
Ne vous laiffe briller que deux ou trois matins.
Ah! confolez vous en, Jonquilles, Tubéreufes,
Vous vivez peu de jours, mais vous vivez heureufes.
M. DE DESHOULIERES.

LONDON:

Printed by STEPHEN COUCHMAN, Throgmorton-Street.
Publifhed at No. 3, St. GEORGE'S-CRESCENT, *Black-Friars-Road;*
And Sold by the principal Bookfellers in Great-Britain and Ireland.
MDCCCVI.

Goodia lotifolia Salisb.
Hand-coloured copper engraving of a drawing by Sydenham Edwards, engraved by Francis Sansom, and published in *Curtis's Botanical Magazine* in 1806. The genus was named in honour of Peter Good who made the original collection while the gardener on Matthew Flinders' voyage of circumnavigation of New Holland. William Curtis started his *Botanical Magazine* in 1787. It is still being published.
[MEL 20.0 x 12.0] [MEL 23.0 x 14.0]

venture to make money by selling engravings of colourful exotics, because the interest in *Flora Londonensis* was not great. Curtis needed money and his action reflected the growing appeal of exotic plants.

In 1790 James Edward Smith commenced a periodical entitled *English Botany*. It ran for twenty-four years in thirty-six volumes until 1814 and was produced in close collaboration with James Sowerby, the illustrator. William Jackson Hooker prepared a series of supplements to this work between 1829 and 1866. Second and third editions to the original work, complete with supplements of their own, had additional illustrators contributing, including

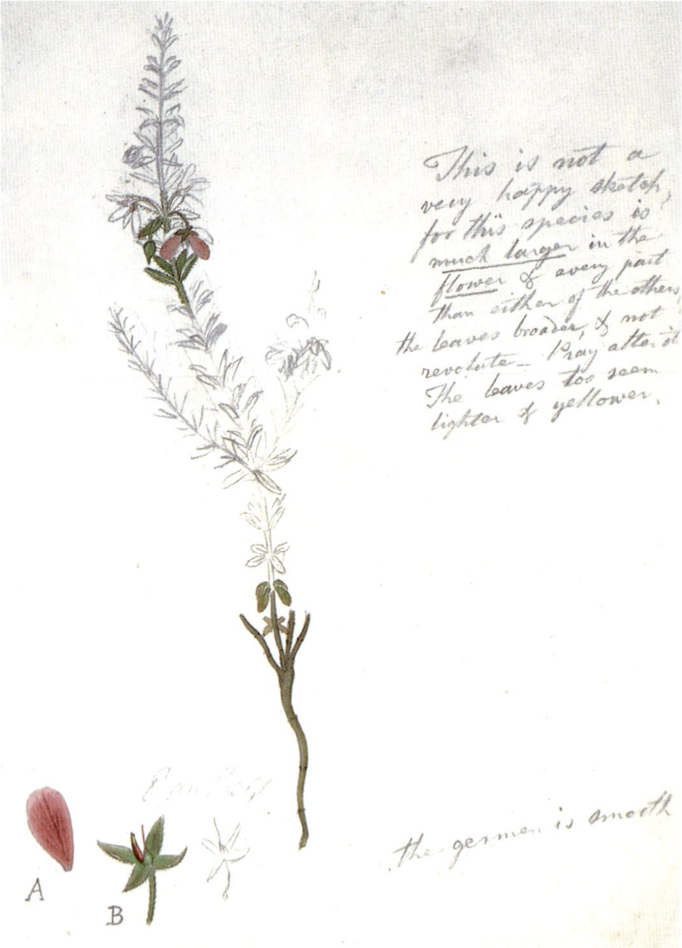

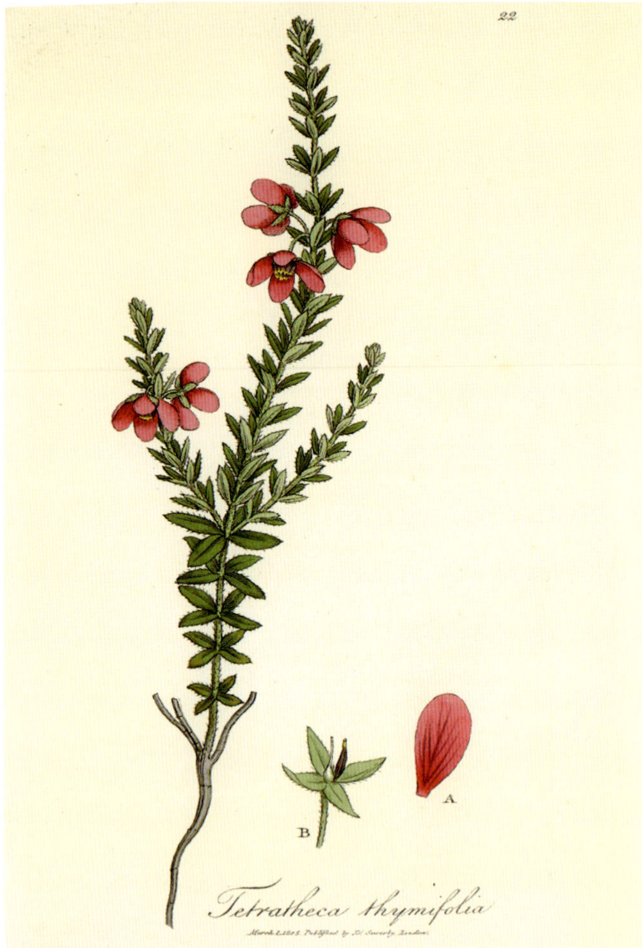

Tetratheca thymifolia Smith
Original drawing (left) by James Sowerby with corrections by James Smith during the preparation for publication. This is an example of the close cooperation between the artist and scientist in the preparation of botanical illustrations. Sowerby's hand-coloured engraving (right) was published in J.E. Smith, *Exotic Botany*, 1804–1806.
[BM] [NLA 19.8 x 11.2]

sons of Sowerby. *English Botany* thus became a very confusing periodical requiring very careful citation.

In 1804 Smith published *Exotic Botany* with the aim of 'introduc[ing] to the curious cultivator plants worthy of his acquisition from all parts of the globe, and to teach those who have correspondents abroad what to enquire for'. This too was illustrated by Sowerby and included many Australian plants.

James Sowerby must have been the most sought-after, busy and prolific artist of the late 1700s and early 1800s, being a major illustrator for a number of very significant publications of the time. He was a scientist in his own right as well as a competent painter and engraver. James also raised a family of natural history artists. For botany his oldest son, James de Carle Sowerby, was particularly significant.

In 1797 Henry Charles Andrews[4] began to publish his *Botanist's Repository*. This ran to nine volumes from 1797 until 1809. Andrews was a botanical painter and engraver and specialised in heaths, roses and geraniums. He painted from live material but did not keep an herbarium. The *Botanist's Repository* was entirely engraved and hand-coloured by Andrews. The descriptions in the first five volumes of the *Repository* were probably done by his father-in-law, John Kennedy, the nurseryman.

These grand periodicals, which provided a plate with a page or two of text describing the plant depicted, were produced more for horticulturists and wealthy owners of exotic gardens than for taxonomic botanists. As plants from other countries, including New Holland, were being returned to England and successfully cultivated, they became the subjects for such publications. Inevitably, however, the two disciplines of horticulture and botany ran together and became indistinguishable. In several cases species were formally named and described in these periodicals for the first time, rather than in the more scientific journals.

Plant material arriving in England from Australia came through a number of initiatives, but in the early days the bulk of it was due to the interest and patronage of Joseph Banks.

Banks became central to a growing network of collectors, scientists and artists. Moreover, he had connections in the halls of the aristocracy and the government. From this position of power Banks advised on what was needed to further the scientific interests of Britain in the southern hemisphere, and he selected appropriately qualified and experienced naturalists, artists, gardeners and other scientists for numerous voyages of exploration. This of course included the James Cook voyages. The Forsters, David Nelson, Archibald Menzies and others were among the collectors before the famous team of Robert Brown and Ferdinand Bauer in the early 1800s. All of these men and others, less notable, maintained a steady flow of material back to England.

Surprisingly, naturalists were not sent with the first fleets to help in establishing the penal colony in New Holland. Any such work was carried out by the medical men and senior officers who had an interest in natural history, surgeon John White being the first. Governor Phillip complained about the lack of naturalists and ultimately Joseph Banks arranged for George Caley to be assigned to collect plants in Australia, but not until 1800.

Meanwhile, Denis Considen,[5] assistant-surgeon on the *Scarborough* of the First Fleet, devoted some of his energies to making use of native plants for medicinal purposes. He claimed to be the discoverer of eucalyptus oil. He wrote to Joseph Banks in 1788:[6]

> We have a large peppermint tree which is equal to, if not superior, to our English peppermint. I have sent you a specimen of it. If there is any merit in applying these and many other simples to the benefit of the poor wretches here, I certainly claim it, being the first who discovered and recommended them.

Considen's involvement was brief because ill health forced him to leave the colony in 1793.

David Burton[7] arrived in the colony in 1791 on the *Gorgon* as a Superintendent of Convicts. Burton was a competent gardener and became the public gardener at Parramatta, collecting and returning plants to England. Regrettably his time in the colony was short too,

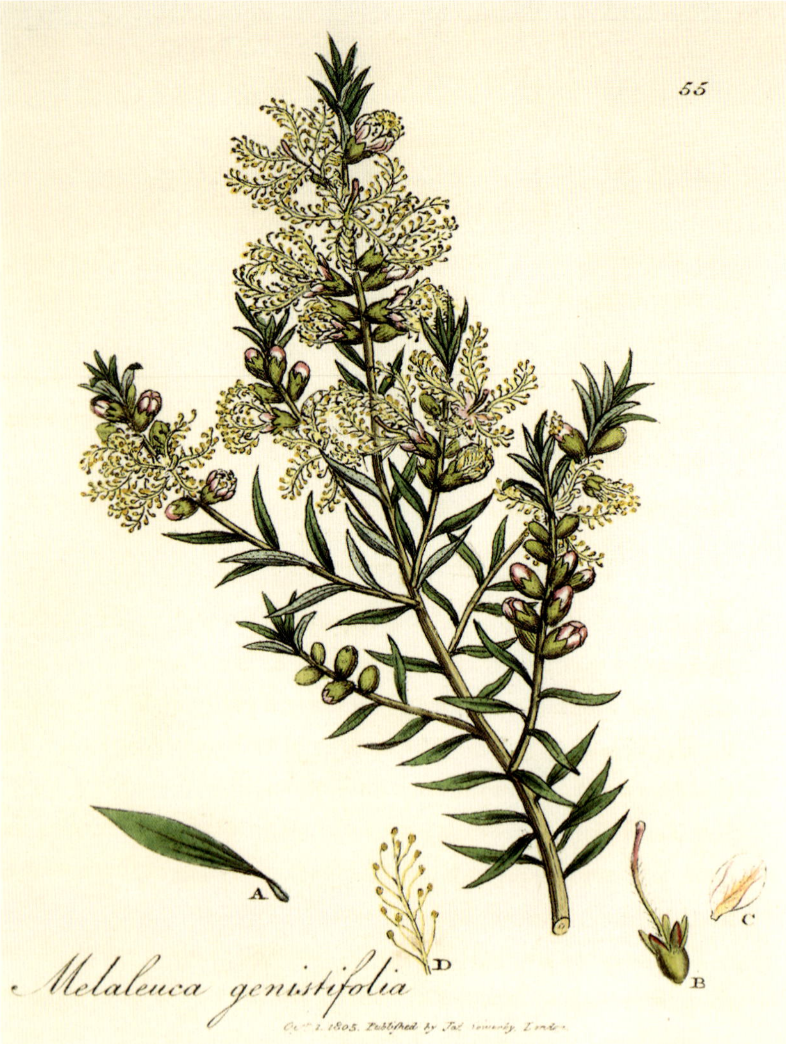

Melaleuca decora (Salisb.) Britten
(= *Melaleuca genistifolia* Smith)
Hand-coloured copper engraving by James Sowerby published in J.E. Smith *Exotic Botany* in 1804. The plants for the description and illustration were collected by David Burton.
[NLA 14.2 x 11.2]

because he accidentally shot himself while shooting ducks on the Nepean River, late in 1792.

Lieutenant-Governor Colonel William Paterson was born in 1755 at Montrose, Scotland, birthplace of Robert Brown. Paterson came to the colony in 1791 as a Captain in the newly formed New South Wales Corps. Paterson was very interested in natural history and actively collected for Joseph Banks on Norfolk Island, around Port Jackson and in the Blue Mountains. In 1794 he became Acting Governor, between the postings of Phillip and Hunter. His botanical interests, with support from Joseph Banks, led him to become a Fellow of both the Linnean Society and Royal Society of London.

In 1804 he was chosen to establish a settlement in northern Van Diemen's Land. Robert Brown accompanied him on the voyage to do so. Paterson chose Port Dalrymple on the River Tamar for the settlement and from there he also sent material back to Banks in England. In 1808 Paterson returned to Port Jackson to take over the running of the colony following the failure of Governor Bligh to maintain control. Paterson kept the position of Acting Governor until Governor Macquarie's arrival in 1809. Paterson was suffering ill health and decided to join the convoy taking Bligh back to England for his court marshall emanating from the Rum Rebellion. Bligh was on the *Hindostan* and Paterson was on the *Dromedary*. Paterson did not make it, dying at sea near Cape Horn on 21 June 1810.

In 1798 Joseph Banks gave George Suttor[8] the responsibility of bringing a consignment of 'useful' plants to the colony. Suttor (1774–1859) was required to tend plants destined for the orchards of the colony and was promised a parcel of land upon their safe arrival. He was to sail with these on the *Porpoise* together with his new wife and several significant individuals, including the newly appointed Governor Philip Gidley King and his wife. They were due to

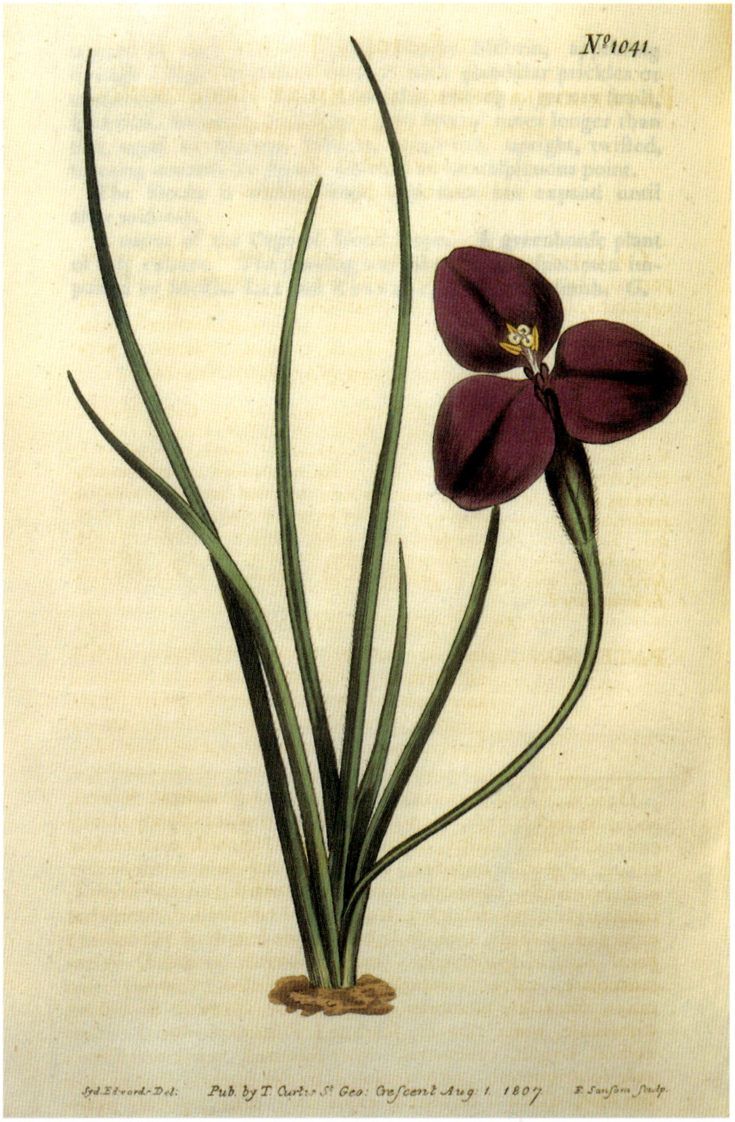

Patersonia sericea R.Br.
Hand-coloured copper engraving of a drawing by Sydenham Edwards, published in *Curtis's Botanical Magazine* in 1807. This plate is part of the original description of this genus and species, named in honour of William Paterson.
[MEL 20.0 x 11.7]

sail at the end of 1798. A sequence of hitches delayed the *Porpoise* until September 1799. By this time a fellow passenger, Mungo Park, a surgeon interested in natural history and chosen to collect for Banks, withdrew from his posting to the colony. This allowed Banks to send George Caley instead. Shortly after they finally set sail, the ship was damaged in a storm and forced to return to England. The passengers, except for the hapless Suttor, were transferred to the *Speedy*. Suttor was later put on a refitted and renamed Spanish corvette, *Porpoise* (2). The second *Porpoise* sailed early in 1800.

There is every indication that Suttor's attempts to tend the plants were only partially successful with the failure of hops, in particular, being significant. Hops were needed for brewing beer in the colony in an attempt to break down the addiction to rum. Suttor reported to Banks late in 1800 on a few of each of several fruit trees and grape vines arriving with him, and went on to confess: 'It is with painful concern I inform you, Sir, that I have not been able to bring the hops and many other valuable plants that I have on board.'[9]

Suttor was granted a parcel of land, but there is not much evidence that he was a prolific collector. His interests were very clearly horticultural, but he also kept in touch with the scientific community and offered hospitality to several visiting scientists, including Allan Cunningham.

George Caley[10] (1770–1829) developed an interest in plants as a youngster, which led him to London and Joseph Banks. Caley, it seemed, desperately wanted to go to New Holland to collect plants. His desires were frustrated by Banks who insisted that he work as a gardener at the Royal botanic gardens to learn as much as possible about the Australian plants growing there. Banks and Caley aggravated each other — the one with his wisdom and the other with his impatience. Caley's aspirations were finally met through Mungo Park's withdrawal from his posting. Through the delay, Caley met George Suttor, but they were soon separated with Caley ultimately sailing in 1800 on the *Speedy* together with the new Governor, Philip Gidley King, and Francis Barrallier, chart-maker. Just as Caley's youthful impatience had irritated Banks, so too it irritated King.

Nevertheless, Caley seems to have been very conscientious. Initially he set up a botanical garden at Parramatta to house plants while awaiting their return to England. He also seems to have accepted every possible opportunity to collect in areas previously unvisited. He returned volumes of material, both botanical and zoological, to England on suitable vessels over the ten years he spent in the colony. For botany, he not only collected material but also prepared detailed descriptions. Like Suttor he seems to have cooperated generously with other visiting scientists, especially Robert Brown. After King left the colony Caley got on better with Governor Bligh. Indeed when Caley returned to England, he and George Suttor provided character references for Bligh at the court marshal.

In 1800 Joseph Banks had Lieutenant James Grant appointed to the command of the *Lady Nelson* with a view to carrying out inshore survey work in New Holland. Grant was to collect biological and mineralogical material for return to England. The *Lady Nelson* arrived at Port Jackson in December 1800. In March 1801 Grant set off to survey the southern and south-western coastline of New Holland. George Caley went on board as naturalist together with Francis Barrallier. They sailed to Bass Strait collecting in the vicinity of Jervis Bay, Bass Strait islands and Western Port Bay.

In 1802 Baudin's ships put into Port Jackson and François Péron was active in collecting around Port Jackson and environs. Support by Caley seems to have been willingly given but hampered by language problems. Matthew Flinders also arrived in 1802 bringing Robert Brown and Ferdinand Bauer. There is no doubt that Caley helped Brown and Bauer greatly over the next three years.

Early in 1804, Caley collected in the vicinity of the Hunter River. In 1805 he found the opportunity to visit both Norfolk Island and Van Diemen's Land. In between these trips further afield, Caley penetrated north and south of the Sydney plain and west into the Blue Mountains. He proved himself to be an able explorer and had the wisdom to seek the assistance of the Aborigines. It seems that he came to the brink of finding a passage through the Blue Mountains to the plains beyond early in 1810, shortly before he left the colony. All round he was a capable bushman as well as an effective naturalist. He returned with Bligh on the *Hindostan*.

The explosion of periodicals was not limited to England. Jules Paul Benjamin Delessert (1773–1847), described as 'a baron, French industrialist, banker, philanthropist and amateur botanist',[11] commissioned and published *Icones Selectae Plantarum* in five volumes between 1820 and 1846. Delessert is renowned for his private herbarium which embraced collections from significant scientists or collectors, including several who came to Australia. Delessert was generous with the use of his herbarium, much as Joseph Banks was in England, and certainly made it available to Augustin Pyramus de Candolle and his son, Alphonse Louis, for their *Prodromus* (1823–1873). The collection ultimately formed the nucleus of the Conservatoire Botanique, Geneva.

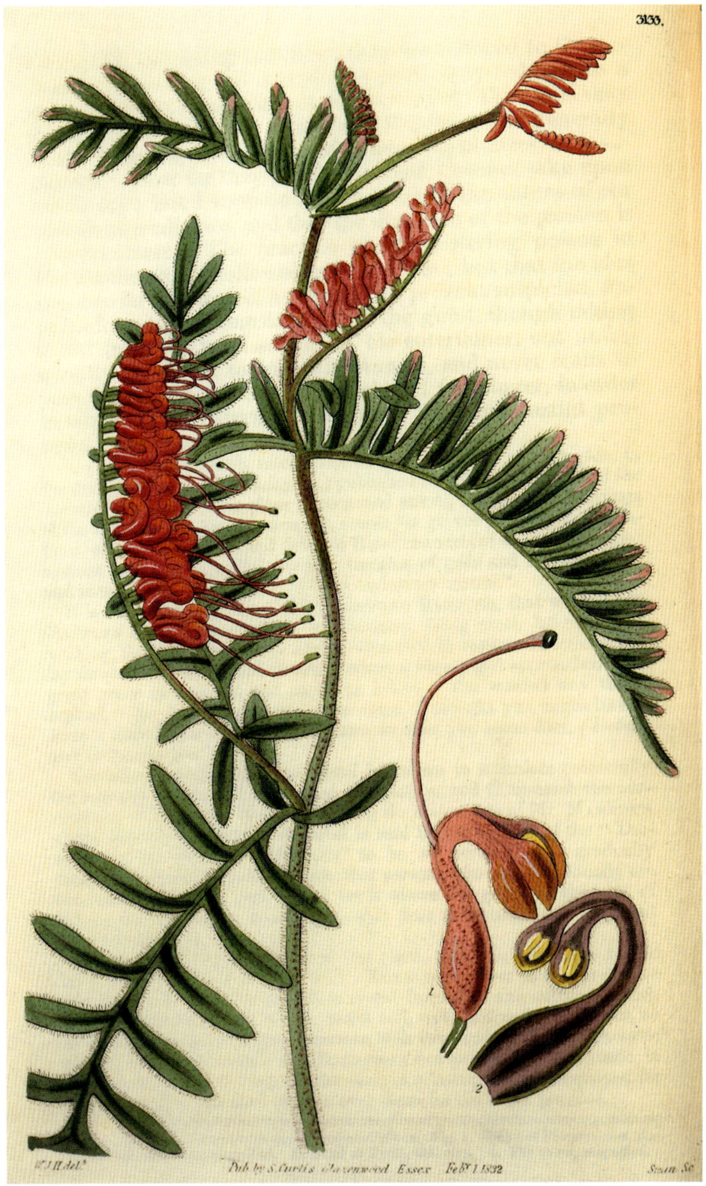

Grevillea caleyi R.Br.
Hand-coloured engraving of a drawing by William Jackson Hooker, engraved by Joseph Swan and published in *Curtis's Botanical Magazine* in 1832. The plant was named in honour of the Banksian collector, George Caley.
[MEL 20.0 x 12.0]

Meanwhile, Delessert arranged for Augustin de Candolle to provide descriptions and for Pierre-Jean-François Turpin to illustrate selected items from the herbarium for publication in the *Icones*. With some exceptions in Volume 3, all the description is provided by de Candolle, even though some volumes were published after his death in 1841. Turpin provided the illustration for the first two volumes.

Jean Christoph Heyland became the major illustrator for the next three volumes. Heyland had an artistic potential which was recognised and encouraged by de Candolle. Once Heyland proved himself, he became de Candolle's illustrator for nearly a quarter of a century and provided quality illustrations as exemplified in the *Icones*. One of Heyland's signatures was detailed shading for a portion of the illustration, while relying on line for the rest of the figure. Eulalia Delile was another illustrator, especially for the third volume. Fifteen engravers were used for the work, including most of the high profile engravers of the times such as Auguste Plée, Poiret and Sauvage.

Other European botanists were also describing and illustrating Australian plants, though with lesser impetus than that prevailing in England and France.

Concurrently, a number of dictionaries and encyclopaedias were also published, in particular by the French. Notably Jean Baptiste Antoine Pierre Monnet de Lamarck and Jean Louis Marie Poiret produced *Encyclopédie méthodique. Botanique* (1783–1817) in eight volumes and five supplements, and *Tableau Encyclopédique et méthodique. Botanique* (1791–1834) in six volumes. The former was not illustrated, but the latter had a thousand botanical plates each depicting a genus of plants. A number of illustrators and engravers were used, with L. Fossier and Jacques Eustache de Sève making major contributions to the drawing.

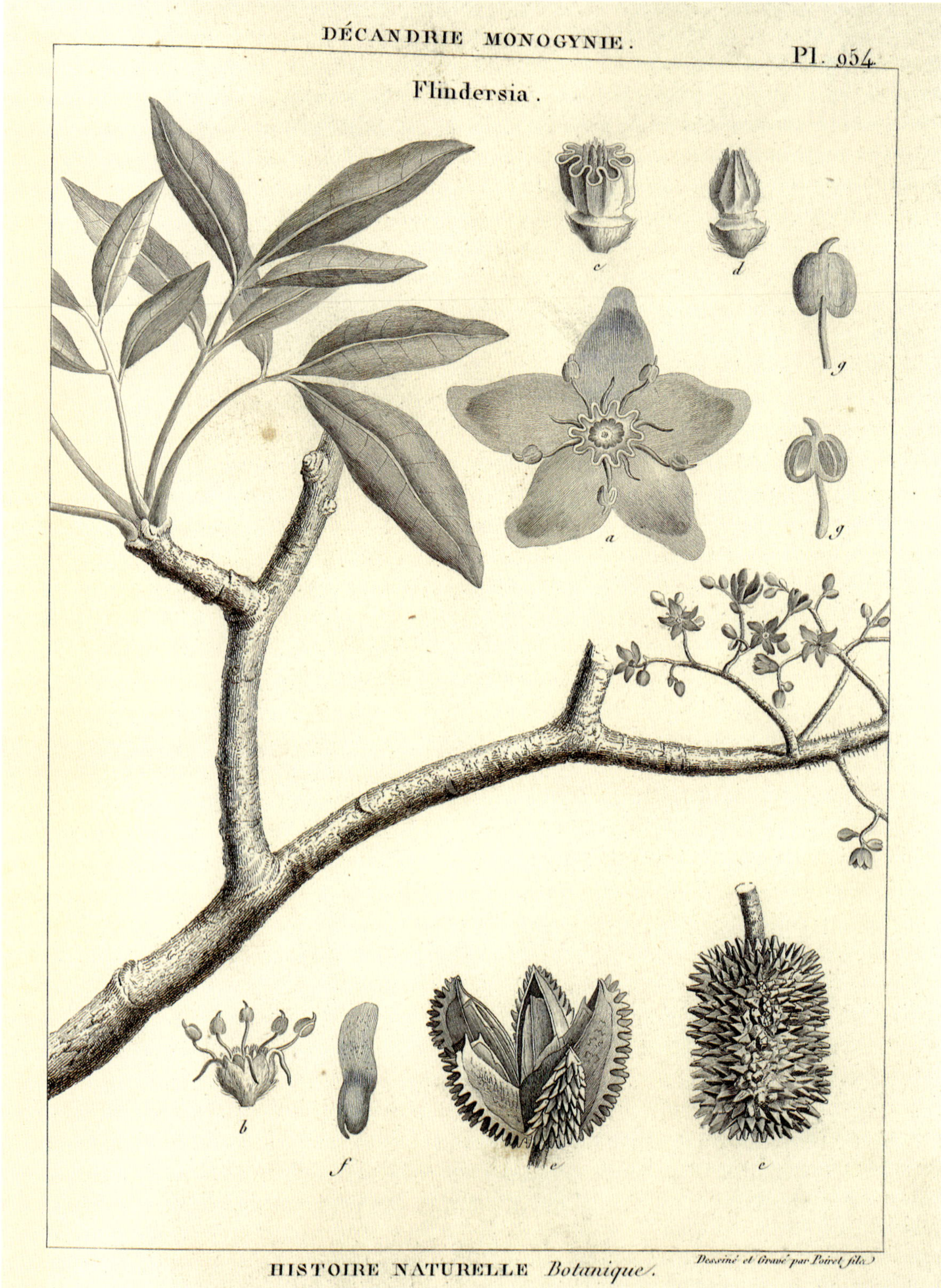

Flindersia australis R.Br.
Copper engraving by Poiret, published in J.B.A.P.M. de Lamarck & J.L.M. Poiret, *Tableau Encyclopédique et Méthodique. Botanique* in 1823. The six volumes of this work contained a thousand plates, each depicting a genus of plants.
[MEL 26.0 x 19.5]

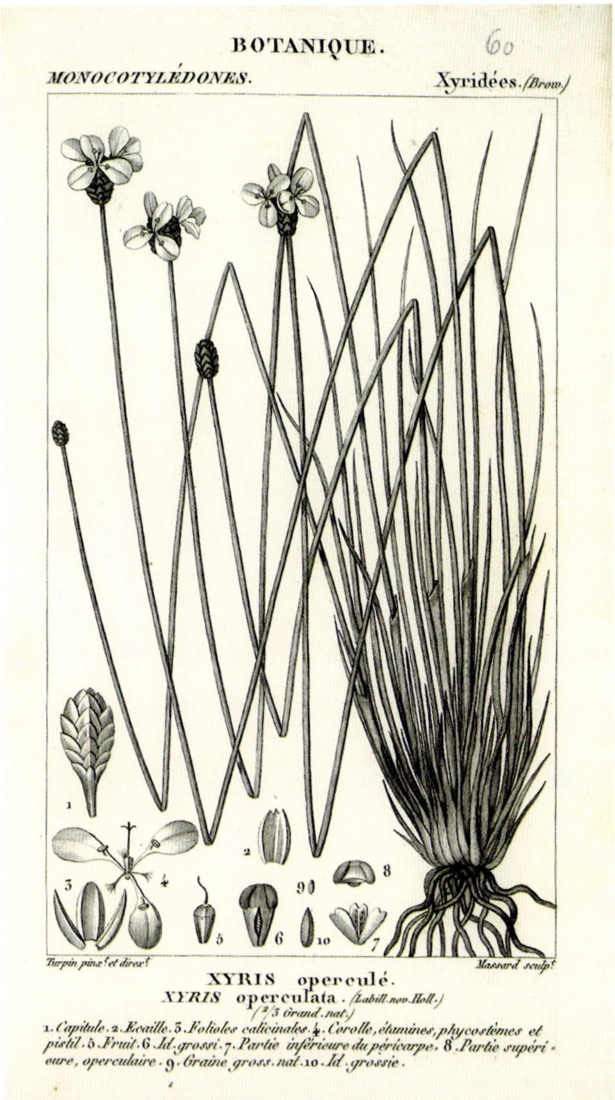

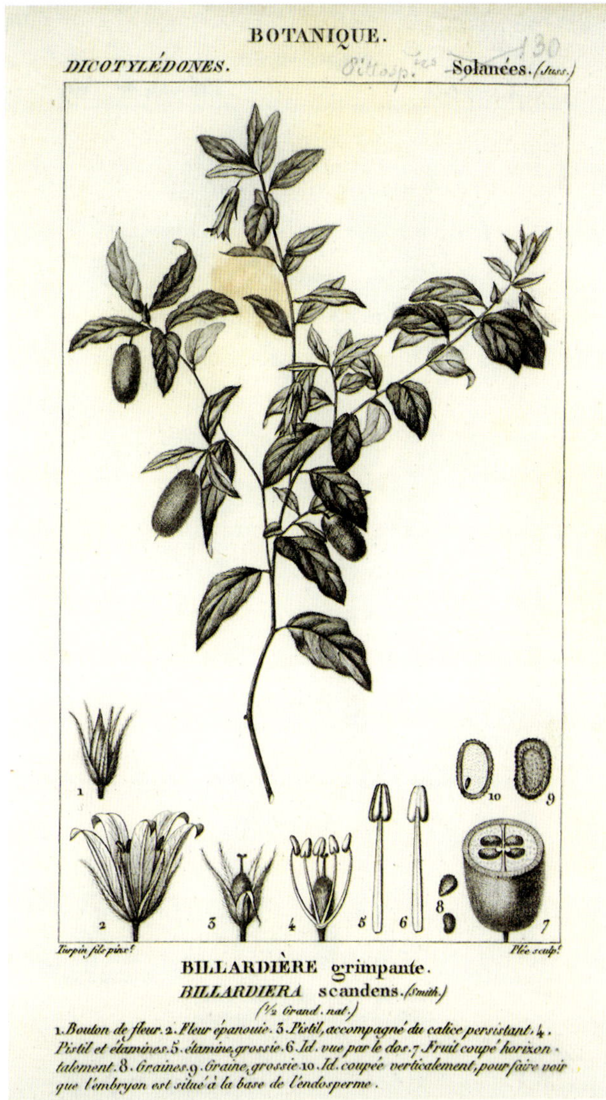

Xyris operculata Labill.; Copper engraving of a drawing by Pierre Turpin, engraved by Massard (left). *Billardiera scandens* Smith; Copper engraving of a drawing by a son of Pierre Turpin, engraved by Auguste Plée (right). Both are published in F.G. Levrault, *Dictionnaire des Sciences Naturelles*, 1816–1830. Pierre Turpin's son has not previously been recorded as a botanical artist.
{CANB 19.5 x 11.5} {CANB 19.5 x 11.5}

F.G. Levrault edited *Dictionnaire des Sciences Naturelles* (1816–1830) in sixty volumes with twelve *Planche* volumes, five of which were devoted to botanical illustration. The drawings were drafted and their engraving supervised by Turpin. An interesting exception is found in this work where one plate of an Australian plant was drafted by Turpin's son. 'Turpin fils' has not been recorded as a botanical illustrator, so it is assumed that he pursued a different career. A number of engravers were used to prepare the plates for the *Dictionnaire*. They employed a mix of line and stipple to good effect.

In England, Abraham Rees complemented the French works with *The Cyclopedia* (1802–1820) in thirty-nine volumes. Sydenham Teast Edwards did the drawings which Milton engraved.

Sydenham Edwards[12] (1769–1819) contributed illustrations to a number of works, notably *Curtis's Botanical Magazine*. Edwards was a protégé of Curtis who had him trained in the skills of botanical painting and the two became firm friends. Edwards began illustrating for *Curtis's Botanical Magazine* in its second volume in 1788. He went on to be its most prolific illustrator for many years.

When Edwards discontinued his association with the *Magazine*, reportedly due to a 'misunderstanding', he was astute enough to realise that there was room for another periodical depicting exotics in Britain. In 1815 he began his *Botanical Register*, 'consisting of coloured figures of exotic plants cultivated in British gardens with their history and mode of treatment'. This became another in the mode of the grand horticultural periodicals. Edwards did the paintings, others effected the engraving and hand colouring. Initially John Bellenden Ker (-Gawler)[13] provided the text. Edwards illustrated the magazine until his death in 1819. *The Botanical Register* was continued and in 1829, when John Lindley took over, it became known as *Edwards's Botanical Register*. Lindley retained editorship until the periodical was discontinued in 1847. It ran to thirty-three volumes over thirty-two years.

In 1817 Conrad Loddiges began *The Botanical Cabinet*. He was a Dutch horticulturalist with a renowned and richly stocked nursery in Hackney, England. The *Cabinet* included twenty volumes until 1833.

George Loddiges,[14] son of Conrad, followed in his father's footsteps as a nurseryman and kept the *Cabinet* in production after his father's death. George Loddiges was a principal illustrator for the *Cabinet* and George Cooke was the engraver. In reality several members of both the Loddiges and Cooke families were involved in the illustration of the work. Many of the illustrations are vignette-like, and the descriptions are generally brief with some being highly romantic and scarcely scientific. George Cooke seems to have done botanical engraving work only for the Loddiges family. Otherwise he was a prolific topographical engraver.

While these grand publishing enterprises were flourishing, Allan Cunningham[15] (1791–1839) became a gardener at Kew in 1808 and came under the influence of William Aiton. Aiton desired to expand the collection at Kew and appealed to Joseph Banks to have collectors sent to unexplored parts of the world. Cunningham was chosen to explore and collect in Brazil in 1814. He was then sent to New South Wales in 1816 to become another of the valued Banksian collectors. This appointment proved to be particularly valuable for Australian botany, Cunningham becoming an admirable successor to George Caley. Cunningham was also both an able botanist and competent explorer. The similarity did not end there as Cunningham also had difficult relations with the resident Governor, Lachlan Macquarie.

Cunningham arrived just as inland and coastal exploration was intensifying, enabling collection in areas previously unvisited. Shortly after his arrival he joined General Lieutenant John Oxley[16] on an expedition to inland New South Wales during which they explored the Lachlan and Macquarie Rivers. Oxley was Surveyor General from 1812 until

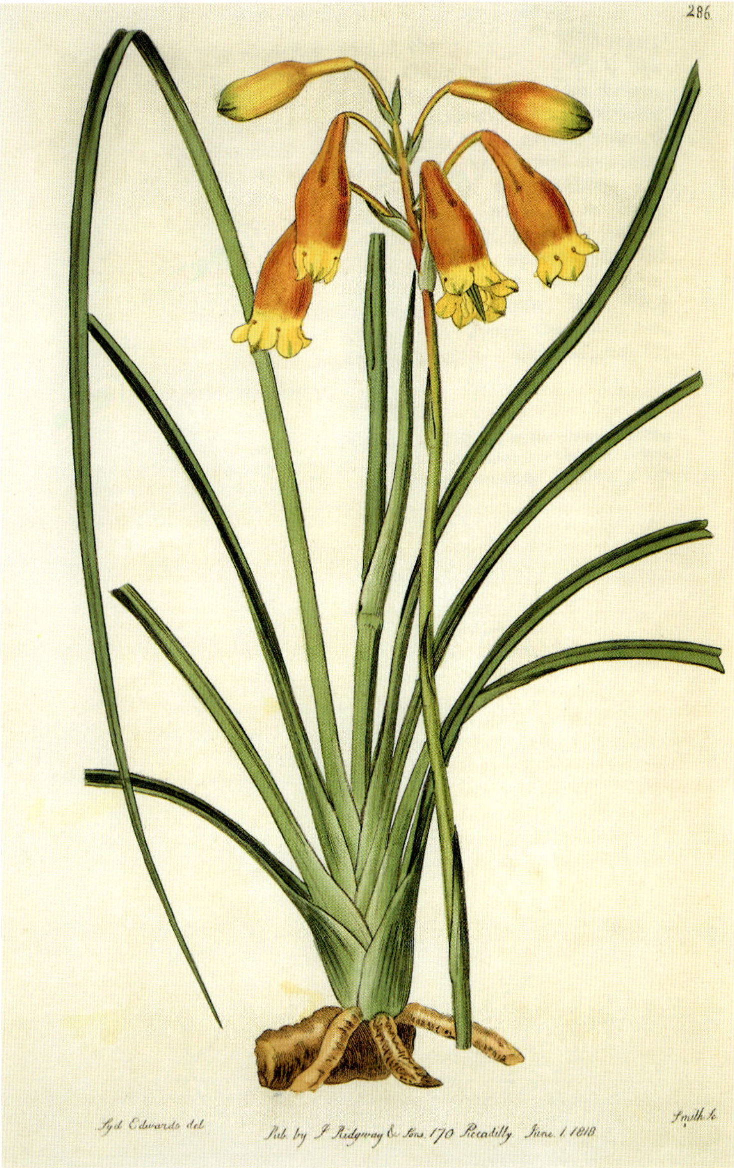

Blandfordia nobilis Smith
Hand-coloured copper engraving of a drawing by Sydenham Edwards and engraved by Smith. Published in Edwards' *Botanical Register* in 1817.
[CANB 20.7 x 12.7]

1828. Charles Fraser, of the 46th Regiment, also joined this expedition. The expedition left Bathurst in April 1817 and returned there four months later. Cunningham estimated that he collected about four hundred and fifty plant species, all new to science.

Charles Fraser[17] arrived in the colony in 1816 and, having some horticultural and botanical knowledge, was made Superintendent of the Sydney Botanic Gardens. Five years later he was officially appointed Colonial Botanist. Fraser and Cunningham cooperated over numerous activities, including collecting trips. However, there is evidence that Cunningham was envious of Fraser, particularly over ventures in which Fraser participated while Cunningham was away on others.

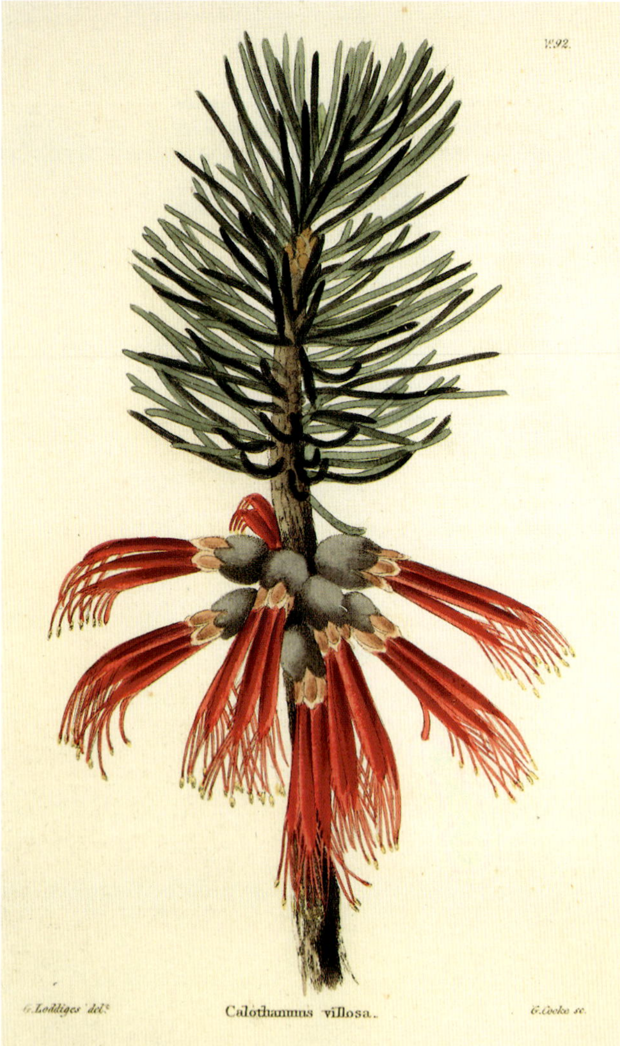

Calothamnus villosus R.Br.
Hand-coloured copper engraving of a drawing by George Loddiges and engraved by George Cooke. Published in C. Loddiges, *Botanical Cabinet,* in 1817.
[CANB 16.3 x 10.4]

In 1817 a constructive alliance began for Cunningham with Captain Phillip Parker King,[18] son of Philip Gidley King, former Governor of the Colony. They sailed in the *Mermaid* late in December 1817 to survey the north-west coast. The *Mermaid* was ideal for working close to shore but less than ideal for those on board because she was very small. They spent nearly a month collecting in the region of Cape Leeuwin, before sailing north towards North West Cape. From there they worked more closely along the coast and nearby islands. Port Essington was investigated as a possible location for a future settlement. When supplies ran out in May they sailed for Timor and thence back to Port Jackson, arriving late in July. This journey was not judged to be particularly successful because it did not find any major rivers draining from the inland.

While they were away, Charles Fraser accompanied John Oxley on another expedition to survey the Macquarie River. By all accounts, the trip was successful botanically.

Cunningham completed a trip to the Illawarra district south of Sydney before the *Mermaid* sailed again, this time for Tasmania. Cunningham accompanied King on this voyage, leaving late in December and arriving back in mid-February 1818. In May the *Mermaid* sailed for the north-west again, this time with the *Lady Nelson*. Oxley was on the latter and was to survey Port Macquarie and the Hastings River. When that was completed they separated and the *Mermaid* continued north. Cunningham seized every opportunity to collect until he became ill near the Liverpool River. From there on the explorers began the survey again but made little progress due to general ill health and lack of fresh water. They returned to Port Jackson early in 1820, Cunningham armed with a considerable collection of plants, fruits and bulbs.

The third voyage of the *Mermaid* to the north-west sailed in July 1820. This trip was hampered by damage from a grounding in the Barrier Reef, which gradually resulted in increased leakage. The explorers put ashore for careening at Prince Frederick Sound which gave Cunningham three weeks of concentrated collecting. Unfortunately they were unable to mend the ship effectively and had to return to Port Jackson, arriving early in December.

The *Mermaid* was declared unseaworthy and replaced with the *Bathurst*. King and Cunningham sailed for the fourth time in May 1821. Cunningham became ill again on this voyage and was unable to take best advantage of opportunities to collect as the surveying progressed. In September they sailed to Mauritius to replenish supplies. Cunningham had six good weeks there. The explorers sailed for Australia in November arriving in King George Sound in December. Early in January they sailed west and north to Rottnest Island, the Abrolhos Islands and further north to North West Cape. The voyage was soon abandoned and they arrived back in Port Jackson in April 1822. This and the previous three voyages were not regarded as successful because they failed to find the hoped-for major river. Botanically, however, they had to be regarded as successful. On the last one alone Cunningham collected over twelve hundred species, more than half of which were new to science.

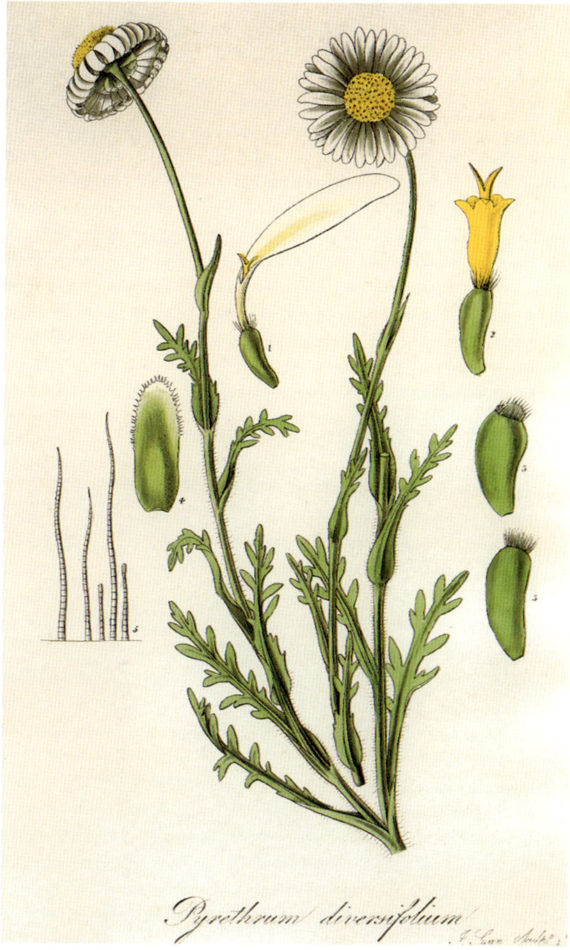

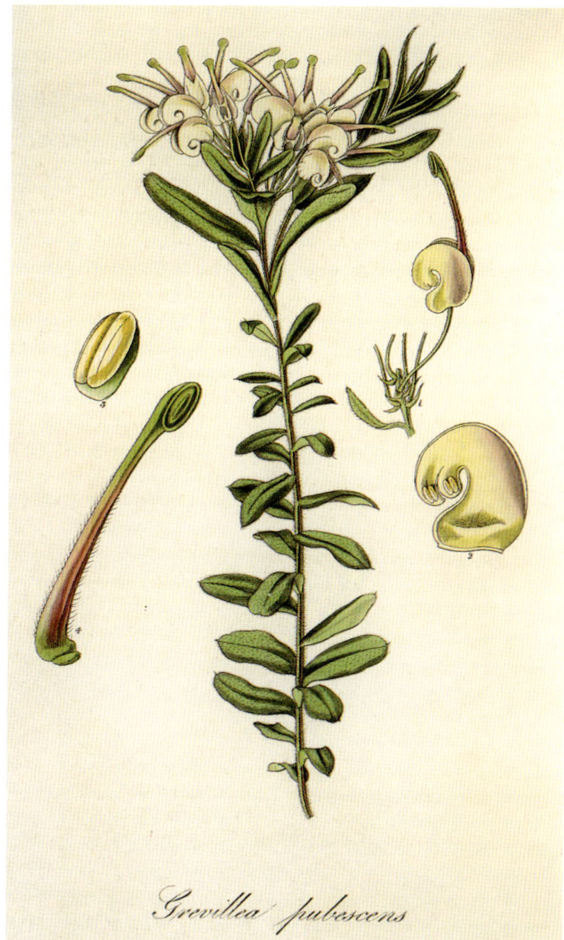

Brachycome diversifolia (Hook.) Fisch & C.A. Mey. (= *Pyrethrum diversifolium* Hook.) (left); *Grevillea baueri* R.Br (= *Grevillea pubescens* Hook.) (right). Both engravings were published in W.J. Hooker's *Exotic Flora* in 1827. The plants were cultivated in Edinburgh from seed collected by Charles Fraser on his numerous collecting trips. He became Colonial Botanist in 1821.
[CANB 24.0 x 15.0] [CANB 24.0 x 15.0]

At Port Jackson, Cunningham continued his field work south, west and north. To the south he ventured to the headwaters of the Murrumbidgee River; to the west he penetrated the Liverpool Range to the Liverpool Plains; and to the north he joined Oxley on an expedition to Moreton Bay to investigate its suitability for the establishment of a new settlement. Cunningham was continually active in his collecting work as well as being available for visiting botanists, such as Jules Sébastien Cesar Dumont D'Urville from the French expedition of Louis Isidore Duperry[19] on the *Coquille* in 1824. The *Coquille* was renamed *Astrolabe* and d'Urville made another voyage into the Pacific in 1826. On this occasion he found evidence of the fate of La Pérouse at Vanikorao. On this voyage and a third, in the *Astrolabe* and the *Zélée*, Dumont d'Urville put into Hobart. Journals of these voyages were published, with *Voyage autour du monde ... Coquille* (1826–1830) by Duperry having most relevance to Australia.

Sir Joseph Banks died in 1820. Whilst there were no more Banksian collectors Banks, through his estate, continued to support some activites which he had initiated during his lifetime. His estate, for example, continued to provide the salary of Franz Bauer until he died in 1840.

Collections made by the Banksian collectors continued to enter the scientific record long after the duration of the collecting trips and long after the death of Joseph Banks. Banks' legacy to the science of taxonomy and to the horticultural industry is inestimable.

Of the periodicals, James Smith wrote:[20]

> ... several periodical publications, illustrative of exotic plants, have been at different times undertaken in this country, and some have been continued with great success; yet none of them can keep pace with the botanical riches daily flowing in upon us.

They certainly served a vital role for both science and horticulture.

6. Hooker, Lindley and the Private Collectors

It is a subject of surprise and regret that so little anxiety should have been shown by geographers, and even by men of science in general, to increase our knowledge of the interior of the Australian continent. But so it is — that land of anomalies may still be said to be almost a terra incognita.

Allan Cunningham [1]

Joseph Banks made an important contribution to Australia's discovery, colonisation, early exploration and scientific record over a period of fifty years. For about twenty years after his death in 1820, this contribution continued through the many successors he had helped in their chosen career.

William Jackson Hooker [2] (1785–1865) was one of those successors. Hooker was born in Norwich and developed an interest in natural history early in his childhood. His discovery of an unusual moss led Hooker to corresponding with the botanist, Dawson Turner. This in turn led him to London and to Joseph Banks. In 1806 Hooker became a Fellow of the Linnean Society and so began his networking with famous botanists of the time. In 1812 he became a Fellow of the Royal Society. By 1816 he had completed his first book, the *British Jungermanniae*, a fine illustrated study of English bryophytes. He married in 1815 and was anxious to settle into a career in botany.

With the support of Joseph Banks, Hooker was appointed to the Chair of Botany at Glasgow University in 1820. There he developed a reputation as an excellent teacher, botanist and gardener. Judging by his publication record Hooker must have been something of a workaholic. He retained the position at Glasgow for twenty years.

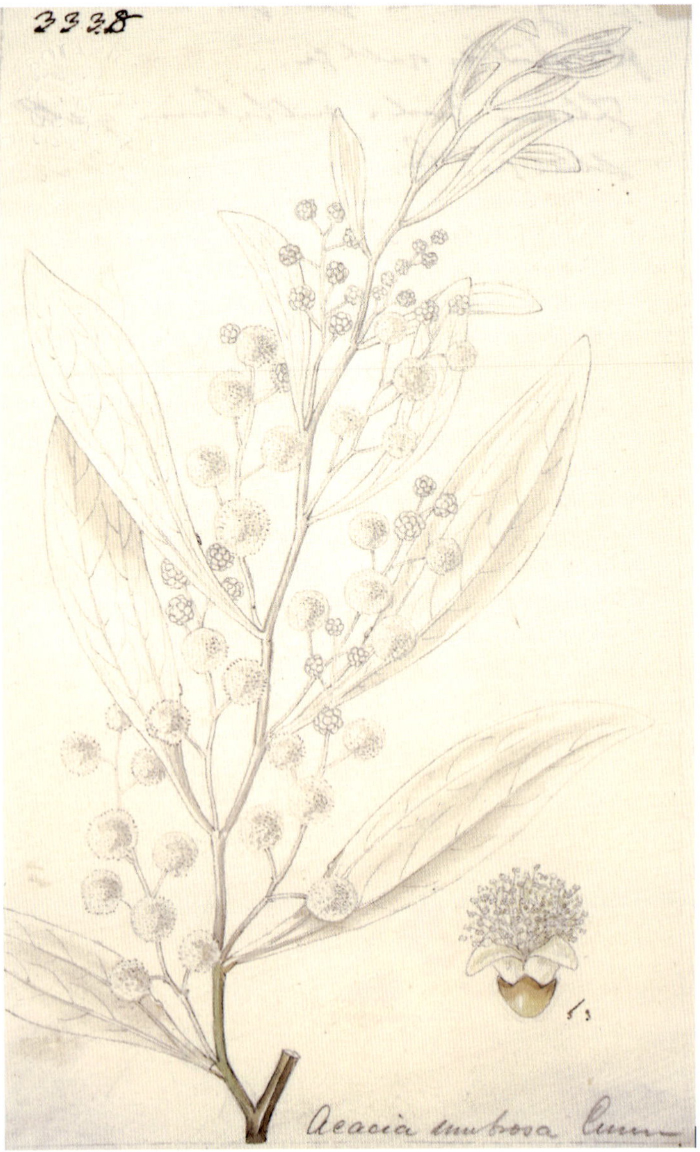

Acacia umbrosa G.Don
Original pencil and wash drawing by William Jackson Hooker. This was published in *Curtis's Botanical Magazine* in 1834.
[KEW 23.5 x 14.6]

In 1841 Hooker was appointed Director of the Royal Botanic Gardens at Kew. Following the death of Joseph Banks it is generally conceded that the Gardens at Kew declined. As a result the government conducted an inquiry into their management in 1838. Subsequently it officially took over the Gardens from the Crown and created the position of Director. Sir William, who had been knighted in 1836, was appointed and set about laying the foundations of the Gardens that one sees when visiting today. Concurrently the scientific credibility of the Gardens was established, not just with Hooker's credentials and activity, but also with the establishment of a library and herbarium. Hooker must have had great stamina because he retained the position of Director until he was eighty, in 1865.

In 1821 Hooker began a sequence of publications, some of which were in the style of the contemporary entrepreneurial periodicals. Others were more in the scientific periodical, or journal, style. The former tended to rely on living material for their subjects and the latter tended to use dried herbarium specimens.

Hooker's first periodical, *Exotic Flora*, ran to three volumes between 1821 and 1827. The drawings, hand-coloured copper engravings, were by several artists. However, acknowledgement on the plates is either rare, or cropped off. The plates are larger than the text and it seems that the bookbinder took his measurements to accommodate the letterpress. However, most of the plates seem to have been drawn by Hooker himself. He was a competent draftsman and is not to be confused with the William Hooker who specialised in painting fruits. All of the plates are engraved by Joseph Swan, who flourished as a botanical engraver during Hooker's term as Professor at Glasgow University.

Joseph Swan was a local and established engraver and copperplate printer when Hooker arrived in Glasgow. The two teamed up with Swan providing all of Hooker's engraving work until Hooker moved to Kew in 1840, Swan choosing not to follow.

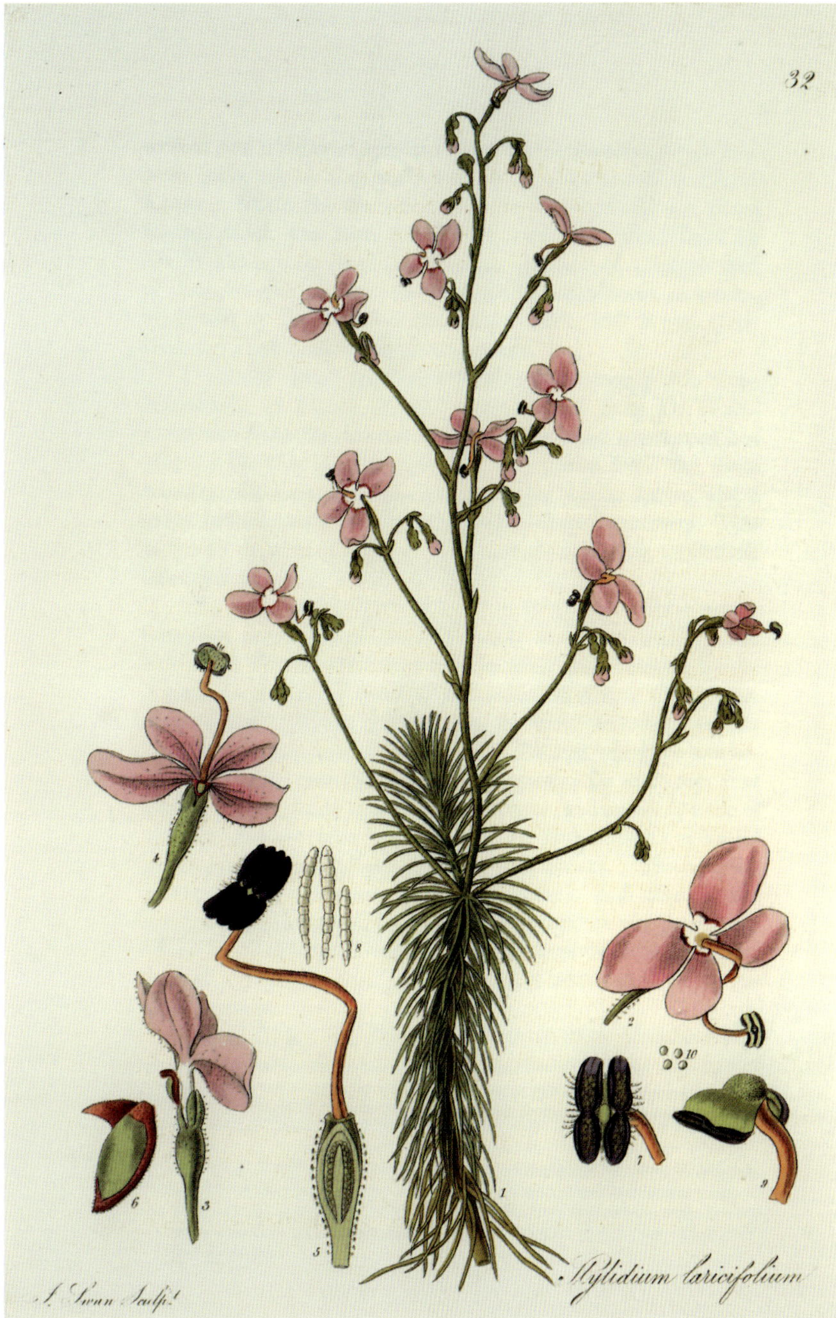

Stylidium laricifolium A.Rich.
Hand-coloured copper engraving drawn by William Hooker, engraved by Joseph Swan and published in Hooker's *Exotic Flora* in 1823.
[CANB 22.3 x 14.5]

By the early 1820s plants collected by Charles Fraser and Allan Cunningham in Australia were becoming established for horticulture and coming to Hooker's attention. Hooker treated some of these in *Exotic Flora.*

Exotic Flora was followed by *Botanical Miscellany*, which ran to three volumes (1830–1833), and the *Journal of Botany* comprising four volumes (1834–1842). The latter

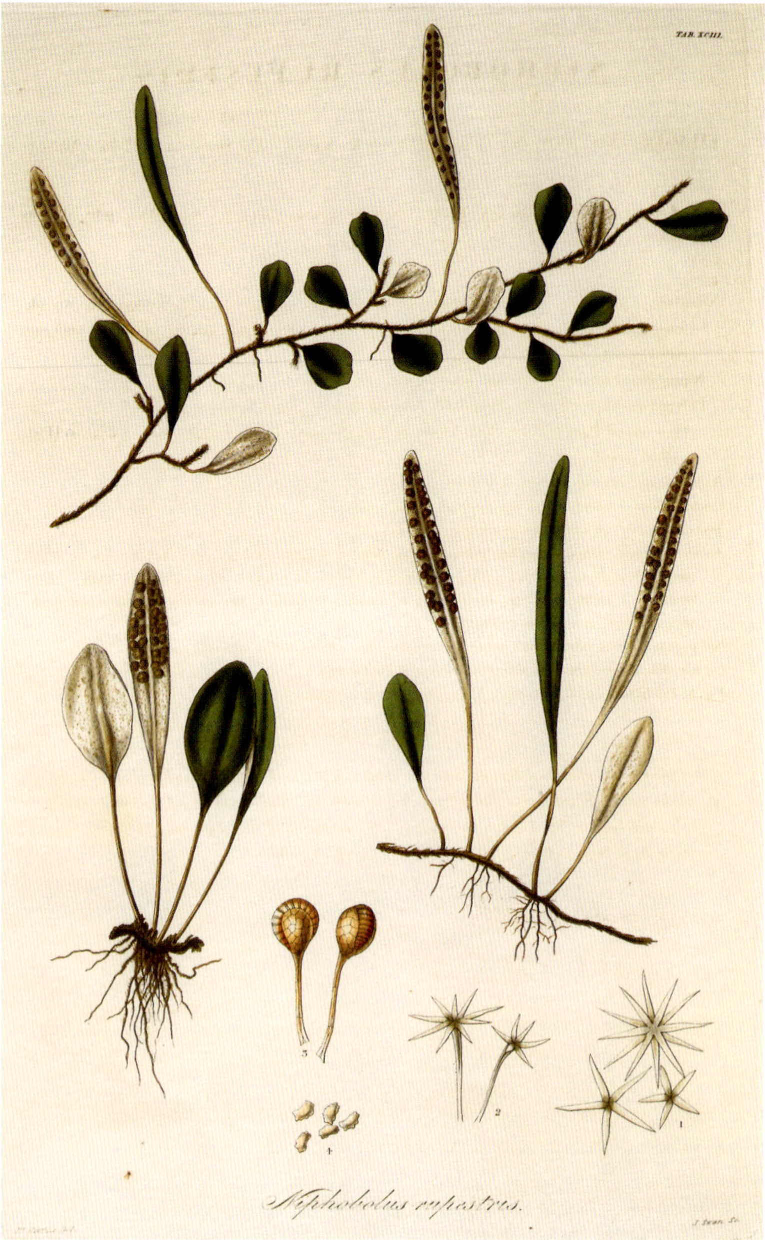

Pyrrosia rupestris (R.Br.) Ching (= *Niphobolus rupestris* R.Br.)
Hand-coloured copper engraving by Joseph Swan from a drawing by Robert Greville and published in Hooker's *Icones Filicum* in 1831.
[CANB 30.2 x 18.4]

two were journal-like, went through a series of name changes, and were finally published as *Hooker's Journal of Botany and Kew Garden Miscellany* (1849–1857) in nine volumes. The illustrations in these were uncoloured engravings and fewer than in *Exotic Flora*. In 1837 Hooker commenced the *Icones Plantarum* which was in the entrepreneurial-periodical style, except that the drawings were mostly based on herbarium specimens and are uncoloured. This periodical has endured to the present day (albeit issued only occasionally in the past decade). From 1867, following Hooker's death, it has been titled *Hooker's Icones Plantarum*.

Diplazium australe (R.Br) N.A.Wakef.
(= *Athyrium australe* R.Br.)
Hand-coloured lithograph from a drawing by Franz Bauer, executed by Walter Fitch using zinc plates, published in Hooker's *Genera Filicum* in 1842.
[CANB 23.5 x 15.3]

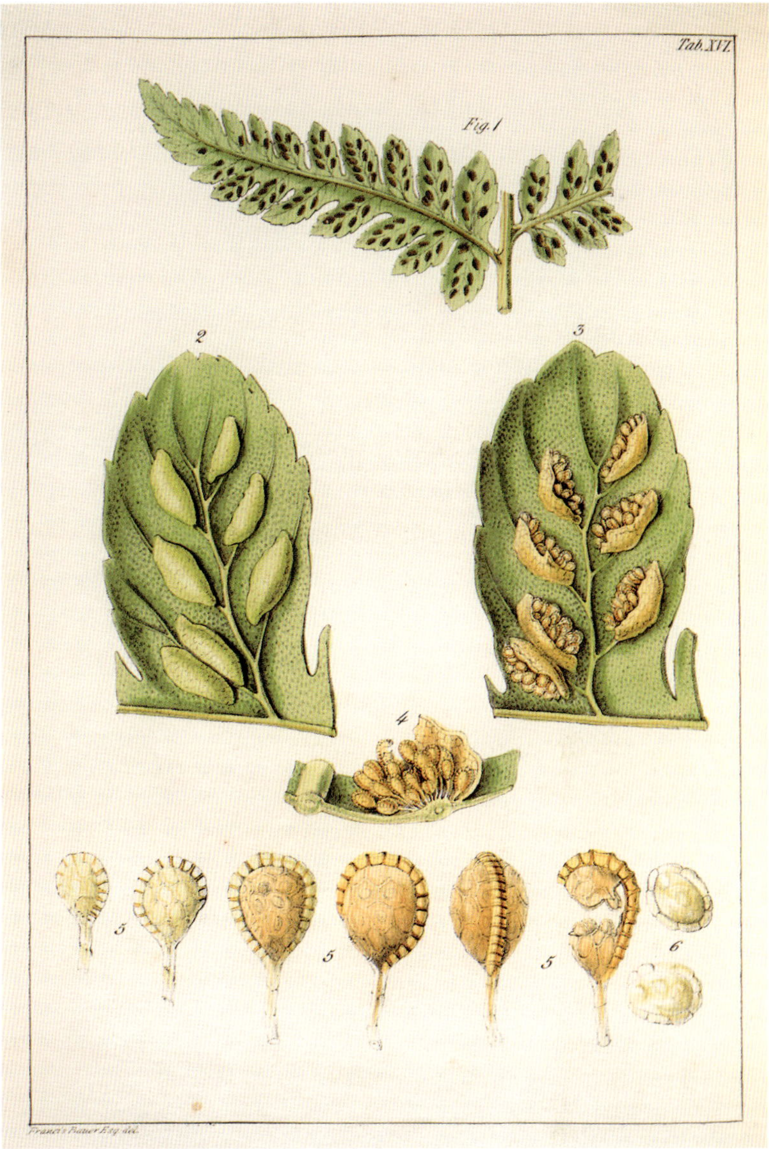

Hooker was quite a prolific publisher and worked on other concurrent illustrated publications such as *Curtis's Botanical Magazine*, which he edited from 1827 till 1865.

Hooker began his career with bryophytes (mosses and liverworts) and, while his name is associated with so much that is at the advanced end of the evolutionary scale, he maintained an academic and scientific interest in the more primitive plants. Amongst these are the ferns. He teamed up with Robert Kaye Greville [3] to produce *Icones Filicum* (1827–1832). For this work Greville did the drawing and Joseph Swan the engraving. Greville specialised in lower plant botany, especially algae (seaweeds) and ferns, and was also a competent illustrator.

Later, in 1842, Hooker published *Genera Filicum; or illustrations of the ferns, and other allied genera; from the original coloured drawings of the late Francis Bauer Esq., botanic painter to Her Majesty; with additions and descriptive letterpress, by Sir William Jackson Hooker*. The lithographs for this work were executed by William Fitch using zinc plates.

Hooker's significance for Australian botany began while he was at Glasgow, but increased greatly when he took up the directorship at the Royal Botanic Gardens. He, and the herbarium at Kew, became the major recipient of collections from all around the world including, of course, Australia. This replacement of Sir Joseph Banks and his Soho herbarium (now absorbed into the British Museum) as the focus of taxonomic botany took twenty years to come about. Once it did Hooker, his network, his artists, and the accumulated collections and publications combined to found one of the greatest herbaria of all time.

John Lindley [4] (1799–1865) was an active academic and colleague of William Hooker. He was born into the nursery industry, his father owning a nursery in Norwich. Lindley was self-taught and yet another workaholic. He came under the notice of both William Hooker and Joseph Banks early in his career and Banks employed him as an assistant in his library and herbarium. This certainly meant that he also came into contact with Robert Brown. In 1822, following the death of Banks, Lindley became an assistant secretary for the Royal Horticultural Society and retained a life-long association with it. Lindley was appointed the inaugural Professor of Botany at University College, London, in 1829, a post which he held until 1860. In 1838 he prepared a report on the Royal gardens at Kew which led to their becoming the Royal Botanic Gardens.

Lindley's publication record is enormous, the more so given that he illustrated many of his publications himself. He was blind in one eye which must have been something of a handicap. In illustration he collaborated with Augusta Innes (Baker) Withers and Sarah Ann Drake. Sarah Drake, in particular, provided many excellent drawings for his works and for the works of others with whom he was associated.

Botanically Lindley collaborated with other workers, especially horticulturists, such as Joseph Paxton. Academically he furthered the interest of taxonomic botany by coding and standardising the terminology and by preparing several educational texts so competently that his work on botanical terminology is still used as a fundamental source. Lindley specialised in the taxonomy of orchids. As an artist he was quite competent, though not in the league of those he employed, such as Drake.

Sarah Drake, who came from the same area in Norfolk as John Lindley and his wife, was invited to join the Lindley family in London around 1830. Drake seems to have filled several roles, including governess, being affectionately known as Ducky by Lindley's children. Clearly she was inspired by Lindley to take up botanical art and soon took over from him as the major illustrator of his works. Drake returned to Norfolk in 1847 and looked after some elderly relatives. She married John Sutton Hastings in 1852. Drake died in 1857, reputedly of diabetes, but there is speculation that she may have absorbed cumulative poisons from her painting materials.

Drakea, a Western Australian genus of orchid, was named for Sarah Drake by Lindley in 1840. Lindley's concept of the genus required resolution. His type description of *Drakea elastica* actually embraced several species, though he thought he was dealing with only one. His words were too brief to help the scientists decide which specimen should be selected as the type of the *Drakea elastica*. The use of the woodcut in *A Sketch of the Vegetation of the Swan River Colony* (1839) helped resolve the definition of *Drakea elastica* and the naming of the other species.

Lindley became associated with Australian material in several ways, including his passion for orchids and through his involvement with some of the periodicals. Lindley was editor of Sydneham Teast Edwards' *Botanical Register* (renamed *Edwards's Botanical Register*) from 1829 until it was discontinued in 1847. This work treated numerous Australian species. Lindley also worked on the collections made during Thomas Mitchell's expeditions surveying inland regions of eastern Australia.

Independently, Francis Bauer also developed a passion for orchids. The two men collaborated to publish *Illustrations of Orchidaceous Plants by Francis Bauer Esq. F.R.S., L.S. & H.S.*

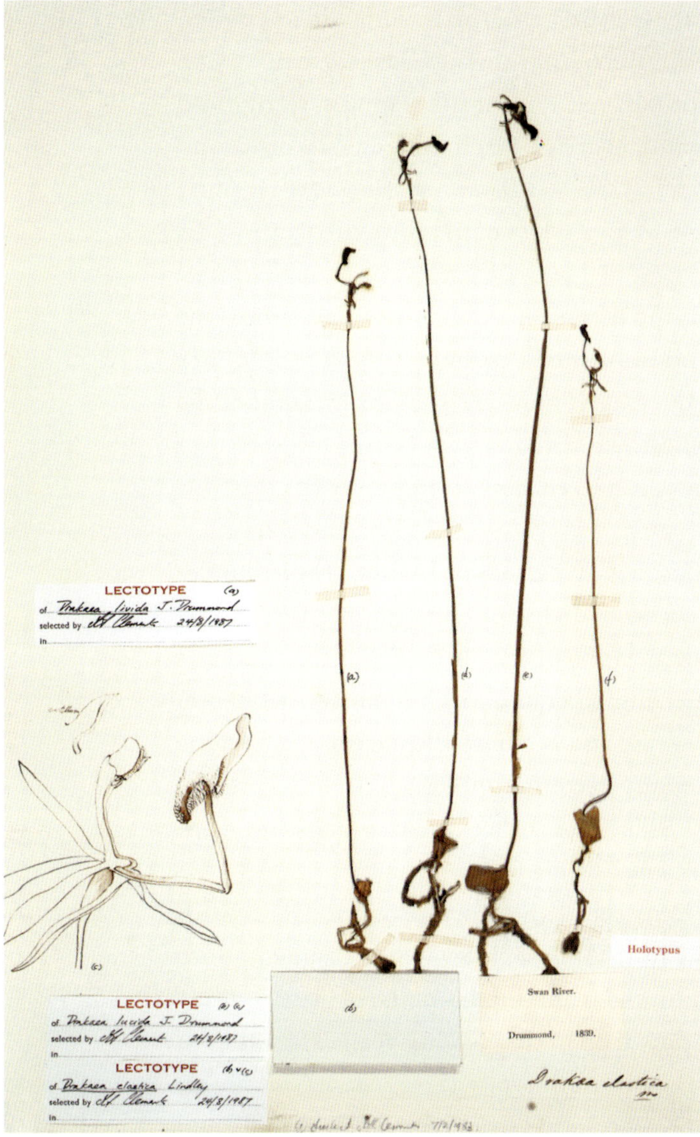

Drakea elastica Lindley
Herbarium specimen of *Drakea elastica* (left) with a pen sketch by John Lindley.
Scientists have used this sketch to define Lindley's concept of the species and have chosen the specimen as its lectotype.
The engraver of the woodcut (below), published in *Edwards's Botanical Register* in 1839, has based his engraving on Lindley's sketch.
[KEW] [CBG 6.0 x 5.0]

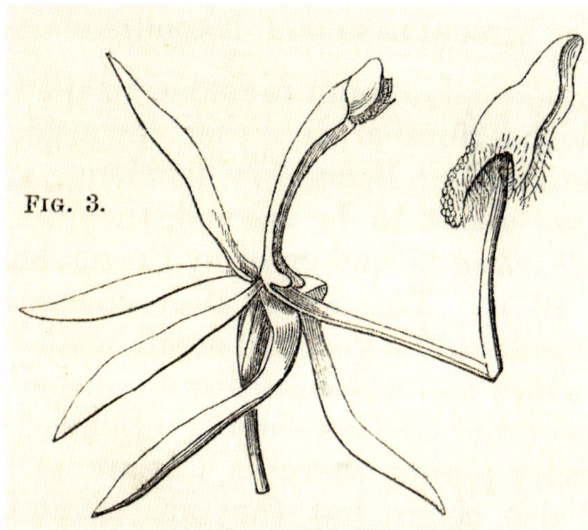

with Notes and prefatory Remarks by John Lindley. This appeared in fascicles between 1830 and 1838, making a total of thirty-five plates, twenty-eight of which were by Francis Bauer and seven by John Lindley. The plates were lithographed by Maxim Gauci.

In 1993 Joyce Stewart and William Stearn published *The Orchid Paintings of Franz Bauer*. This is an excellent tribute to Bauer, being a comprehensive attempt to expose his development as an artist, his outstanding expertise, and his remarkable attention to minute detail through the use of the microscope.

Lindley published his own illustrated orchid treatment, *Sertum Orchidaceum: a wreath of the most beautiful orchidaceous flowers*, as fascicles between 1837 and 1841. It included forty-nine plates which were hand-coloured lithographs based on drawings predominantly by Sarah Drake. Gauci was the lithographer for this work also.

Lindley served to complement and enhance the focus of taxonomic botany being brought about by Hooker. Even though a little to one side, Lindley made a major contribution to the record of Australian flora, literally, scientifically and illustratively.

Walter Hood Fitch [5] (1817–1892) began a career in Glasgow as an artist for a calico-printer. In his spare time he worked for William Hooker in the herbarium at Glasgow University. It is probable that he was seeking inspiration for designs and motifs. Hooker recognised his ability and employed and trained him as a botanical artist. Fitch joined Hooker around 1833 and his work began to be published in 1834 in *Curtis's Botanical Magazine*. It must have been an immense relief for Hooker to be able to spend less of his time illustrating his own publications.

Fitch went on to become one of the most prolific botanical illustrators of all time. One estimate is that he achieved over twelve thousand published drawings. This productivity must have been due largely to his own abilities and tenacity; however, the employment of the lithographic technique must also have made an enormous difference. A couple of years after his first scientific drawings were published, Fitch began to employ lithography to create his images.

Fitch learned lithography around 1829 in Glasgow. It is thought that he studied under Hugh Wilson. Thus, he came to Hooker with the knowledge of lithographic technique. In 1837 Hooker began his *Icones Plantarum*, the periodical in which Fitch had his first lithographic work published. It is considered that this liberated Fitch in an artistic sense. Whether that is so or not, Fitch was an extremely fast and accurate illustrator and became renowned for his added ability of being able to mentally 'reconstitute' a plant from its pressed state on an herbarium sheet into a realistic 3D illusion. Most other illustrators had to work from live plant material to achieve a satisfactory image.

When Sir William Hooker was appointed Director of the Royal Botanic Gardens in 1841 and moved to London, Fitch chose to follow him. As Joseph Swan had decided not to join them as engraver, Hooker was left with a gap in the team. Fitch's range of artistic abilities, including lithography, was therefore heavily drawn on in the ensuing years. He was even a competent woodcut engraver, an ability drawn upon to illustrate the *Popular Guide to the Royal Botanic Gardens at Kew* (1844). The move to Kew resulted in an increase in the rate and range of Hooker's publications and William's son, Joseph Dalton Hooker, was beginning his publishing career too. Fitch had to work hard to accommodate all the demands of illustrating periodicals, journals and one-off books.

While these eminent scientists, Hooker and Lindley, were establishing themselves in their careers, the scientific record continued to proliferate, appearing in both scientific and grandly illustrated entrepreneurial periodicals.

In 1825 Benjamin Maund began *The Botanic Garden*, publishing thirteen volumes until 1850. Maund also published *The Botanist*, a five-volume work, from 1836 to 1842. Edwin Dalton Smith did the paintings for the early volumes of *The Botanic Garden* and the latter volumes and *The Botanist* were illustrated by several women artists including Augusta Innes Withers, Jane Taylor and S. Maund. S. Watts did the engraving for *The Botanic Garden*, and both he and S. Nevitt were the engravers for *The Botanist*. A feature of Benjamin Maund's works is the inclusion of miniatures and vignettes. Miniature woodcuts, which

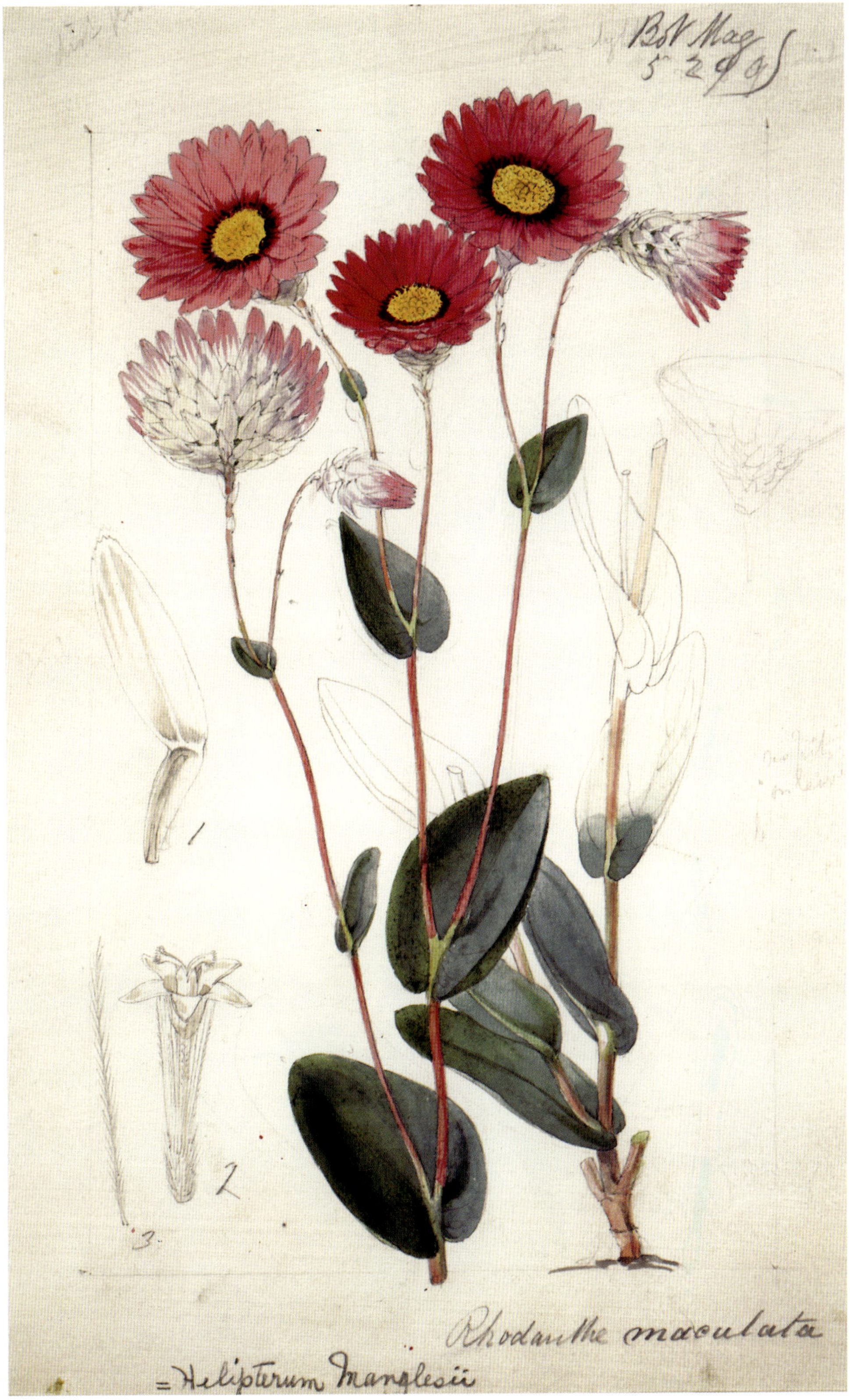

Helipterum manglesii (Lindley) Benth. (= *Rhodanthe manglesii* Lindley)
Watercolour painting by Walter Fitch, one of the most prolific botanical illustators of all time. Fitch was renowned for his ability to mentally 'reconstitute' a plant from its pressed state into a realistic 3D illusion.
[KEW 26.0 x 17.0]

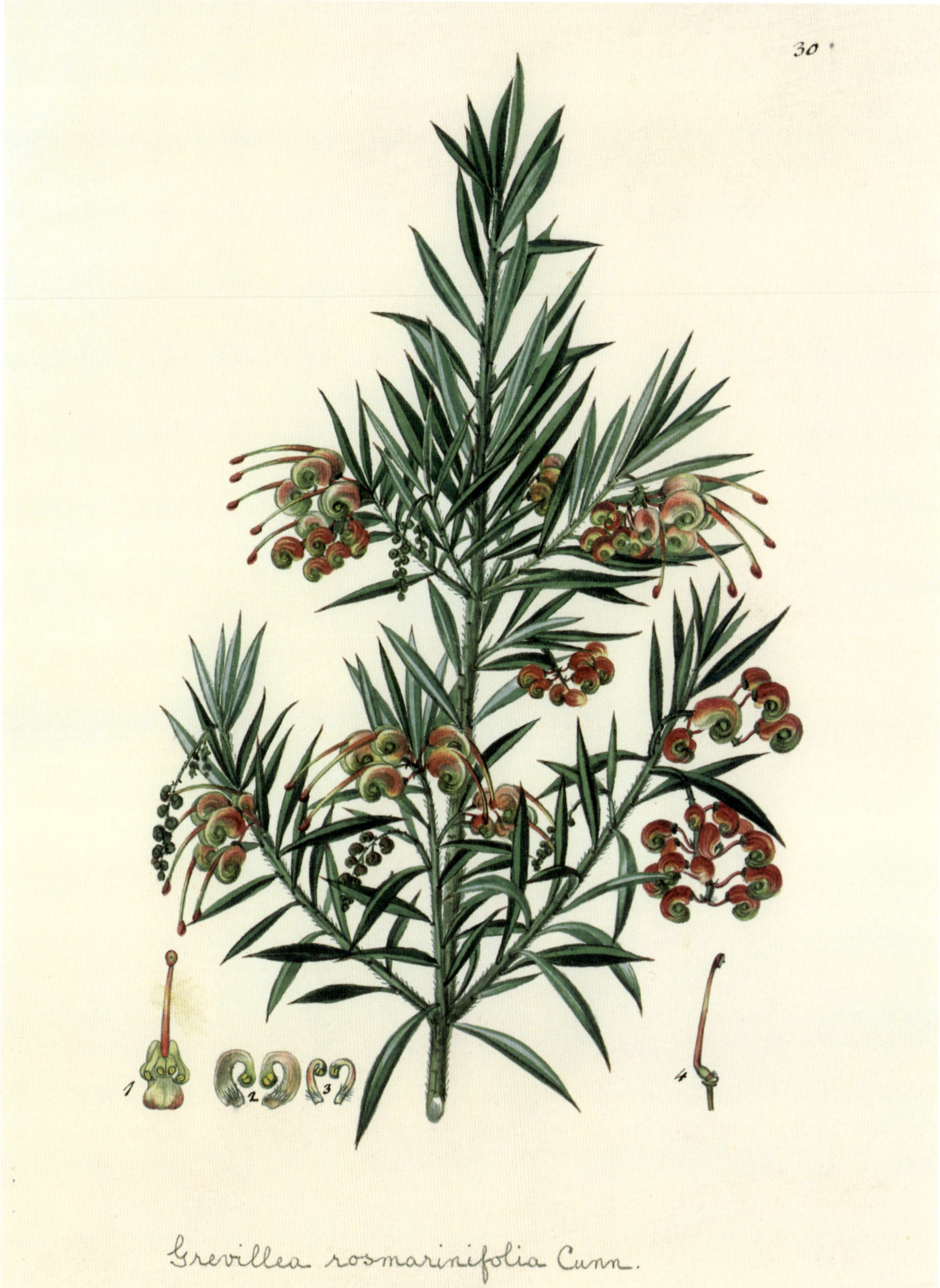

Grevillea rosmarinifolia A.Cunn
Watercolour painting by Edwin Smith for publication in Robert Sweet's *Flora Australasica* in 1827–1828.
[BM]

have no attribution, were used to facilitate an understanding of the classification of the plants. In later volumes of *The Botanic Garden*, miniature woodcuts provide a Floral Register, or index, of the species illustrated in the volume.

On a much less voluminous scale, Robert Sweet [6] published his *Flora Australasica* in 1827–1828. Sweet was born in Devonshire and took up employment as a gardener at the age of sixteen. This led to his association with a series of nurseries at Stockwell, Fulham and Chelsea, and to an interest in plants beyond the garden. By 1818 he was publishing horticultural and botanical works. His *Flora Australasica*, by its title, is a fourth Flora of Australia, though it contains only fifty-six species and plates. It is illustrated with hand-coloured copper engravings, the drawings by Edwin Smith and the engraving by S.Watts.

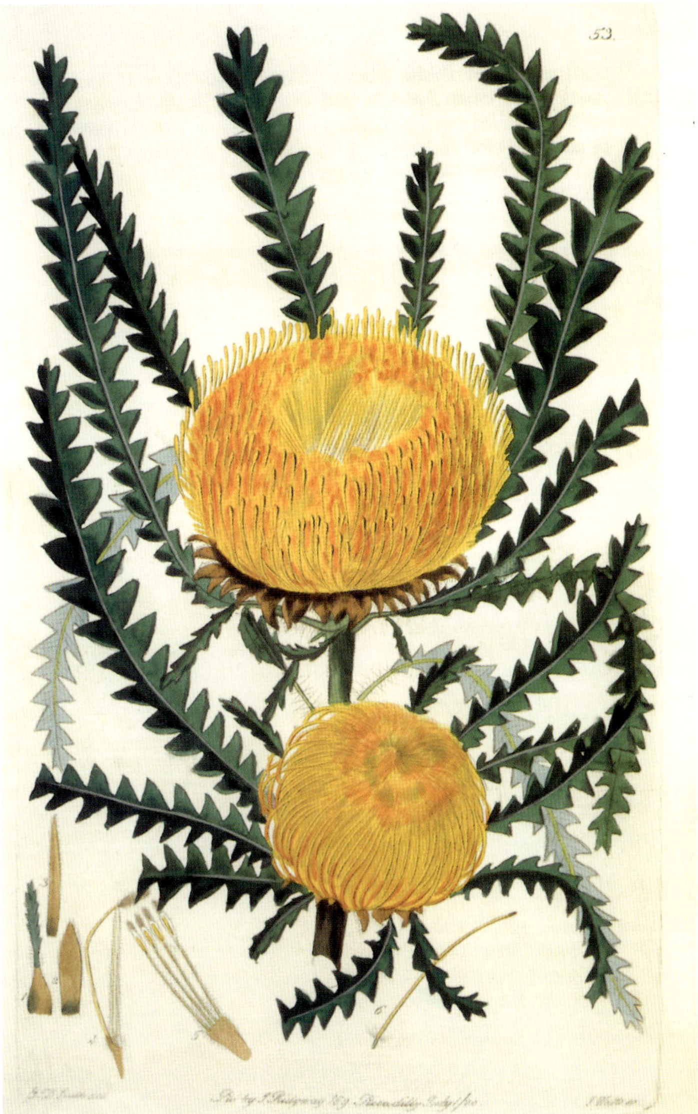

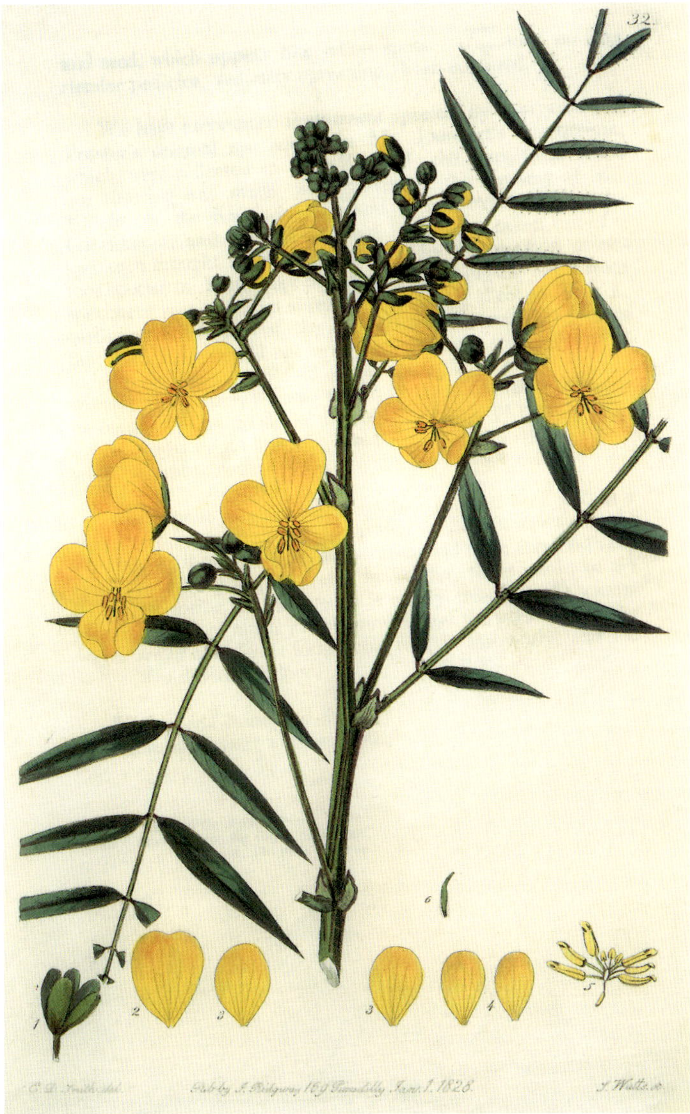

Dryandra formosa R.Br. and *Senna barklayana* (Sweet) Randell (= *Cassia barklayana* Sweet)
Hand-coloured copper engravings, engraved by S. Watts from paintings by Edwin Smith and published in Robert Sweet's *Flora Australasica*, 1827–1828.
[CANB 20.3 x 11.7] [CANB 20.2 x 12.4]

Edwin Smith started his career as a fruit and flower painter. Sweet recognised Smith's ability and employed him to provide scientific illustration. Smith seems to have begun formal botanical work around 1820, and was elected a Fellow of the Linnean Society by 1823. However, he seems to have supported himself more as a miniature portrait painter. He applied the skill of miniature work to botanical illustration too. This is especially exemplified in Maund's *The Botanic Garden*.

In 1834 Joseph Paxton began *Paxton's Magazine of Botany*. This ran to sixteen volumes until 1849. *Paxton's Flower Garden* scarcely qualifies as a periodical, having only three volumes published between 1850 and 1853. John Lindley was a co-author of this. The team of Paxton and Lindley resulted in an illustration style characteristic of Lindley's educational publications, where the text has illustrative drawings within the letterpress of the page. The drawings were reproduced as woodcut engravings.

Both of Paxton's works, however, are grand, being illustrated with colour plates and woodcut engravings. In the *Magazine* there were four artists and artisans. Frederick William Smith painted and engraved his own subjects, and Samuel Holden painted and engraved or lithographed his own subjects. The woodcuts are not signed, but C.J. Flemin and Thomas Orlando Sheldon Jewitt are are thought to have done them.[7] The woodcuts are superior to the other artwork.

In the *Flower Garden* the painting and lithography were done solely by L. Constans. Constans' work is superior to that in the *Magazine*. The woodcut engraver remains anonymous.

Not a great deal is recorded of these artists. Frederick Smith is reportedly a brother of Edwin Smith; Samuel Holden was a botanical artist who specialised in painting orchids; and Thomas Jewitt was both a botanist and wood-engraver. Jewitt illustrated with both woodcut and lithograph, mainly for horticulturalists. He also worked for the algologist, W.H. Harvey, illustrating his *Phycologia Britannica* (1846–1851). Little more than names is recorded for both Flemin and Constans.

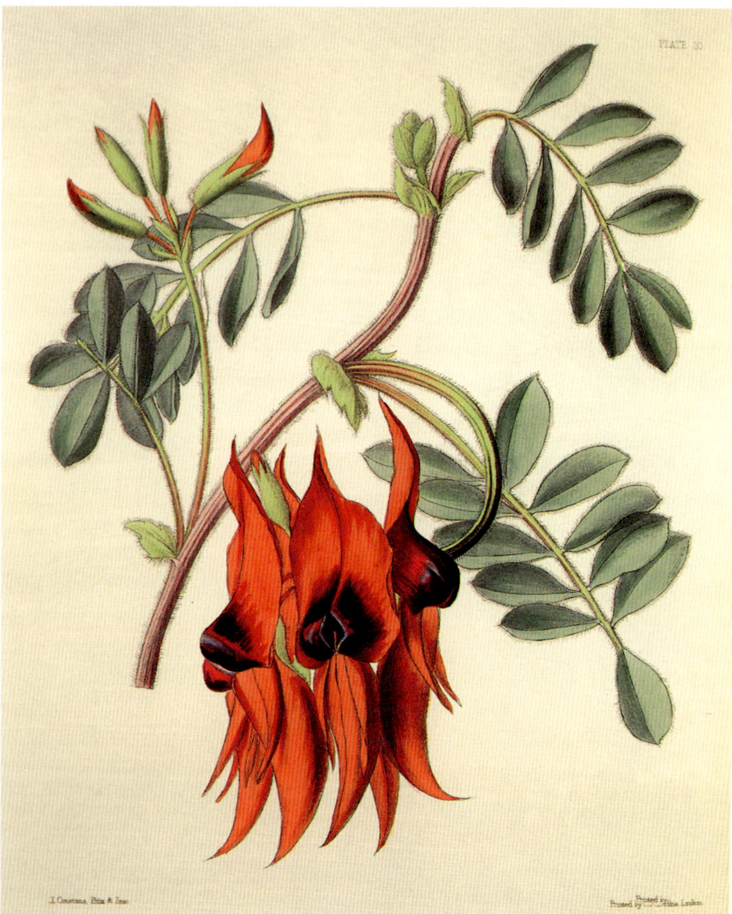

Swainsona formosa (G.Don) J. Thompson
(= *Clianthus dampieri* Lindley)
Hand-coloured lithograph of Sturt's Desert Pea drawn and lithographed by L. Constans, of whom little is known. It was published in Paxton's *Flower Garden* in 1850–1851.
[CANB 24.5 x 18.5]

Acacia oxycedrus DC. and *Acacia genistifolia* Link (= *Acacia diffusa* Ker Gawler)
Woodcut engravings were included within the letterpress of the page in Joseph Paxton's *Flower Garden* published in 1850–1851. The artist and engraver are not acknowledged — the artist was possibly L. Constans. [CANB]

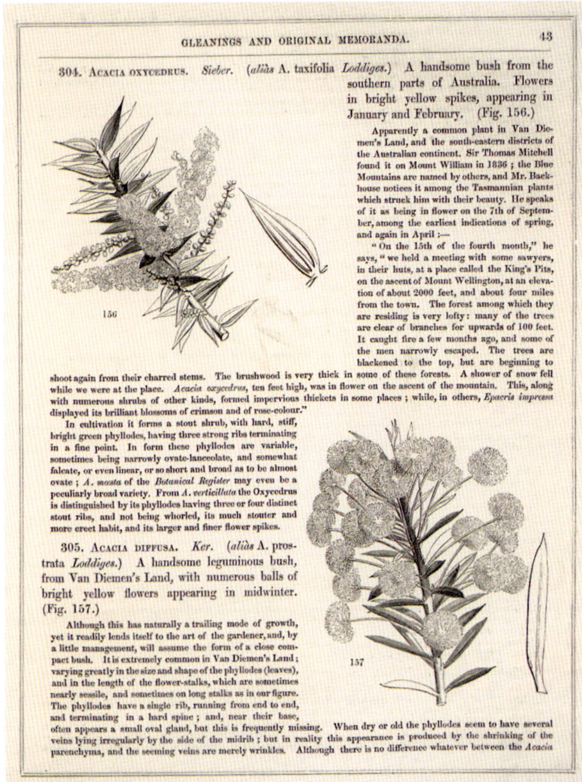

GLEANINGS AND ORIGINAL MEMORANDA. 43

304. ACACIA OXYCEDRUS. *Sieber.* (*aliàs* A. taxifolia *Loddiges.*) A handsome bush from the southern parts of Australia. Flowers in bright yellow spikes, appearing in January and February. (Fig. 156.)

Apparently a common plant in Van Diemen's Land, and the south-eastern districts of the Australian continent. Sir Thomas Mitchell found it on Mount William in 1836 ; the Blue Mountains are named by others, and Mr. Backhouse notices it among the Tasmannian plants which struck him with their beauty. He speaks of it as being in flower on the 7th of September, among the earliest indications of spring, and again in April :—

"On the 15th of the fourth month," he says, "we held a meeting with some sawyers, in their huts, at a place called the King's Pits, on the ascent of Mount Wellington, at an elevation of about 2000 feet, and about four miles from the town. The forest among which they are residing is very lofty: many of the trees are clear of branches for upwards of 100 feet. It caught fire a few months ago, and some of the men narrowly escaped. The trees are blackened to the top, but are beginning to shoot again from their charred stems. The brushwood is very thick in some of these forests. A shower of snow fell while we were at the place. *Acacia oxycedrus*, ten feet high, was in flower on the ascent of the mountain. This, along with numerous shrubs of other kinds, formed impervious thickets in some places ; while, in others, *Epacris impressa* displayed its brilliant blossoms of crimson and of rose-colour."

In cultivation it forms a stout shrub, with hard, stiff, bright green phyllodes, having three strong ribs terminating in a fine point. In form these phyllodes are variable, sometimes being narrowly ovate-lanceolate, and somewhat falcate, or even linear, or so short and broad as to be almost ovate ; *A. mœsta* of the *Botanical Register* may even be a peculiarly broad variety. From *A. verticillata* the Oxycedrus is distinguished by its phyllodes having three or four distinct stout ribs, and not being whorled, its much stouter and more erect habit, and its larger and finer flower spikes.

305. ACACIA DIFFUSA. *Ker.* (*aliàs* A. prostrata *Loddiges.*) A handsome leguminous bush, from Van Diemen's Land, with numerous balls of bright yellow flowers appearing in midwinter. (Fig. 157.)

Although this has naturally a trailing mode of growth, yet it readily lends itself to the art of the gardener, and, by a little management, will assume the form of a close compact bush. It is extremely common in Van Diemen's Land ; varying greatly in the size and shape of the phyllodes (leaves), and in the length of the flower-stalks, which are sometimes nearly sessile, and sometimes on long stalks as in our figure. The phyllodes have a single rib, running from end to end, and terminating in a hard spine ; and, near their base, often appears a small oval gland, but this is frequently missing. When dry or old the phyllodes seem to have several veins lying irregularly by the side of the midrib ; but in reality this appearance is produced by the shrinking of the parenchyma, and the seeming veins are merely wrinkles. Although there is no difference whatever between the *Acacia*

While these scientists and artists were busy describing the plants of the world, yet more plants were arriving from afar. In Australia, collection continued at a pace, sometimes by private residents, sometimes by officials and explorers, and sometimes by intrepid Europeans with backing from a variety of sources.

Among the first in the latter category was William Baxter [8] who appears to have been employed by a number of nurserymen and seed merchants, John Mackay of Clapton being one of them and Francis Henchman another. Baxter made two visits to Australia, one between 1823 and 1825 and another between 1828 and 1829. On the first trip he is known to have collected from King George Sound east to Kangaroo Island. On the second trip he sailed on the *Lucy Ann* and collected at Wilsons Promontory in Victoria and Cape Arid, Lucky Bay and King George Sound in Western Australia. On this trip Charles Fraser arranged for Baxter to supply material to the Sydney Botanic Gardens as well as to English employers. The Governor of the day, Governor Ralph Darling, approved this arrangement. When Baxter returned from the west in 1829, he tried to cheat Fraser out of his due portion of the collections for the Sydney Gardens but Fraser called in the Collector of Customs and forced the honouring of the agreement.

For science, Baxter's collections made their way into the hands of Robert Brown, Robert Sweet and William Hooker.

Franz Wilhelm Sieber,[9] a Czechoslovakian, collected in New South Wales for approximately nine months in 1823. Sieber's backing is not clear, however he was a professional collector and made numbered sets for sale in Europe. One of these sets was obtained by Alphonse Louis Pierre Pyramus de Candolle, and many species in it were described in the *Prodromus* (1823–1873). Unfortunately the *Prodromus* is not illustrated.

Allan Cunningham returned to Australia from New Zealand in 1827 to continue his collecting and exploring. In 1828 Charles Fraser and Allan Cunningham set off for the recently established Moreton Bay Settlement in the *Lucy Ann*. In the intervening year, Cunningham had constrained himself to report-writing and one geographical exploration to the Darling Downs. Now that he was on the other side of the range he wanted to find a passage through to the Darling Downs from Moreton Bay. At Moreton Bay the party explored and collected locally and Fraser completed his mission to establish a public garden. Fraser then returned to Port Jackson and Cunningham went on to find the pass. This, now known as Cunningham's Gap, provided the way for agriculture to extend beyond the confines of the Great Dividing Range, west of the Moreton Bay settlement.

In England, Robert Brown published his Supplement to the *Prodromus Florae Novae Hollandiae*, the *Proteaceas Novas*, in 1830. Brown had capitalised on the collections of Baxter, Caley, Cunningham, Fraser and Sieber, which had been incorporated into the collection at the British Museum. The work of these collectors was being recognised both in science and in horticulture but, like the *Prodromus* in 1810, this Supplement was not illustrated and yet another opportunity was missed to publish relevant illustrations of Ferdinand Bauer.

Meanwhile Cunningham continued working in the colony. In 1831, however, he was recalled to England due to a need to economise at the Kew gardens. Once there, Cunningham set about putting his collection in order.

When Charles Fraser died later in 1831, Allan Cunningham was offered the post of Colonial Botanist. He declined the offer and recommended that his brother Richard would be a suitable recruit and Richard was duly appointed.

Richard Cunningham (1793–1835) was a younger brother of Allan, and both had gone to Kew in 1808 and worked for William Aiton. Here they came under the notice of Robert Brown and Joseph Banks. While Allan was away on his extended collecting trip, Richard was employed as a clerk at Kew. With this experience and the support of Brown, Richard was successful in gaining the post of Colonial Botanist and Superintendent of the Sydney Botanic Gardens. He arrived at Port Jackson early in 1833, together with a collection of ornamental and useful plants. Of particular significance were fruit trees and grape vines for cultivation by local orchardists and farmers.

Richard immediately worked on establishing his stock before setting off to expand the collection with more local plants. Late in 1833 he travelled to New Zealand. Back in the colony he joined Mitchell's expedition to survey the Darling River in 1835. Richard seems not to have been a good bushman. He strayed from Mitchell's party in the vicinity of the Bogan River and was lost. Three days later he was found dead, killed by Aborigines.

Following the death of Richard, Allan was again entreated to take up the posting of Colonial Botanist. He accepted and was back in the colony early in 1837. Meanwhile, James Anderson was made acting Superintendent of the Sydney Botanic Gardens.

When Allan arrived, he was not impressed with the volume of market gardening he was required to do and resigned. He wrote:[10]

> Tell all that I have discharged the Government cabbage-garden in disgust, and am now to enter with all my might, mental and corporeal, on a more legitimate occupation for a few months.

That, regrettably, was all he had. Seventeen months later, in June 1839, he died of the tuberculosis which racked his body. His friend, James Anderson, now Superintendent of the Gardens, cared for him in the last few months.

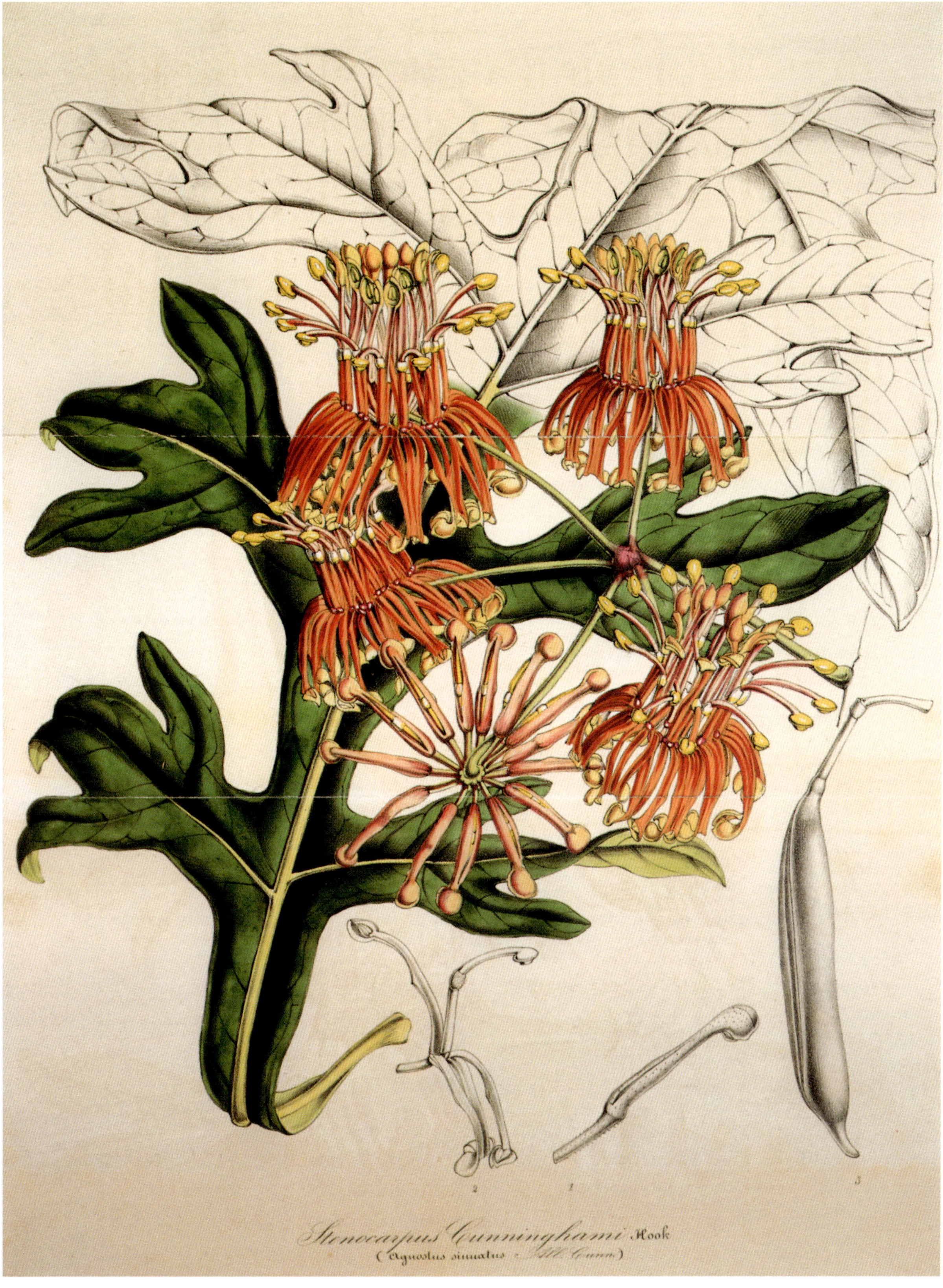

Stenocarpus sinuatus Endl. (= *Stenocarpus cunninghamii* Hook.)
Unacknowledged hand-coloured lithograph based on a drawing in *Curtis's Botanical Magazine*, published in the French periodical, *Flore*, in 1847.
[CANB 33.0 x 23.2]

Thomas Livingstone Mitchell[11] (1792–1855) was appointed to the General Survey Department, New South Wales, in 1827. He came to the posting with the extremely good grounding in sciences necessary to make an excellent explorer. Although he did not, however, have a particularly good knowledge of botany, this was not a disadvantage. In 1828 he became Surveyor General, succeeding John Oxley.

During the twenty-seven years he held the post, Mitchell led four major expeditions of exploration. In 1831 he explored western New South Wales in search of a large river flowing to the north-west. In 1835 he set out to prove whether the Darling River flowed into the Murray River. Both expeditions were abandoned due to confrontations with Aborigines. Richard Cunningham died on the second expedition, though his death did not precipitate the abandonment.

In 1836 Governor Bourke sent Mitchell to survey the course of the Darling from where he had stopped the year before. This expedition was very successful and led him to both the Murrumbidgee and Murray Rivers and beyond into Australia Felix, now Victoria. He even reached the south coast in the vicinity of Portland. For this journey he was knighted, given an honorary degree at Oxford University, and made a Fellow of the Royal Society of London.

In 1845 Mitchell set out to find an overland route to the Gulf of Carpentaria. He discovered a north-westerly flowing river which he believed would run into the Gulf. He named it the Victoria. His Victoria River is now known as the Barcoo River. In reality the water in his river evaporated on its way to the Centre. Mitchell did not continue his journey to the Gulf.

Mitchell published three journals reporting on these four expeditions. He proved to be a skilled draftsman and made several landscape and human studies. These were published as lithographs and a few engravings. Plants were not included as illustrations other than as obvious components of the landscape or as studies of a whole tree. There is a pleasing drawing of the tree and fruit of *Eucarya murrayanum* Mitchell, the Bitter Quandong. The drawing and description were both provided by Mitchell. Edmund Besley Court Kennedy,[12] second in command on the northern journey, provided a drawing of an unusual *Brachychiton* or Bottle Tree.

The plants collected on these expeditions were sent back to England for study by Lindley who provided appropriate editing of the text together with very brief descriptions of new species. These were published as footnotes in each of the journals, *Three expeditions into the interior of eastern Australia* (1838), and *Journal of an expedition into the interior of tropical Australia* (1848).

Numerous expeditions were being undertaken to explore the inland at this time. Geology and natural history benefited from most of these in some degree.

Charles Sturt[13] was born in India and joined the British Army. In 1826 he came to Australia in charge of convicts. His interest in Australia lead to his appointment to lead an expedition to follow the Macquarie River. This expedition discovered the Darling River. In 1829 Sturt carried out a similar expedition, this time following the Murrumbidgee River. The expedition 'discovered' the Murray River, which Hume had previously discovered in 1824, and traced it to its termination in Lake Alexandrina.

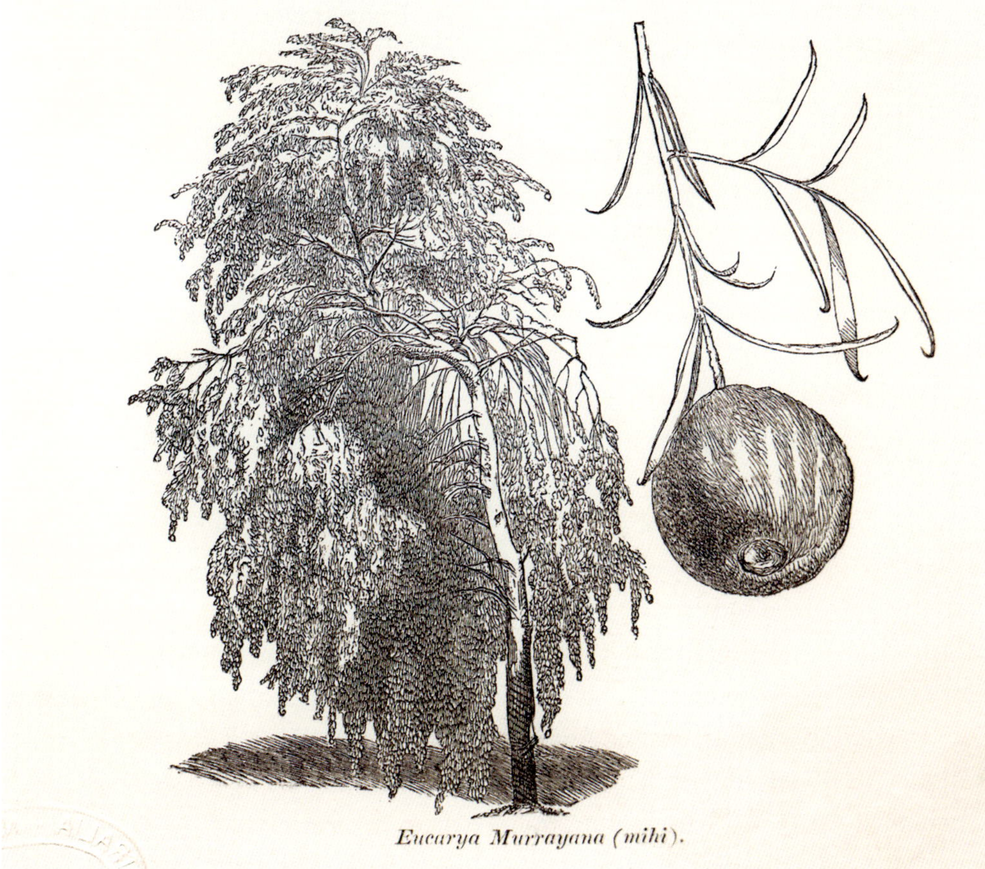

Santalum murrayanum (Mitchell) C.Gardn. (= *Eucarya murrayana* Mitchell)
Woodcut engraving of Bitter Quandong based on a painting by the explorer Thomas Mitchell and published in his journal, *Three Expeditions into the interior of Eastern Australia,* in 1838.
[NLA 9.5 x 11.0]

Following the 1829 expedition, Sturt returned to England on sick leave. When his health was restored he returned to Australia in 1834 to settle, first in New South Wales and then in South Australia where he became Surveyor General. In this capacity he explored central Australia. In 1844 he set out to reach the Centre by following the Murray and then the Darling Rivers to the north and then north-west to the Grey Range. Drought and extreme heat thwarted successive attempts to explore further, the Simpson Desert forming an unsurmountable barrier. By all accounts the explorers endured great hardship, even suffering the sailor's disease, scurvy. Two men lost their lives. Sturt published journals reporting on his expeditions.[14]

Botanist John Carne Bidwill[15] came to Australia in 1838 and worked around Sydney and briefly in New Zealand. He became Government Botanist and Director of the Botanic Gardens for a few months in 1847 before being replaced by Charles Moore, an English appointee. Bidwill was later appointed Commissioner of Crown Lands in Queensland. This posting lead to his demise. While working alone to blaze a new road between Moreton Bay and Wide Bay, he became ill and starved for eight days. Even though he was ultimately rescued, he died of complications.

Friedrich Wilhelm Ludwig Leichhardt[16] emigrated to Australia from Germany in 1842. He was a competent natural historian and wanted to explore, collect and reveal Australia's plant and animal wealth for science. Leichhardt undertook several exploratory expeditions between Sydney and Moreton Bay in the first couple of years after arrival. Then in 1844 he set out from Moreton Bay to travel across land to Port Essington. The expedition was successful, but not without loss and hardship. John Gilbert,[17] former collector for John Gould, lost his life.

Science was a loser too. Leichhardt wrote of this to a French colleague, Gaetano Durando, in 1846:[18]

> You will easily imagine, that even allowing I did my best, a man entirely given to one occupation could have done much more. This refers however almost entirely to the sketching of my plants, and to the putting down more explicit notes on the different espèces we found. For the time came, when I had to open all my fine green hide boxes, to make a poor choice of the dried plants, and to throw the greatest number of them away unable to carry them any further, as 4 of my pack horses drowned, and the means of carrying my collection of plants and geological specimens were consequently destroyed. I fully lost by this 4,000–5,000 specimens. There are however still some very interesting remnants, which I shall send to you, to have them determined, allowing you the duplicates and triplicates. I shall do the same with my Moreton Bay plants, for I wish very much to establish a good well-named herbarium in the Museum in Sydney, that we have some means of ready comparison.

Many collections found their way back to Germany and France. Then in 1848 Leichhardt set off from Moreton Bay to cross overland to the west coast. The party perished and their fate has still not been resolved.

Other explorers of the time include Paul Edmund de Strzelecki,[19] George Grey, John Eyre,[20] and Edmund Kennedy.

Gradually the *terra incognita* view of Australia was diminishing due to the efforts of the explorers. Most of them contributed to early scientific knowledge in a multitude of ways including collections of plants which were transported back to Europe. William Hooker and John Lindley, and their respective teams of artists, then played their role in bringing these into the scientific record.

7. Establishment

In Australia itself, in advancing from the tropics to the coldest latitudes, or from the driest to the most humid districts, or from the interior to the seashore, or in ascending the mountains, the changes in vegetation are in every aspect analogous to what occur in other parts of the globe.

Joseph Hooker [1]

The eastern mainland of Australia attracted great interest among natural historians during the first forty or fifty years with Van Diemen's Land and Western Australia gaining attention later. New South Wales, the site of the first settlement, became the major centre of British activity in Australia. Van Diemen's Land and Western Australia, on the other hand, are geographically better placed to have attracted landings, especially to obtain fresh water, and to result in opportunist collection. And collection in those regions remained largely opportunist, or linked to survey journeys, until settlements were established.

In Van Diemen's Land, once the original settlement was established at Risdon Cove in 1804, collectors were attracted there. William Paterson also established a settlement at Port Dalrymple later in the same year.

Robert Brown was the first significant collector in Van Diemen's Land, visiting during its first year of settlement. A later one was Robert William Lawrence.[2] Lawrence's father arrived from England in 1823 as a settler, for health reasons. Robert stayed in England to complete his education before joining the family in 1825. Robert had developed a significant interest in natural history and horticulture and soon became an enthusiastic plant collector. He established a correspondence with William Hooker and began returning collections to

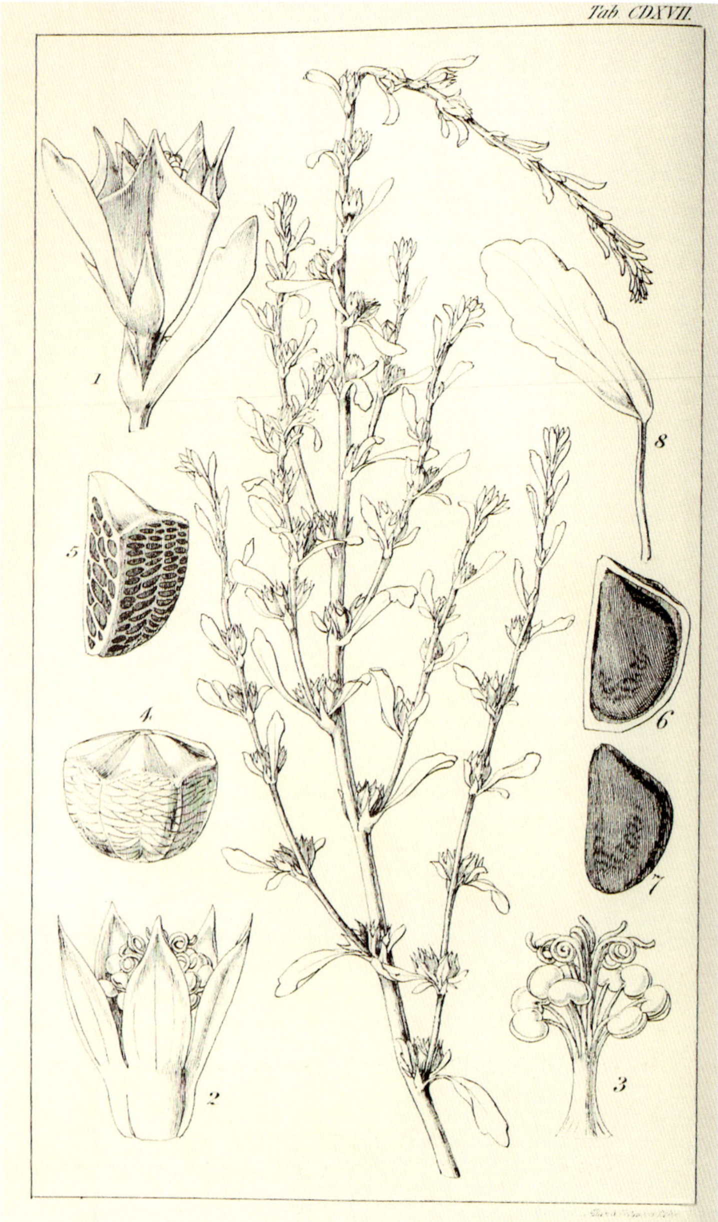

Lawrencia glomerata Hook.
Uncoloured lithograph drawn and lithographed by Walter Fitch and published in William Hooker's *Icones Plantarum* in 1842. Hooker named the genus in honour of Robert Lawrence.
[MEL 19.3 x 11.0]

Hooker in Glasgow. Lawrence also made acquaintance with Ronald Gunn and so stimulated active collection in Van Diemen's Land for many years to come. Unfortunately he was not destined to be active for long, dying in 1833, apparently from an epileptic fit.

Ronald Campbell Gunn[3] came to Hobart in 1830 as a Superintendent of Convicts and was transferred to Launceston in 1833. There he befriended Robert Lawrence and developed an interest in botany and collecting plants. In a sense, Gunn took up from where Lawrence left off, by continuing the liaison with William Hooker and by supplying collections to Hooker. In 1835 Gunn was posted to the less onerous position of Police Magistrate at Circular Head in the north-west. Here he had more time for collecting. His work apparently impressed Hooker sufficiently for him to speak highly of Gunn to John Franklin who became Lieutenant Governor of Van Diemen's Land in 1837. This in turn lead to Gunn being transferred to Hobart where he became an authority on natural history and met all the scientists visiting the settlement. From a botanical point of view, Joseph Hooker[4] was probably the most important of such visitors. Other contemporary collectors in Van Diemen's Land were Joseph Milligan,[5] James Backhouse,[6] and Johann Lhotsky.[7]

Joseph Dalton Hooker (1817–1911), second son of William Hooker, became another of the giants in the science of botany in Britain whose career was influenced by travel and Australia. Joseph grew up in a world of botany and natural history as his father pursued his teaching at Glasgow University. During that time Joseph had a classical education and went on to study medicine. Through his father's contacts, Joseph joined the James Clark Ross expedition in the *Erebus* and *Terror* to the Antarctic. Ross was a renowned Arctic explorer and he was an obvious choice by the British Government to lead an expedition to the Antarctic to

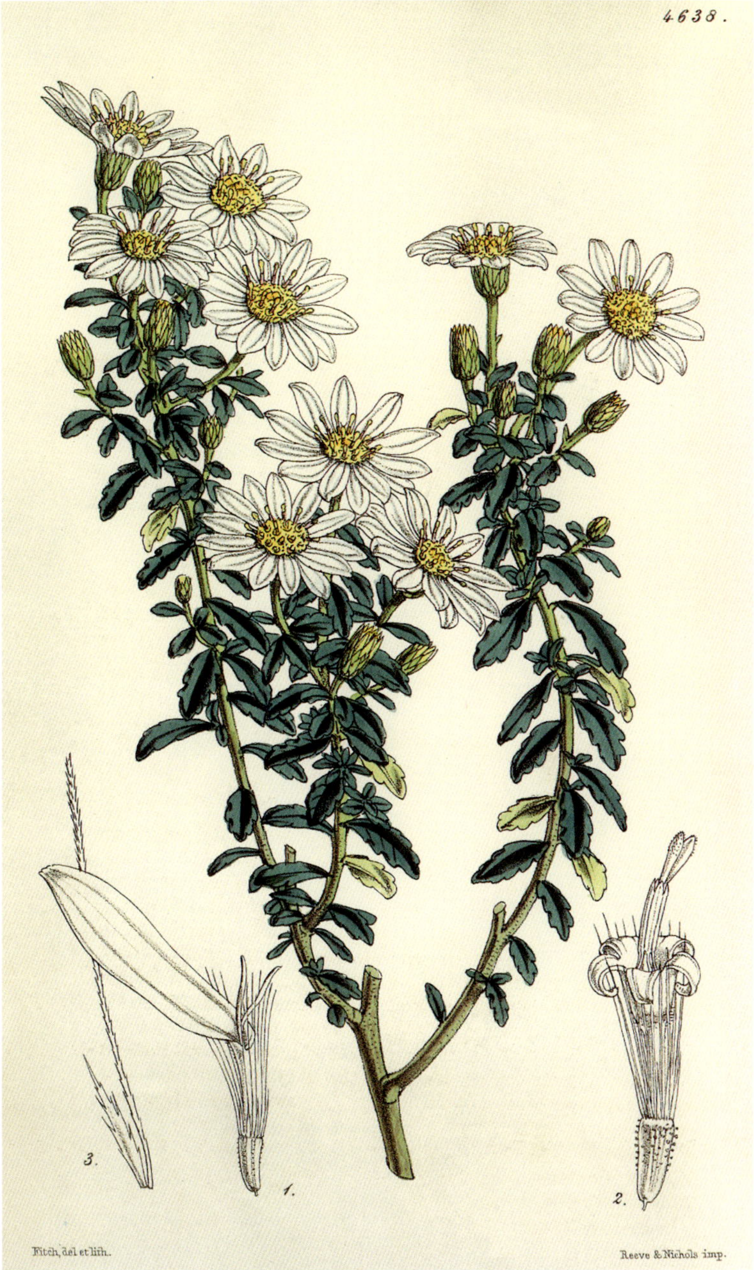

Olearia phlogopappa (Labill.) DC.
(= *Eurybia gunniana* DC.)
Hand-coloured lithograph drawn and lithographed by Walter Fitch and published in *Curtis's Botanical Magazine* in 1852. De Candolle named this species in honour of Ronald Gunn.
[MEL 23.5 x 14.5]

determine the position of the South Magnetic Pole. Hooker was appointed Assistant Surgeon and Naturalist.

The expedition sailed in September 1839 and returned almost exactly four years later. Ross sailed south for the Cape of Good Hope and then on to Kerguelen's Island. The explorers spent some time on Kerguelen before setting sail for Van Diemen's Land in mid-1840. In November Ross sailed south to Antarctica reaching 78 degrees S in the region of South Victoria Land and the Ross Sea. In 1841 Ross returned to Van Diemen's Land and then went north to Sydney. In Sydney, John Bidwill assisted Hooker with collections in the general vicinity.

The expedition then sailed east to New Zealand before trying to approach nearer to the South Pole for a second time. However the explorers came perilously close to disaster and made little further progress south. Ross then sailed north and east to the Falkland Islands in 1842. Following a short trip to Cape Horn the explorers set sail again for the Antarctic late in December. From there Ross sailed north to Capetown and back to England in September 1843.

Joseph seized every opportunity to collect and record whatever was available. Together with what was already known of the botany of the regions visited, and his own observations, Joseph was beginning to formulate what are now known as phytogeographic concepts — concepts of relatedness and distribution patterns of plants. He was witnessing, and must have been formulating, an image of what we now know as the Gondwanan flora.

This, of course, links with the observations made by Charles Darwin a few years earlier on the voyage of the *Beagle* between 1831 and 1836. Darwin was observing geological history

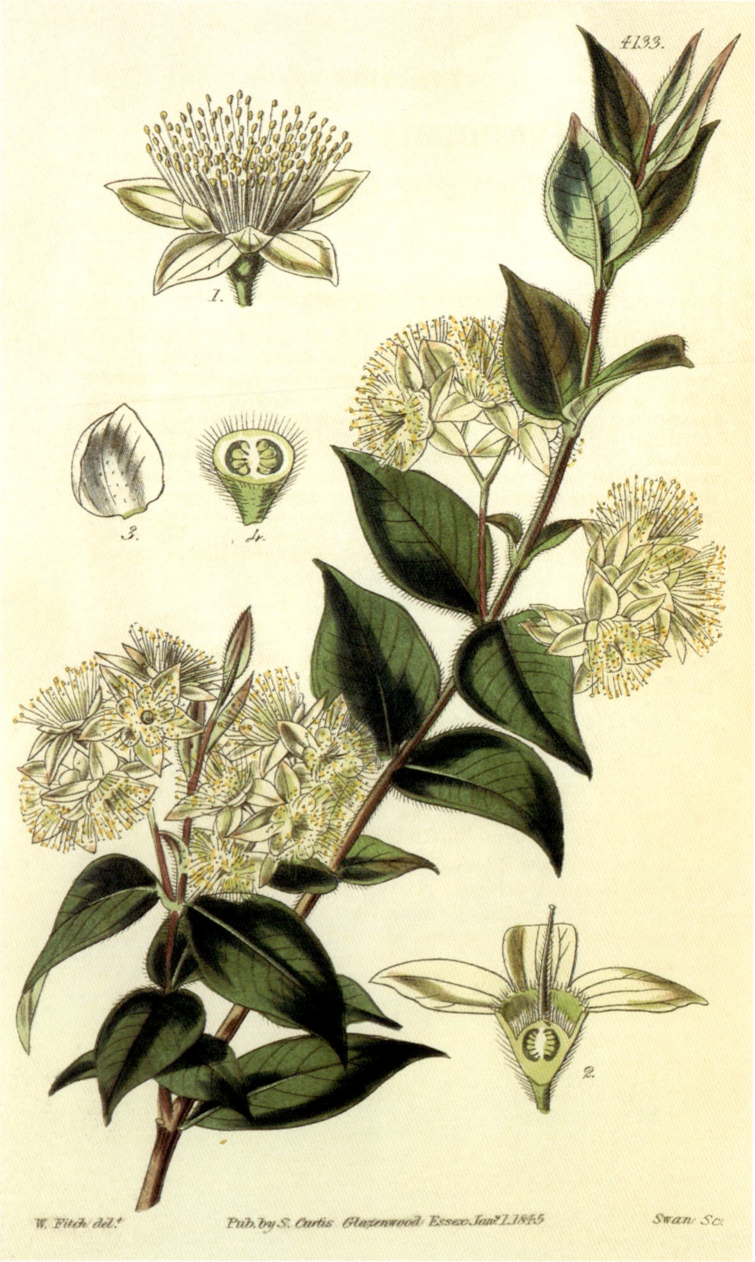

Backhousia myrtifolia Hook.
Hand-coloured engraving drawn by Walter Fitch, engraved by Joseph Swan, and published in *Curtis's Botanical Magazine* in 1845. Hooker named this species in honour of James Backhouse.
[MEL 20.0 x 12.0]

leading to his formulation of the theory of evolution.

Hooker spent a total of six months in Tasmania, three in the spring of 1840 and three in the autumn of 1841. During that time he was able to accompany Ronald Gunn, or a guide provided by Gunn, into a range of localities and habitats. Hooker wrote of this:[8]

> I can recall no happier weeks of my wanderings over the globe, than those spent with Mr Gunn, collecting in the Tasmanian mountains and forests, or studying our plants in his library, with the works of our predecessors Labillardière and Brown.

The mutual advantage of these few months for each man is inestimable. It certainly shows in Hooker's *Flora Tasmaniae.*

Joseph inherited many qualities from his father which worked to his advantage during his arduous voyage into some of the most dangerous waters on the earth's surface. One of course was his stamina and good health. From a botanical point of view, another was his ability to draw. Some of Hooker's drawings were used for the volume of plates depicting plants collected on the voyage; others were drawn and lithographed by Walter Fitch.

In England, while Joseph was away, Sir William Hooker took up his post as Director of the Royal Botanic Gardens. On his return, Joseph installed himself at Kew and started writing up the results of the Antarctic voyage. This was published in three volumes as *The Botany of the Antarctic Voyage of H.M. discovery ships Erebus and Terror in the years 1839–43*. The network which Joseph had set up in Van Diemen's Land continued to contribute to the material from which he drew for the third volume, *Flora Tasmaniae*, which was not published until 1860.

Calytrix acutifolia (Lindley) Craven
(= *Llotskya ericoides* Schauer)
Hand-coloured lithograph drawn by Matilda Smith, lithographed by John Fitch, and published in *Curtis's Botanical Magazine* in 1901. The genus was named in honour of Johann Lhotsky by Johannes Schauer.
[MEL 24.0 x 15.0]

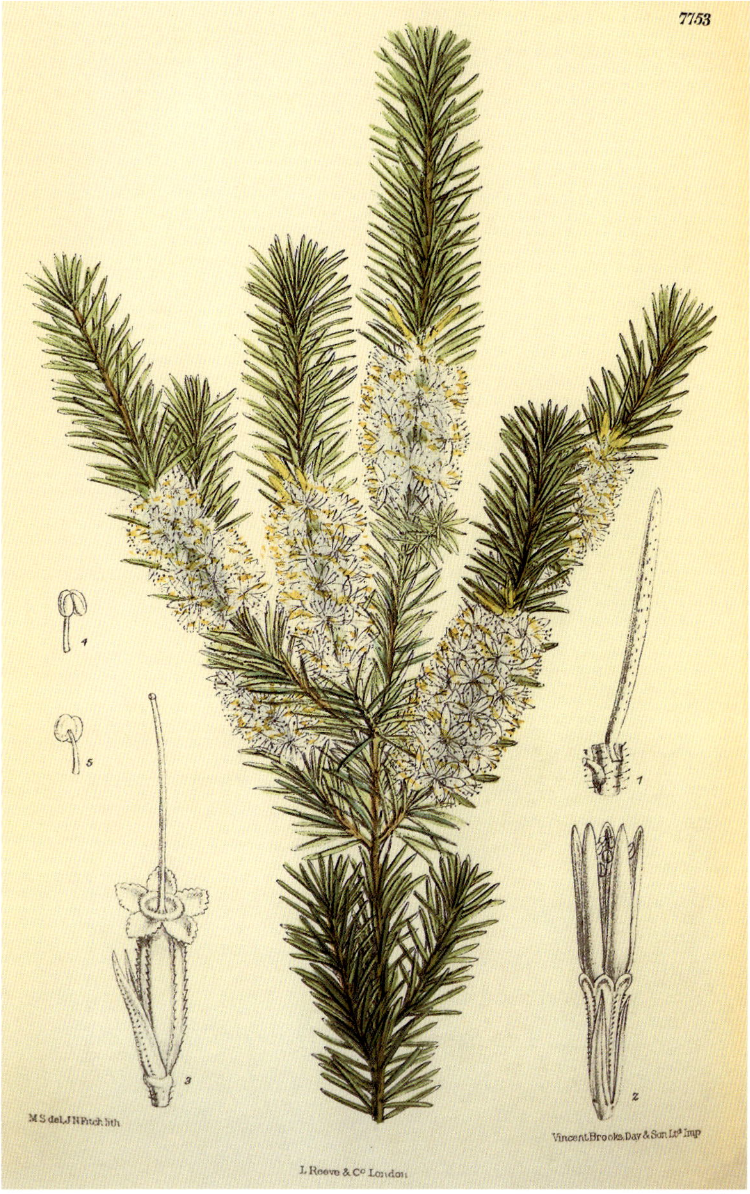

While Ronald Gunn was most notable among the contributing collectors, he was also instrumental in stimulating the interest of others in taking up collection as an active hobby.

One was Augustus Frederick Oldfield,[9] who was described as a 'zealous collector and a careful and acute observer' by Joseph Hooker.[10] Oldfield collected in Tasmania, New South Wales and Western Australia during the 1850s. Details of his education, career and purpose for being in Australia are not known. However, he collected widely and thoroughly while walking from one place to another. He sent his collections to Ferdinand von Mueller in Melbourne and to the Hookers at Kew. Joseph used the Tasmanian material in the compilation of his *Flora Tasmaniae*. Oldfield returned to England in 1863 and also made his collection available to Kew for the compilation of the impending Australian Flora. In this way he made an invaluable contribution to the knowledge of Australian flora, especially that of Western Australia. The tragedy for Oldfield was that his eyesight was failing. Treatment was not successful and he soon became blind. Both Ferdinand von Mueller and Joseph Hooker petitioned Australian governments to provide a pension to compensate Oldfield but, regrettably, their attempts failed.

Joseph Hooker's knowledge of the flora was also greatly helped by William Archer[11] (1829–1874), a Tasmanian-born architect and amateur botanist. Archer collected and studied plants in the district of Deloraine around the family property. His studies included illustration. Between 1856 and 1858, Archer worked at Kew with Joseph, on both his own collection and Tasmanian plants in general. Hooker was extremely grateful for Archer's input. He dedicated the *Flora Tasmaniae* to both Gunn and Archer and said:[12]

> I had the unexpected gratification of receiving from the Governor and Parliament of Tasmania the announcement that they had unanimously awarded me a grant of £350, in consideration of my services in the investigation of the Flora of the Southern Hemisphere, especially that of Tasmania. At the same time I received the most encouraging assistance from my friend William Archer, Esq., of Cheshunt, Tasmania, who forwarded to me a beautiful series of drawings of Tasmanian Orchids, together with £100 to be expended on the Flora; and he soon afterwards arrived himself in England, and rendered me still more valuable aid by his observations and collections, which is duly acknowledged in the body of this work. I have thus been enabled greatly to extend the letterpress and illustrations of this Flora, by putting figures of many more species on the plates; making the descriptions fuller, adding thirty plates, including sixty species, chiefly of *Orchideae* (of many of which Mr. Archer had prepared the drawings), and by appending 130 pages devoted to general considerations on the Botany of Australia and other southern countries ... It remains for me to record my obligations to the unrivalled botanical artist, Mr Fitch, who has, with few exceptions, lithographed the whole of the plates, and drawn the majority of them. The accompanying microscopic analyses (amounting to upwards of 4,500) have been made partly by him, partly by myself, and partly by the distinguished Botanists who have aided me.

Archer seems to have the honour of being the first Australian-born botanical illustrator.

Joseph Hooker had some of his work published in the first parts of *The Botany of the Antarctic Voyage*. However, none of the Tasmanian plants in *Flora Tasmaniae* was drawn by him. They were mostly drawn by and were all lithographed by Fitch. Twenty-seven plates of the orchids were drawn by Archer.

In *Flora Tasmaniae* Hooker's introductory essay on the relationships, phytogeography and other botanical issues is historically one of the most valuable published on Australia's flora. It was written at the time of the debate and controversy over Charles Darwin's theory of evolution of species. Darwin and Hooker collaborated, and Hooker's observations in the essay certainly support Darwin's theory. It is interesting to note that the Tasmanian Government provided the funds to assist the publication of this important essay.

Hobart was slow to get its own science organised with the Tasmanian Botanical Garden being formally established in 1818. William Davidson was the first Superintendent. He was appointed in 1829 and worked until his health failed in 1834. He was succeeded by Martin Tobin until 1840. George Fordyce Story was Director from 1842 until 1847. Independently Lady Franklin, wife of Lieutenant Governor John Franklin, had Ronald Gunn help set up a botanic garden at the end of the 1830s, presumably for the Governor. A Tasmanian herbarium collection was begun by George Story in 1845. This was the first herbarium collection

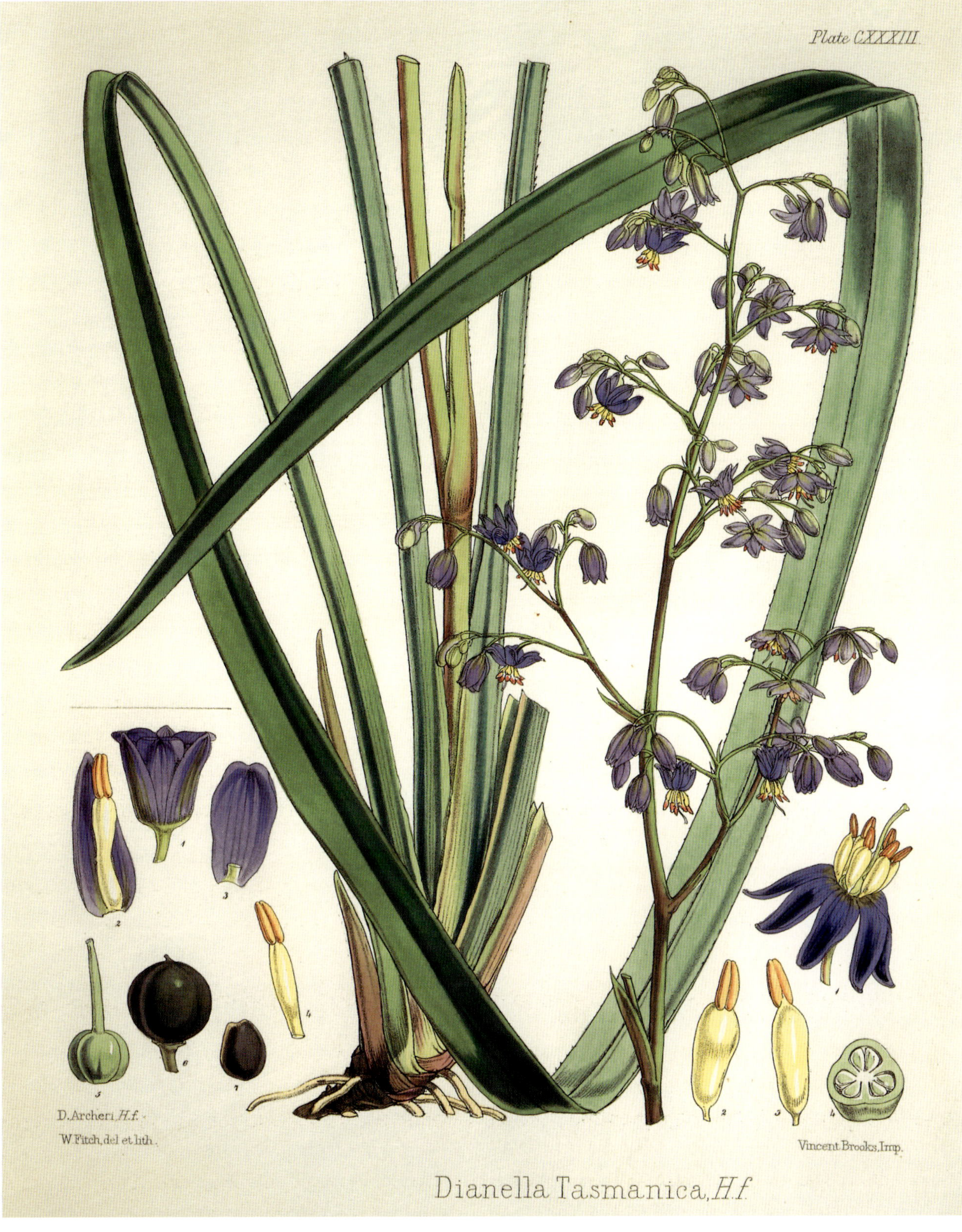

Dianella tasmanica J.D. Hook.
Hand-coloured lithograph drawn and lithographed by Walter Fitch and published in Joseph Hooker's *Flora Tasmaniae* in 1858. The drawing is part of the protologue of the species.
[MEL 30.0 x 22.5]

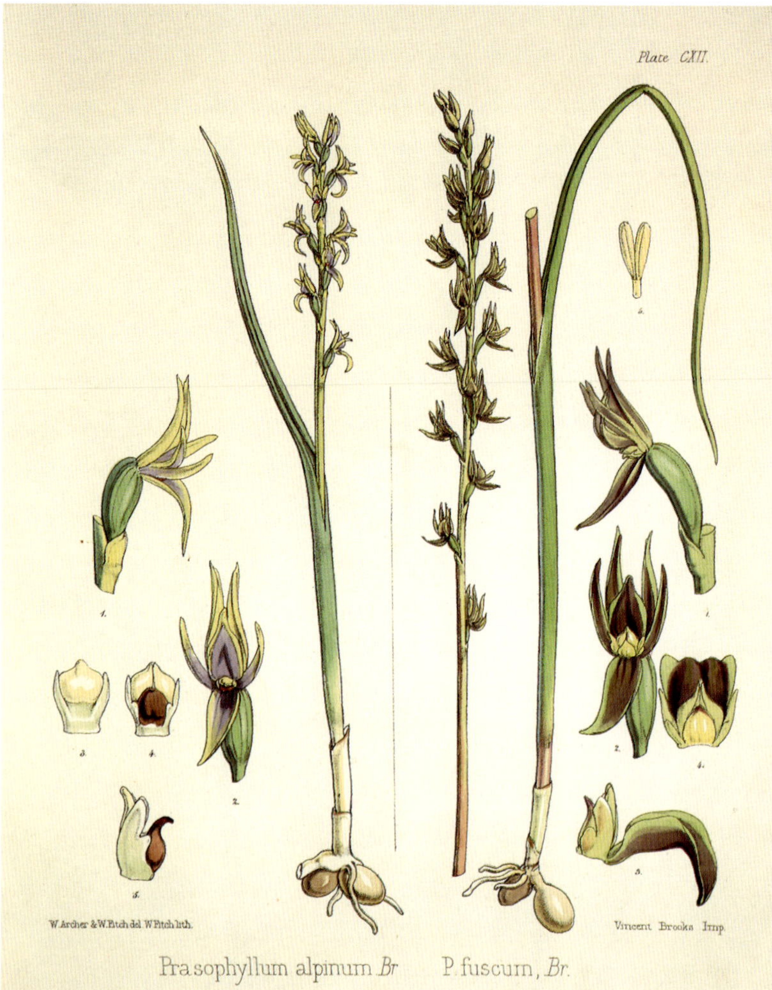

Prasophyllum alpinum R.Br. and *Prasophyllum concinnum* W.H.Nicholls (= *Prasophyllum fuscum* R.Br.)
Hand-coloured lithograph drawn by William Archer and Walter Fitch, lithographed by Walter Fitch, and published in Joseph Hooker's *Flora Tasmaniae* in 1858.
[MEL 30.0 x 22.5]

established in Australia, but it remained relatively obscure. In later years Story sent duplicate material to von Mueller in Melbourne.

Concurrently, the Van Diemen's Land Scientific Society was initiated in the early 1830s in Hobart largely due to the drive of John Henderson, a surgeon. This soon foundered. Through the stimulus of Governor Franklin, the Scientific Society emerged again in 1837 as the Natural History Society of Van Diemen's Land. Out of this the *Tasmanian Journal of Natural Science* appeared in 1841. In 1839 Ronald Gunn had formed the Hobart Town Horticultural Society. Then in 1843 the new Governor, Sir John Eardley-Wilmot, united these two Societies together with the Mechanics Institute into The Royal Tasmanian Society. Faction fighting caused all of the 'good intentions' to limp along and the *Tasmanian Journal* to appear only occasionally. Thus the potential for scientific leadership in Van Diemen's Land was somewhat dissipated.

Richea pandanifolia J.D. Hook.
Uncoloured lithograph drawn by James Grant for the *Tasmanian Journal of Natural Science, Agriculture, Statistics &c.* in 1849 and published in Launceston, Tasmania.
[NLA 21.5 x 13.4]

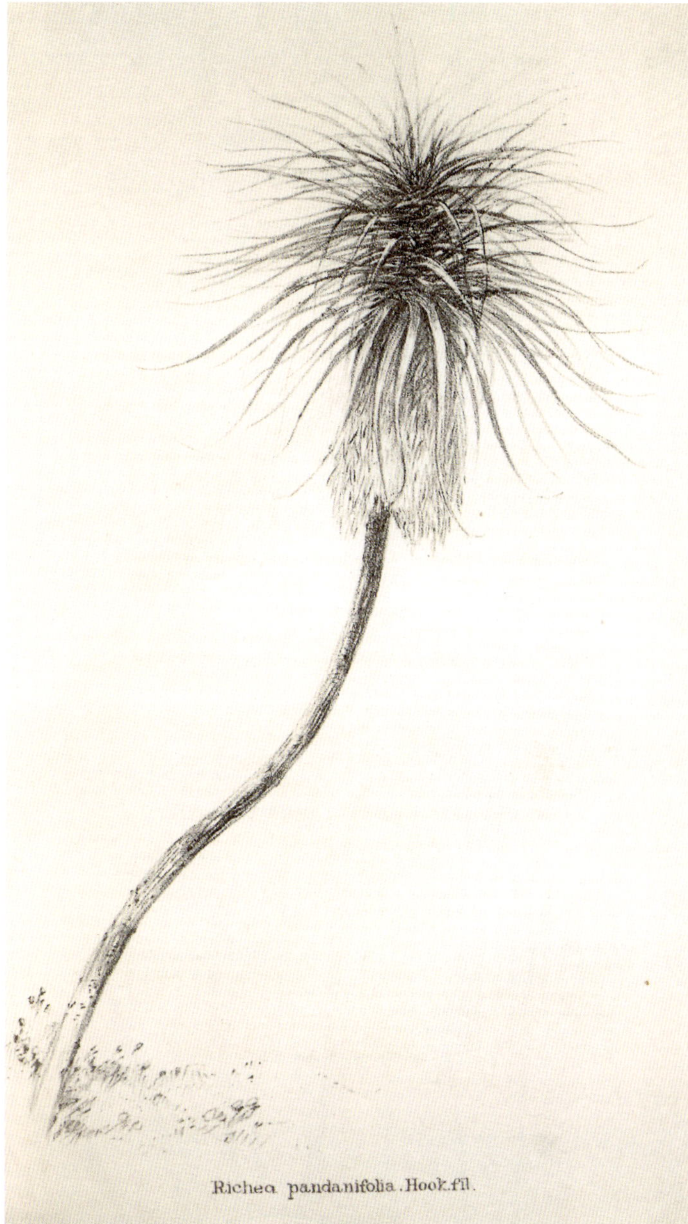

Settlement in Western Australia came twenty-five years after settlement in Van Diemen's Land. In 1827 Captain James Stirling[13] was sent to Australia to establish a settlement at Raffles Bay in the north-west of Western Australia and to advise the British Government on a suitable location for a settlement in the south-west. The British were becoming anxious again over French aspirations and did not have claim to the western half of Australia at the time. In 1828 they sent Captain Charles Fremantle, in the *Challenger*, to take formal possession and to set up buildings on the Swan River, the location recommended by Stirling for a south-western settlement. A few months later, James Stirling was chosen to be the Governor of Western Australia. He was despatched with a contingent of settlers and supplies to occupy Fremantle's buildings and to establish the Swan River settlement. They sailed in the *Sulphur* and the *Parmelia* and arrived in mid-1829.

Just as in the east, so in the west settlement proved to be difficult. This was partly due to the settlers being free and therefore more demanding, less commandable and less united. There was, however, one other significant difference which should have helped greatly in the west. Stirling brought with him a Government Naturalist, James Drummond (1784–1863).[14] Drummond was a Scottish-born horticulturalist who had worked as Curator of the Botanical Gardens at Cork, Ireland. He brought with him a collection of useful plants for establishment at the settlement in a public garden which was originally located on Garden Island. Drummond wasted no time in selecting land of his own and soon had one hundred acres near Guildford, and later much more. There was some altercation over the Government Garden but Drummond secured the salaried position of Superintendent of Government Gardens and the responsibility of setting up the garden adjacent to Government House, doing so in 1831.

Meanwhile, Captain James Mangles[15] fortuitously visited the Swan River colony for three months in 1831. Mangles had a brother working in the settlement from 1829, was a cousin of Lady Stirling, had connections with the East India Company and the F. & C.F. Mangles Shipping Company, and was something of an entrepreneurial traveller. He is also described as being one of a family of 'English gentleman botanists'.[16] James' older brother, George Mangles, had come to the settlement as Superintendent of Government Stock. George had helped James Drummond establish the Government Garden and accompanied him on some collecting expeditions. However George did not stay in Western Australia for long, probably only two years. Their younger brother, Robert Mangles, had a garden in England which was renowned for its Western Australian plants. The Mangles family's botanical network included John Lindley and Conrad Loddiges, in England and George Fletcher Moore, Advocate General to the Settlement, in Western Australia.

James Mangles was so impressed with the flora and its potential that, in 1835, he entreated Moore to collect for him in exchange for useful garden and orchard seeds. In England, James Mangles advised on the development of some of England's great gardens including parks such as Regent's Park and Hyde Park in London. Meanwhile, stocking Robert's garden and the public gardens with exotics was one of James' hobbies.

George Moore arranged for the Drummond family to supply Mangles which started James Drummond on his official plant-collecting career. For Mangles, though, it was not particularly successful, with live plants dying and seed being ruined by insect attack. Drummond's herbarium collection, however, met with greater enthusiasm from John Lindley. New specimens there were described by Lindley in his *Sketch of the Vegetation of the Swan River Colony* (1839).[17] The illustrations

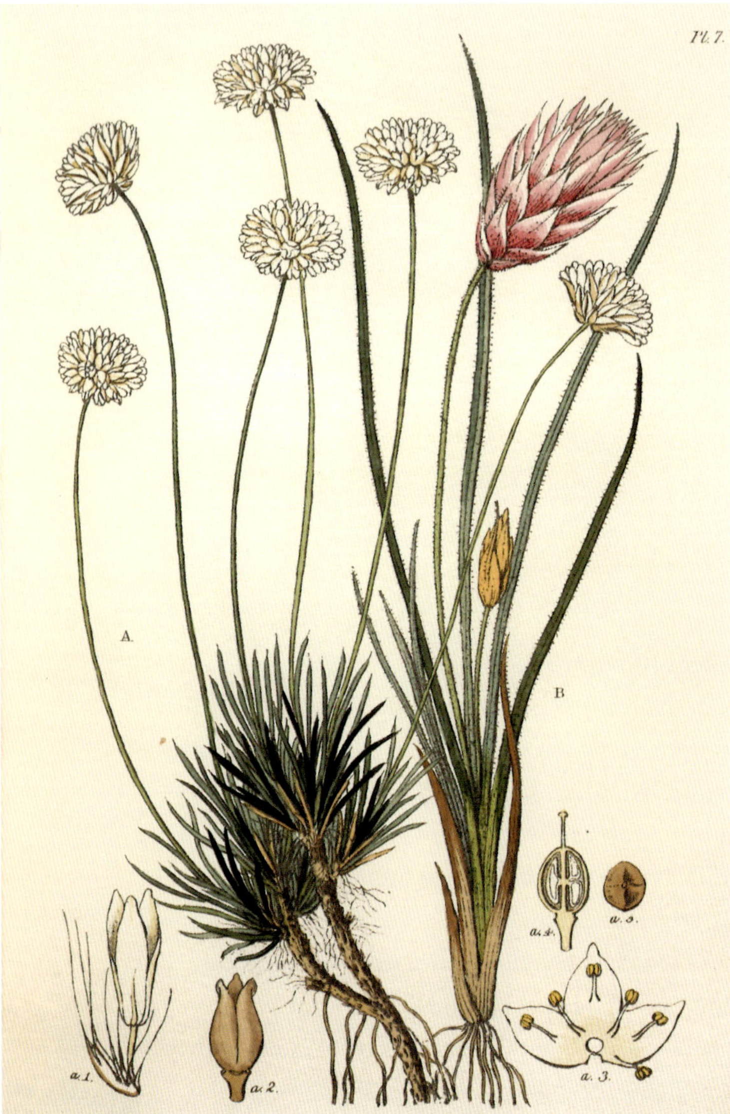

Laxmannia grandiflora Lindley and *Johnsonia pubescens* Lindley (= *Johnsonia hirta* Lindley)
Hand-coloured lithograph published in John Lindley's *A Sketch of the Vegetation of the Swan River* in 1839. The artist and lithographer are not acknowledged.
[CBG 24.0 x 14]

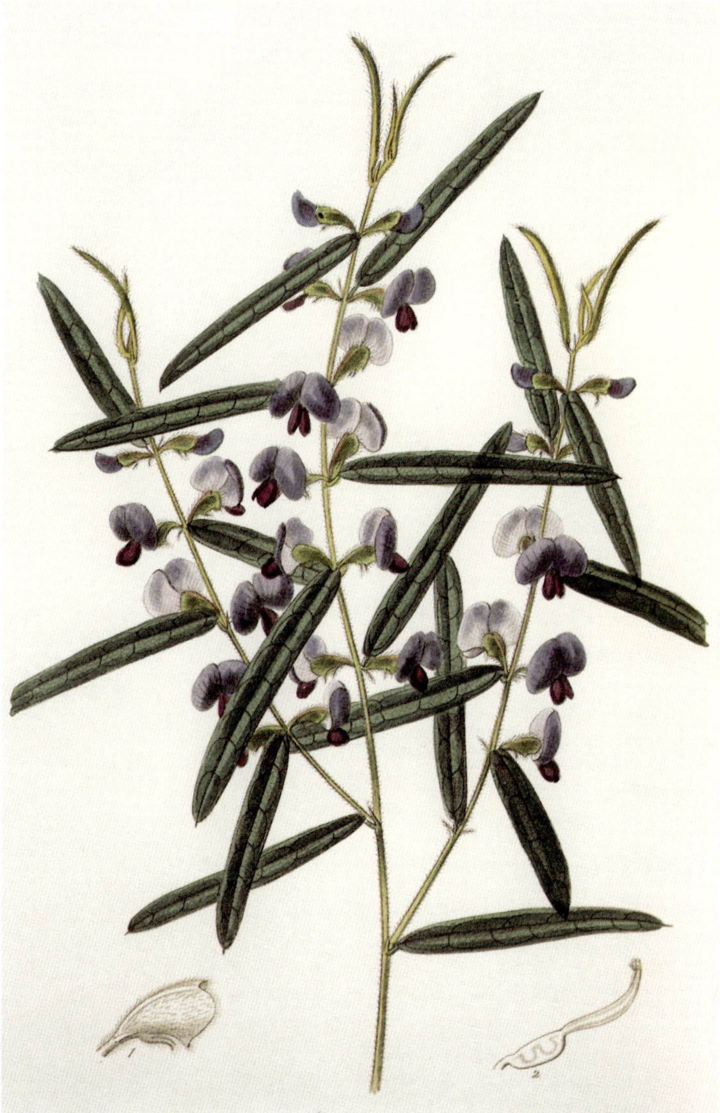

Hovea trisperma Benth. (= *Hovea manglesii* Lindley)
Hand-coloured copper engraving drawn by Sarah Drake, engraved by S. Watts, and published in *Edwards's Botanical Register* in 1838. Lindley named this species in honour of James Mangles.
[CANB 20.2 x 12.2]

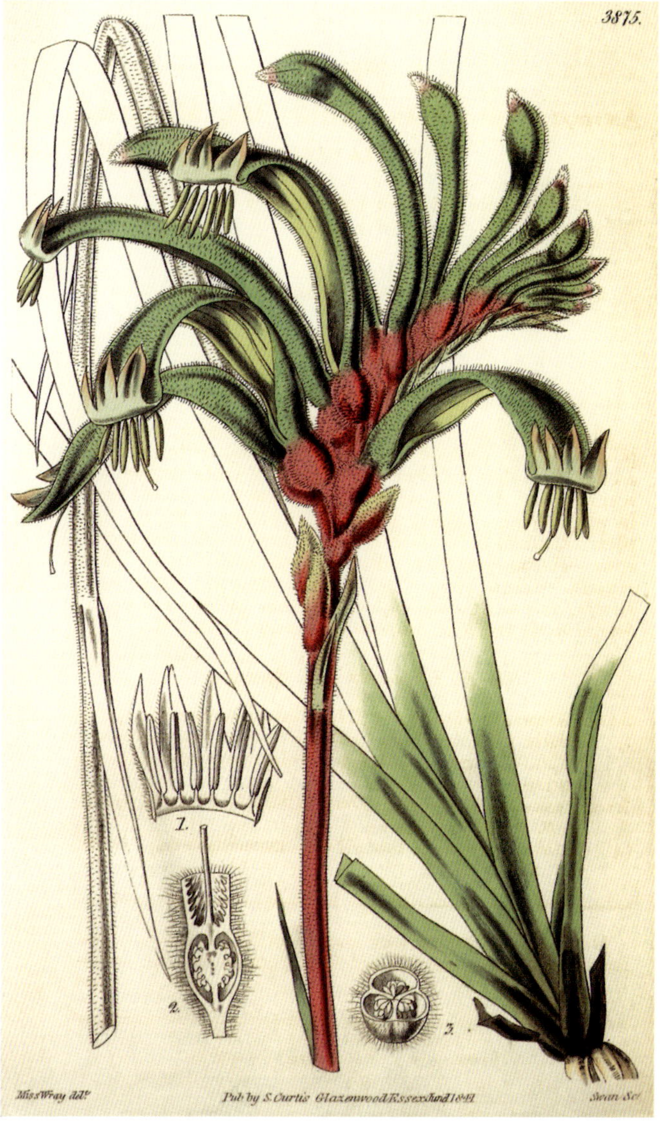

Anigozanthos manglesii D.Don
Hand-coloured engraving by Joseph Swan, based on a drawing by Miss Ray, published in *Curtis's Botanical Magazine* in 1841. The Kangaroo Paw was named for the Mangles brothers, George, James and Robert.
[MEL 21.5 x 12.5]

comprised nine hand-coloured lithographs and four woodcuts. These were unacknowledged and, with the exception of the woodcuts, do not measure up to the standard of contemporary illustration — certainly not to that of Sarah Drake in the same volume.

Georgiana Molloy[18] (1805–1843) settled with her husband at Augusta in 1830, having arrived at Fremantle on the *Warrior*. Molloy developed an interest in the native plants around her and was greatly flattered when asked by James Mangles to collect for him. The contact seems to have been engineered by James' cousin, Lady Stirling. During 1837 Molloy occupied herself greatly with accumulating her 'hortus siccus'. The first collection was enthusiastically received by Mangles and passed on to Lindley and others. So careful and meticulous was Molloy that most of the seed remained viable on arrival in England. It is assumed that material collected by Molloy was also included in Lindley's *Sketch of the Vegetation of the Swan River Colony*. Molloy reported to Mangles after receipt of a copy:[19]

> The notices of the Swan River 'Floral Botanical Register' have quite inflamed my ardour, and instigated me to greater exertion ... The colour of the Patersonia is too blue. The natural colour is more of a violet or Amethystine colour. It really is the colour of Amethyst and there is a yellow species at Augusta.

However, enthusiasm was not enough and Georgiana's on-going contribution was to be short-lived. She succumbed to complications following the birth of her seventh child in 1842.

Coincidentally, Mangles largely lost interest in plants, having turned his attention to exotic birds.

James Drummond was faced with the ever-present hardship of life in a new settlement. His home burnt down, he quarrelled with the Governor over alternative housing and the management of the Government Garden, lost salary and, eventually, lost his post as Colonial Botanist. In spite of it all, Drummond's family continued their passion for finding and collecting new plants and animals.

Eventually the Drummonds settled permanently in the Toodyay Valley in 1836. The later Drummond collections were sent to William Hooker who sought seed and herbarium specimens and written accounts of the botany of the region. In return Hooker assisted in disposing of sets to other subscribers.

One of Drummond's handicaps was an inability to obtain sufficient paper for adequate collection, division and packaging. Nevertheless he persevered, almost obsessively, leaving the farm to his sons to establish and manage during his frequent and extended absences. Drummond collected extensively across the south-west of Western Australia and sent 3500 specimens back to England for study — quite a valuable contribution to the knowledge of the flora, though it must be conceded that many of the collections were scrappy. Perhaps he broke good specimens into too many duplicates.

James Drummond's botanical activity ranged from the time of his arrival in 1829 through to the early 1850s, by which time he was becoming debilitated with age. The highest price he paid for it was the loss of his son, Johnston, to Aboriginal attack (or revenge) on the Moore River in 1845. This loss, however, had more to do with grazing rights than plant collecting.

Even though Drummond had established a Government Garden in 1831 and been Colonial Botanist for a couple of years, not until 1961 was an official Botanic Garden established in King's Park in Perth under the Directorship of John Beard. Three herbaria, however, were established around the turn of the century, the first being with the Department of Agriculture. These were merged in later years.

John Septimus Roe[20] had accompanied Phillip Parker King and Allan Cunningham on surveys of the northern and north-west coast of Australia in 1818 and 1821. Roe returned to the

Swan River settlement in 1829 with Lieutenant-Governor James Stirling as Surveyor General. He surveyed Fremantle and Perth for settlement and received a land grant for his services. Roe led numerous expeditions into Western Australia, mostly in search of suitable pasture country. Later, in 1848–1849, he explored the coast from between Perth and Albany and further east into the Great Australian Bight. Roe was also an enthusiastic amateur naturalist, supporting visiting scientists and certainly assisting Carl Hügel. Roe's own collections formed a nucleus for the Perth Museum and earned him a Fellowship of the Linnean Society of London. He sent duplicates of his plant collections to William Hooker.

Carl Alexander Anselm Freiherr von Hügel[21] embarked on a grand world journey after being jilted by his fiancée. Hügel was an Austrian aristocrat and could afford to indulge in a broken heart in a grand way. He left Austria in 1831 and eventually reached Australia in 1833. He spent seven weeks in Western Australia, three weeks in Van Diemen's Land and less than two weeks in Sydney. As Hügel had an interest in horticulture, it was natural for him to collect plants during his travels. Much of his material ended up in Vienna with Stephan Endlicher who published an *Enumeratio plantarum quas in Novae Hollandiae ... collegit Carolus Liber Baro de Hügel* in 1837, describing plants collected in Australia. This was another unillustrated Australian Flora. Endlicher's *Iconographia Generum Plantarum* (1838) related to his *Genera Plantarum*, not to his *Enumeratio*. The *Iconographia* was dedicated to Allan Cunningham and contained plates prepared from Ferdinand Bauer's drawings, most engraved by A. Gebhardt. Endlicher carefully cross-referenced the figures to collections which he had seen, including those of Robert Brown, Ferdinand Bauer, Franz Seiber, Allan Cunningham and Carl Hügel.

A contemporary German work using material collected by Roe, Hügel and earlier collectors was *Monographia*

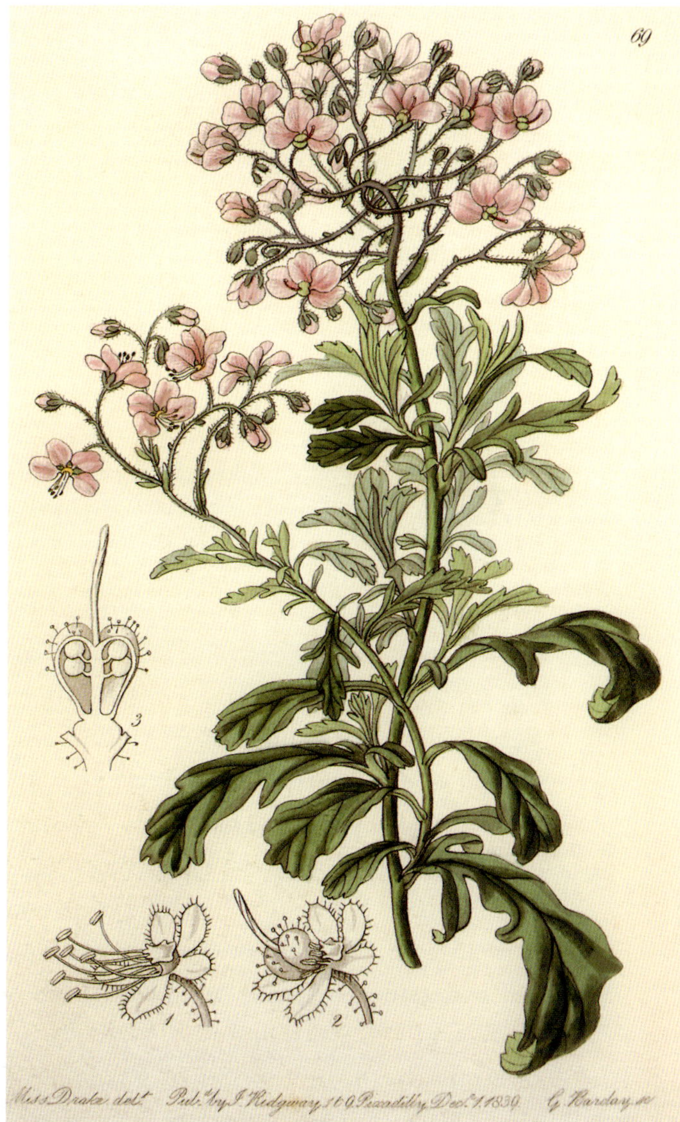

Diplopeltis huegelii Endl.
Hand-coloured engraving drawn by Sarah Drake, engraved by G. Barclay, and published in *Edwards's Botanical Register* in 1839. Named in honour of
Carl Hügel by Stephan Endlicher.
[CANB 20.2 x 12.2]

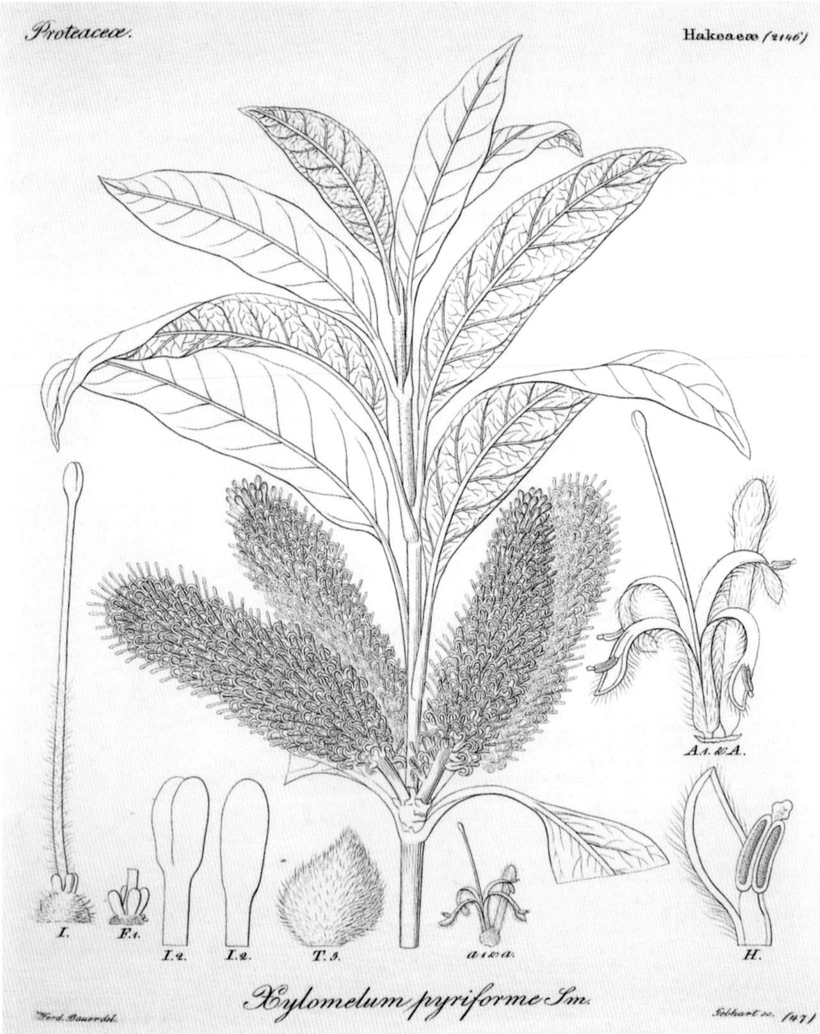

Xylomelum pyriforme (Gaertner) Knight
Uncoloured copper engraving by A. Gebhart based on a drawing by Ferdinand Bauer. This was published in Stephan Endlicher's *Iconographia Generum Plantarum*, 1837–1841, which was dedicated to Allan Cunningham.
[CANB 23.1 x 18.1]

Myrtacearum Xerocarpicarum by Johannes Conrad Schauer in 1841. At the time of publishing this monograph, Schauer was in charge of the Breslau Botanical Garden. The seven plates were drawn from nature by Carl Weitz and lithographed by a firm of lithographers in Bonn with great delicacy of line.

Ludwig Johann August Preiss[22] came to Western Australia in 1838 and worked as a natural history collector for four years. He had studied at university in Hamburg before embarking on a private collecting expedition. During his time in Australia Preiss became a naturalised British subject. He tried to encourage the British Government to buy his extensive collection so that he could finance the emigration of members of his family to settle permanently in Western Australia. The latter enterprise failed. The collection, however, was certainly worth the £3000 that he sought. Preiss collected extensively and meticulously. His records were comprehensive and accurate, adding greatly to the value of each specimen.

He contrasted delightfully with James Drummond with whom he teamed up from time to time — the one a dour, determined bushman, obsessed with finding the new and unusual; and the other, a politely Teutonic, equally determined scientist, obsessed with accurate observation and the careful record of detail.

Ludwig Preiss collected almost as many plant species as did Drummond in a shorter period. The Preiss collection is the more valuable for its quality and detailed records and because the entire collection was described in one work, *Plantae Preissianae* (1844–1848), organised by Johann George Christian Lehmann. This is yet another unillustrated Flora and is somewhat comparable with Robert Brown's *Prodromus* (1810). In contrast, however, Lehmann's work has more extensive descriptions than those provided by Brown, and several authors contributed.

Amongst Preiss' claims to fame is the unsubstantiated belief that he encouraged Ferdinand von Mueller to come to Australia, thereby making an enormous contribution to Australian botany. Preiss also made a significant contribution to zoology through his animal collections.

As each colony was established, botanic gardens and a knowledge of the local botany became central to the success of the settlement. Initially botanic gardens played a two-way role — establishment of plants coming in for use in the colony and the clearing of plants going out, mostly for horticulture and science in Europe. In New South Wales the lack of both, officially, at the beginning of settlement surely provided some lessons. The Government Farm on Farm Cove, however, reached its destiny as the Botanic Garden with the unofficial appointment of Charles Fraser in 1817 as the first Colonial Botanist. The appointment was made official in 1821, thirty-three years after settlement.

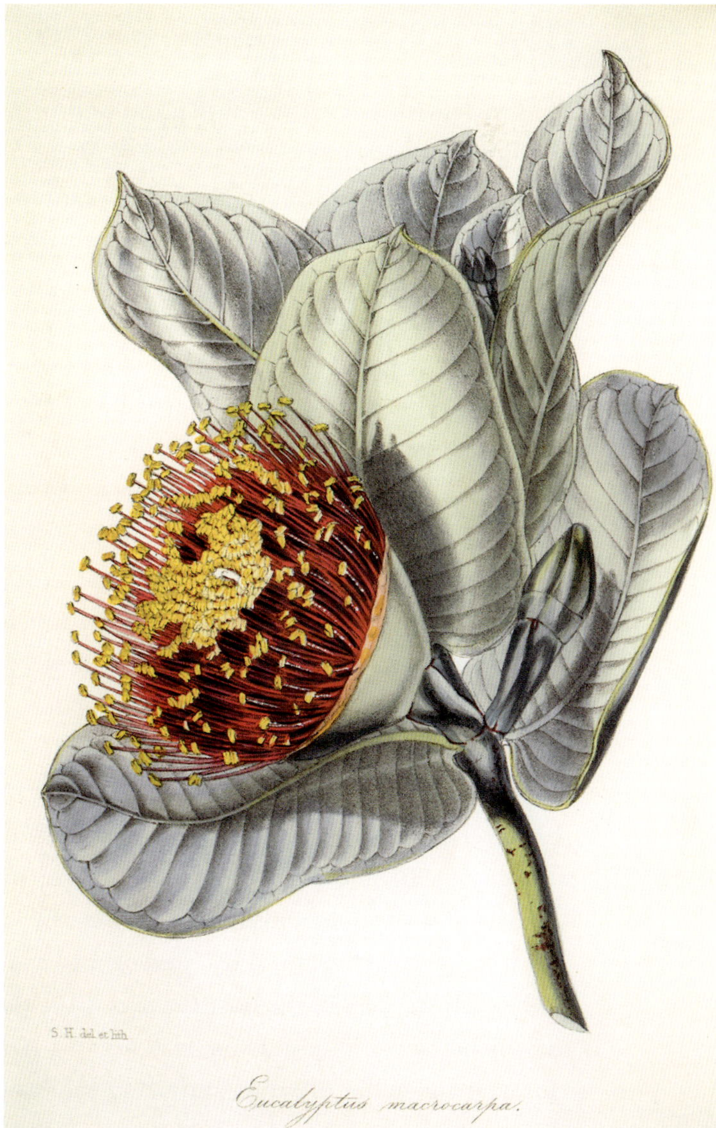

Eucalyptus macrocarpa Hook.
Hand-coloured lithograph drawn and lithographed by Samuel Holden and published in Joseph Paxton's *Magazine of Botany* in 1849.
[NLA 23.7 x 16.0]

Fraser was succeeded by Richard Cunningham, Allan Cunningham (briefly) and James Anderson during the 1830s.

In 1847, following two short-term appointees, John Bidwill was appointed locally as the first Director. Bidwill's term was destined to be even shorter. Charles Moore[23] was appointed from London and, in 1848, replaced Bidwill and served until 1896. The influence of William Hooker and the Royal Botanic Gardens over Australian botanic gardens was significant, but from this example and from an Australian point of view, it was sometimes arrogant and patronising. It will never be known if Bidwill would have served the gardens better than Moore but it is known that a scientific base, other than a small reference herbarium, did not eventuate under Moore. It must, however, be conceded that Moore operated in a difficult, local political climate. An herbarium was built in 1899 under the direction of Joseph Maiden.

Charles Fraser was also given the responsibility of setting up the botanic garden in Queensland. This he effected in 1828, the settlement having been established at Moreton Bay in 1824 by Lieutenant Miller. By 1842 the garden was called the Botanic Gardens, but not until 1855 was a Government Botanist appointed in Walter Hill who held the post until 1881. Frederick Bailey took over as Colonial Botanist that year and soon established an herbarium.

Outlets for scientific publication within Australia were limited. In New South Wales a Philosophical Society of New South Wales tottered briefly in the early 1820s and was followed by a periodical, *Australian Quarterly Journal of Theology, Literature and Science*, in 1828. It too foundered under science-politics over the following few decades. A scientific establishment existed which was based around the Australian Museum, the Sydney Botanic Gardens and the Macleay family and cronies, but the latter were obstructive. By 1850 an Australian Philosophical Society emerged to effect a balance. From this the Entomology Society of New South Wales (1862) and the Linnean Society of New South Wales (1874) were ultimately formed.

In South Australia the first Colonial Botanist was John Bailey, father of Frederick Bailey. Following a series of unsuccessful attempts at establishing a garden, George Francis succeeded in 1855. Richard Schomburgk took over from 1865 until 1890, supervising the establishment of an herbarium during his directorship.

In Victoria the Melbourne Botanic Garden was established by John Arthur in 1846. John Dallachy[24] took over in 1849 as Superintendent of the Melbourne Botanic Gardens. The early establishment of a Mechanics Institute in 1839 in Victoria proved to be a positive action. In 1854 there emerged the Philosophical Society of Victoria and the Victorian Institute for the Advancement of Science. These two amalgamated into the Philosophical Institute of Victoria a year later. In 1860 the Philosophical Insitute of Victoria became the Royal Society of Victoria. Each of these published a *Transactions* titled accordingly.

It is obvious that not a great deal of energy was put into establishing herbaria in Australia in the early decades of the nineteenth century partly because of the slow and dissociated way in which scientific endeavour was organised within Australia in its disunited colonies. However the major cause was probably the very effective role of Joseph Banks and William Hooker and the respective herbaria at the British Museum and at Kew, in England.

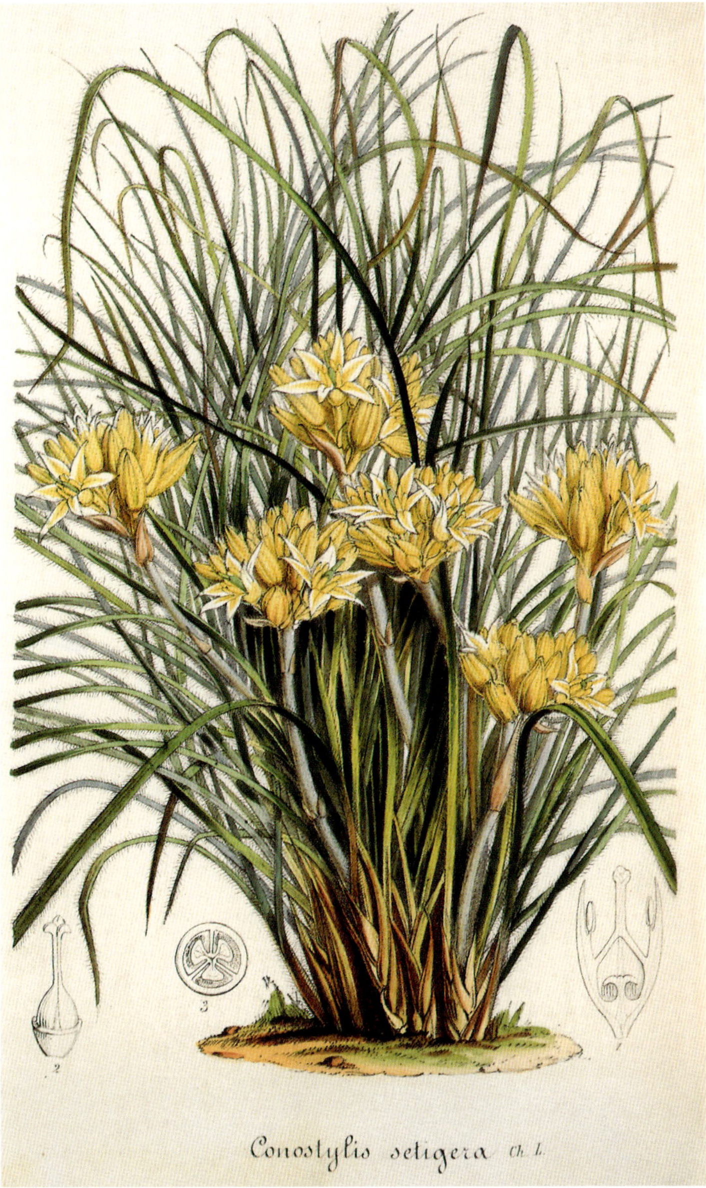

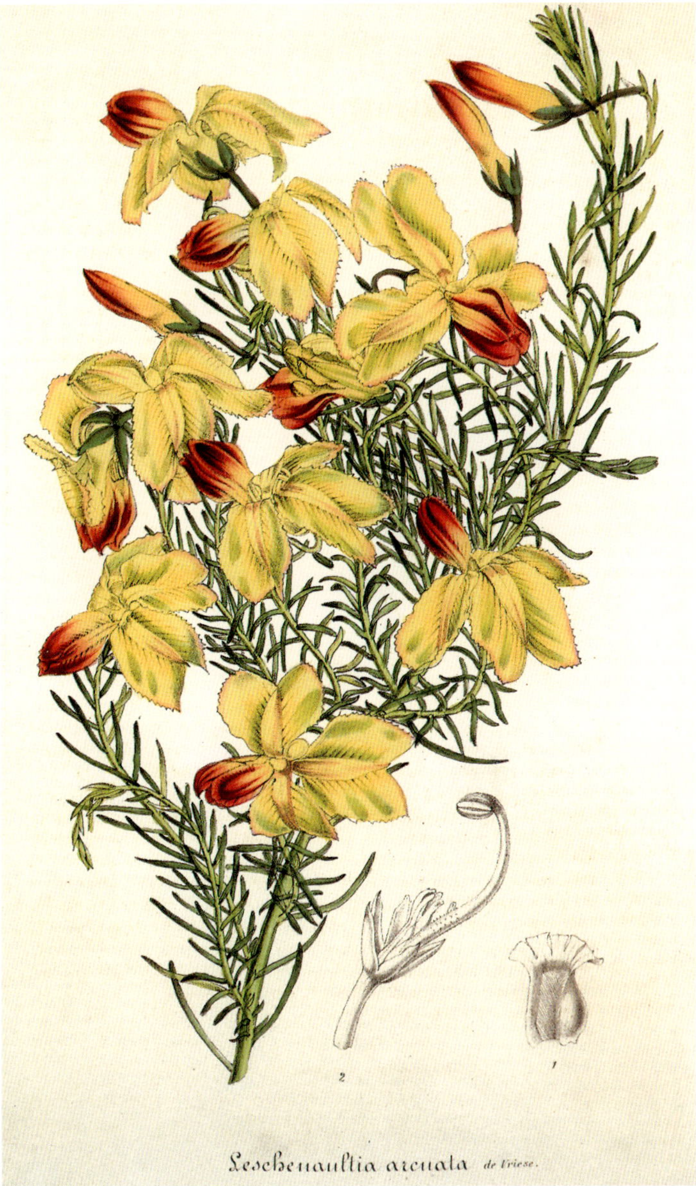

Conostylis setigera R.Br.; *Lechenaultia linarioides* DC. (= *Lechenaultia arcuata* Vriese)
Hand-coloured lithographs published in *Flore des Serres et des Jardins de l'Europe* (1845–1883).
These are copies of illustrations in previous publications, with no attribution to the original artist.
[CANB 23.2 x 15.0] [CANB 23.5 x 14.0]

Inumerable collectors were sending or taking material back from Australia to English and other European herbaria and gardens. The proliferation of this activity is too great to be exhaustively reported here but it certainly led to quite a comprehensive record of the Australian flora, a record in which the Hookers and John Lindley played a major role both with words and with images.

The dominance of William Hooker and John Lindley in making this record, as editors of the grand periodicals, was paid tribute to by a European trio in their publication *Flore des Serres et des Jardins de l'Europe* (1845–1883).[25] In reality, *Flore* was a compilation of

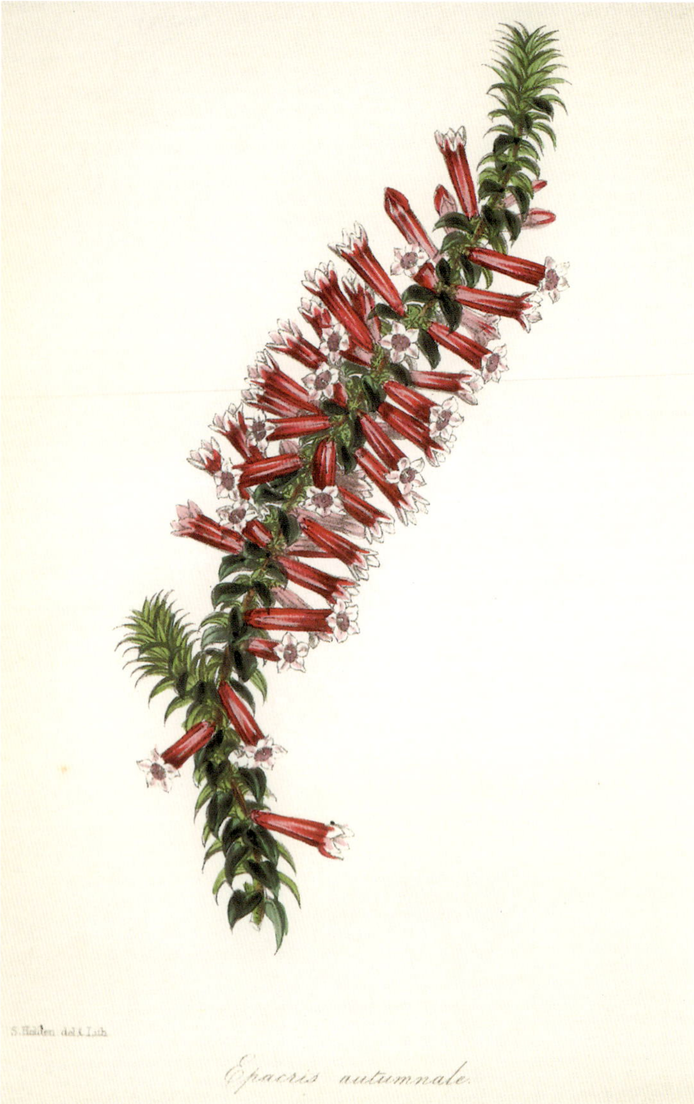

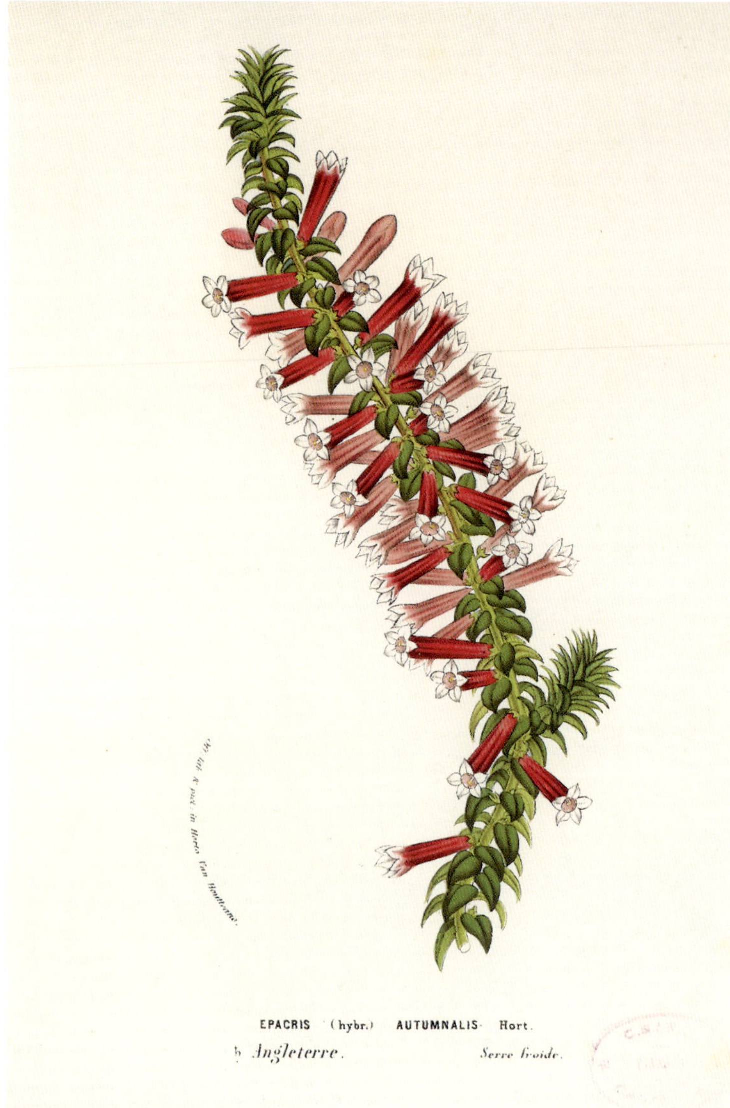

Epacris autumnale Paxton
A hand-coloured lithograph (left), drawn and lithographed by Samuel Holden and published in Joseph Paxton's *Magazine of Botany* in 1844, has been redrawn for *Flore des Serres et des Jardins de l'Europe* and published in 1845 (right). The re-engraving has not been corrected and so is reproduced in mirror image.
{NLA 22.8 x 15.0} {CANB 23.3 x 15.0}

extracts from previous publications, the title page acknowledging the '*Botanical Magazine*, the *Botanical Register* and *Paxton's Magazine of Botany*' as major sources of extracts. In effect, *Flore* became another grand periodical. The illustrations, hand-coloured lithographs, are copies of the original published drawings. They are not attributed to an artist, rather each states 'Off. lith, & pict. in Horto Van Houtteano'.

Breaking the English and European domination in botany was totally dependent upon establishing the science within Australia. That in turn was dependent upon the establishment of botanic gardens and herbaria in Australia. The activities of the first half of the nineteenth century laid the foundations for that to occur during the second half of the century.

8. Ferdinand von Mueller

We have recommended cheap {octavo} works with short descriptions and no plates, or outline lithographs ... to be sold separately & we recommend that expensive illustrations or expensive local Florae, like yours of Victoria, & mine of Tasmania, be left wholly to the Col. Govts. to pay for if they please.

Joseph Hooker[1]

A new era was about to begin for Australian botany — an era marked by the beginnings of political and scientific independence from Europe. Those beginnings, certainly for botanical science and botanical illustration, were pioneered by Ferdinand Jakob Heinrich von Mueller (1825–1896).

In 1847 von Mueller emigrated to Australia, arriving in Adelaide on 15 December with his two sisters, Berta and Clara. By this time a settlement was well established at Adelaide, but not before it had survived several crises. Adelaide was founded in 1836 by Colonel William Light, the South Australia Act of 1834 having provided authority to make land grants in the region.

In Europe, Ferdinand had trained as a pharmacist and studied botany at Keil University. He had a phobia about contracting tuberculosis, because almost all of his immediate family had died of it while he was a young boy. Moving to a country warmer than Germany had appeal and this may have been boosted by enthusiastic reports of Australia from a family friend, Ludwig Preiss, who had visited Western Australia in 1838 for four years. If indeed Preiss influenced von Mueller, Australia owes him a great debt.

Von Mueller began his career in Australia as a chemist, but he soon turned to botany and set about establishing Australian botany firmly in Australia.

From herbarium records it appears that von Mueller began collecting immediately he arrived. In next to no time he explored the general vicinity of Adelaide sufficiently to prepare a useful paper in 1852, documenting the flora of South Australia. This was the first comprehensive record of the plants of that region.

Meanwhile, activity in Port Phillip was becoming significant. Beginning very early in the century, there had been a series of unsuccessful attempts to establish a colony on Port Phillip Bay. During the 1830s the story became quite contorted. Suffice to say here that, in 1836, Governor Richard Bourke sent William Lonsdale as Police Magistrate to supervise a growing, but unofficial, settlement. In 1839 Charles Joseph La Trobe was sent, as the first Superintendent, to replace Lonsdale and soon the settlement began to grow steadily. Then in 1850 New South Wales gained self-government and Port Phillip was separated from New South Wales as a colony in its own right and named Victoria. The following year the discovery of gold at Clunes in Victoria was announced.

Ferdinand von Mueller was attracted to the Victorian goldfields along with hordes of others. Von Mueller had a business head and intended to set up a pharmacy on the goldfields. This, however, did not come to pass. Governor La Trobe appointed von Mueller as Government Botanist to make a scientific survey of the Victorian flora. He took up the post in January 1853 at the Botanic Garden where John Dallachy was Superintendent. These two men worked together at the Garden.

Von Mueller had stepped into an incipient scientific community with a barely developed infrastructure. Undaunted, he set to work immediately and began collecting vigorously in the field. Early in 1853 he and Dallachy set off into the Dividing Range following the Goulburn River to its upper reaches. From there Dallachy returned to keep things going in Melbourne and von Mueller continued on. He came down off the range, crossed Gippsland to the coast and then travelled back to Melbourne. No doubt this was an early revelation of his excellent capacities as a bushman. This trip of approximately 2400 kilometres gained about one thousand collections of plants previously not recorded for Victoria. Later in the year he undertook another trip, this time to the north-west of Victoria through the Mallee and then along the Murray to the vicinity of Albury. In all, the trip was longer but revealed fewer new plants, about five hundred. He made another trip into the alpine region in 1854. Upon his return, William Harvey, the algologist, was in town.

William Henry Harvey[2] (1811–1866) was born in Limerick and had developed an interest in botany while at school. Early in his career, a term of employment in South Africa allowed him time to pursue his hobby. He soon became fascinated by seaweeds. This lead to his appointment as the curator of the Trinity College Herbarium, Dublin, and to his becoming one of the great algologists of all time. Between 1853 and 1856 he travelled widely in the southern hemisphere, arriving in Western Australia in January 1854. He collected extensively along the coast of the settled regions before moving on to Victoria, arriving in April. He spent eight months collecting in several localities. His work was supported by von Mueller and a number of amateur collectors. Harvey was a charming gentleman and it was easy for him to develop a network of willing and supportive collectors.

From Melbourne Harvey travelled to Van Diemen's Land and then to Sydney and other parts of New South Wales. He is reported to have made 20 000 collections of about six hundred species of seaweed during the eighteen months he was in Australia. On his return to Ireland he was appointed Professor of Botany at Dublin University and became a Fellow of both the Linnean and Royal Societies of London. In *The Contented Botanist* (1988) Sophie Charlotte Ducker has documented Harvey's letters associated with his field work in Australia and the Pacific. This compilation demonstrates the extensive and extremely active network among botanists of the time, William Hooker, of course, being one of the central correspondents.

Among Harvey's voluminous publications is *Phycologia Australica* (1858–1863), published in five volumes and illustrated with 300 coloured lithographs. Harvey did all of his own lithography, not only for this work but for several others. He surely joins Hooker and Lindley as a workaholic scientist-artist. The plates in the *Phycologia Australica* were printed in the fundamental colour of the subject and then touched up by hand with watercolour. This is most effective, especially for the Red Algae.

By this time von Mueller's communication and publication record was accelerating. Von Mueller had made contact with William Hooker and despatched to Kew a duplicate set of the specimens he had collected. Von Mueller's first publication appeared in 1850 and by the end of 1854 he had seventeen papers published. Thereby, on all fronts, von Mueller was rapidly establishing himself as the authority on Australian flora. It seems that this was not haphazard. He was aiming to compile a Flora of Australia.

Von Mueller's activity, of course, did not preclude ongoing European enquiry into Australian plants. For example, Willem Hendrik de Vriese, a Dutch scientist, published a monograph, the *Goodenovieae*, in 1854.

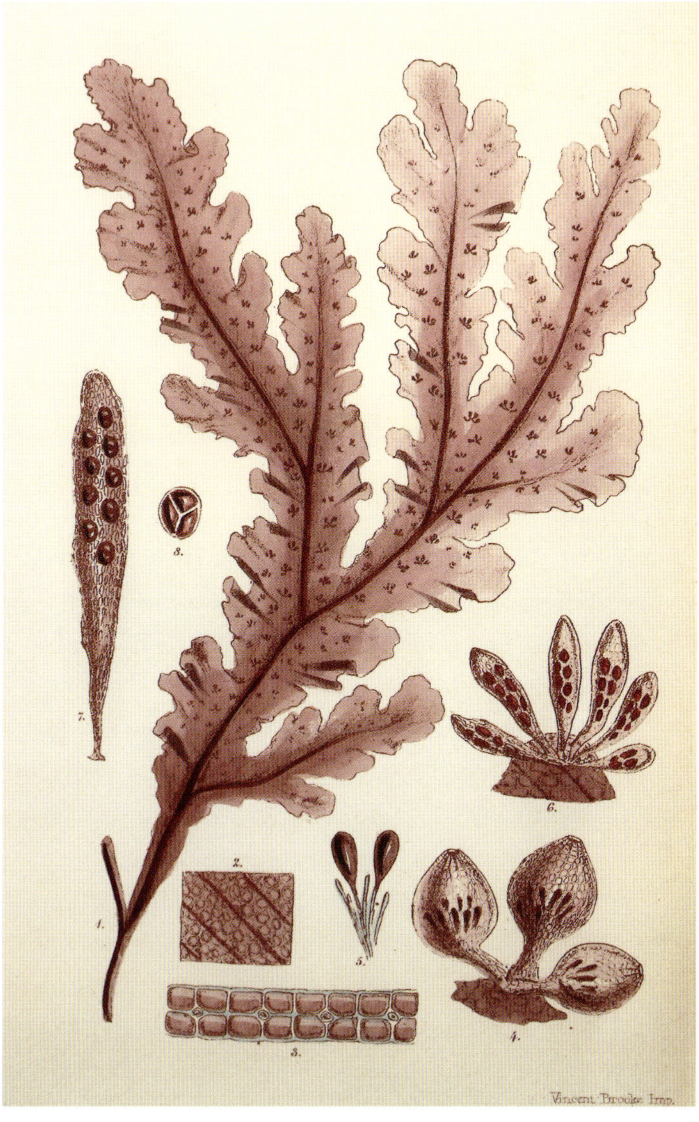

Jeannerettia lobata J.D. Hook. & Harv.
Hand-coloured lithograph of Red Alga, drawn and lithographed by William Harvey for publication in his *Phycologia Australica* in 1858.
[MEL 25.5 x 14.5]

The monograph was largely illustrated with lithographs by Quirijn Maurits Rudolph ver Huell. De Vriese had been Professor of Botany at the University of Amsterdam prior to his retirement to Leiden from where he completed his study of the Goodeniaceae, a family which is predominantly Australian. Of the thirty-eight accompanying plates two are unsigned, one is drawn by Abraham Jacobus Wendel, one by A.J. Kouwels, and thirty-four are drawn by Huell.

To obtain a better knowledge of the northern Australian plants, von Mueller joined the North Australian Expedition of 1855–1856 headed by Augustus Charles Gregory. The expedition sailed from Sydney on the *Monarch* and the *Tom Tough*, calling in at Moreton Bay on the way for supplies and horses. Problems with landing the *Monarch* in the Brisbane River saw the beginning of a sequence of ocurrences of bad luck, which was especially bad once they reached the Victoria River. Suffice to say that a lesser man than Gregory would have allowed the expedition to collapse completely. They explored the Victoria River to its source and then set about crossing inland to the Gulf of Carpentaria.

All the while von Mueller was making plant collections and he reported having discovered twelve new genera and twenty-five new species. He collected representation of about eight hundred species, these being sent to England via Timor where the *Tom Tough* had been sent for supplies. Eventually the expedition reached the Roper River and continued to make steady progress. The explorers had planned to meet up with the *Tom Tough* in the Gulf but, as there was no sign of it, Gregory decided to continue east to the Gilbert River and then south-east to the Burdekin River. Ultimately the expedition returned to Brisbane.

By all accounts von Mueller was a stalwart during times of great privation. Some reports suggest that he carried his determination to the point of being a liability, with a bravado and perhaps lack of concern for the stamina of the horses which could have led to the demise of the expedition. In spite of this, von Mueller seemed to gain self-confidence during the course of this gruelling undertaking.

Von Mueller went on to Sydney to sort out the collection. A duplicate set of the collection from the latter part of the expedition was successfully despatched to Kew but unfortunately the earlier, more valuable, collection deteriorated before arriving in England.

Hooker was doing his best to encourage von Mueller to go to England to write a Flora of Australia. During the expedition von Mueller seemed to have convinced himself that this was to be his next move. He knew he had to go to England to do the job properly, but it was a move he did not make.

Soon after his return to Melbourne in 1857, von Mueller was appointed Director of the Melbourne Botanic Gardens and John Dallachy his Assistant. He immediately set about organising the building of an herbarium which was to become, arguably, the most significant in the southern hemisphere.

The year 1858 proved to be challenging — the Herbarium was completed; von Mueller began the publication of his magnum opus, the *Fragmenta Phytographiae Australiae* (1858–1882);[3] and contemplated preparation of the proposed Flora of Australia. By now, von Mueller was determined to write the Flora in Australia. However there was the huge problem that the resources — the plant collections — were firmly housed in European

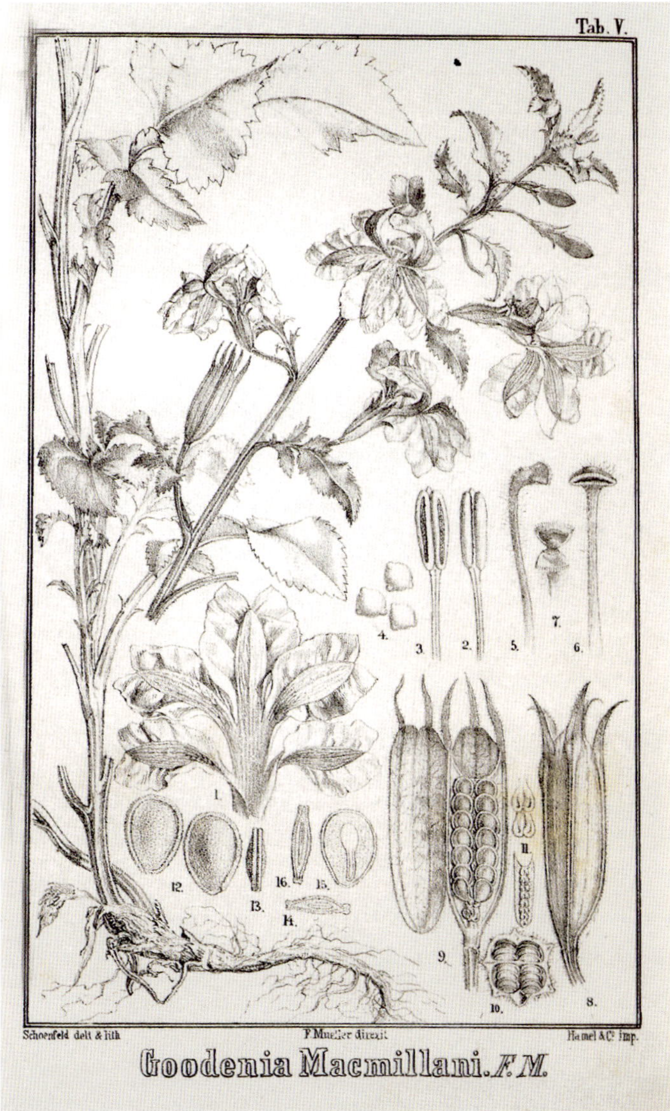

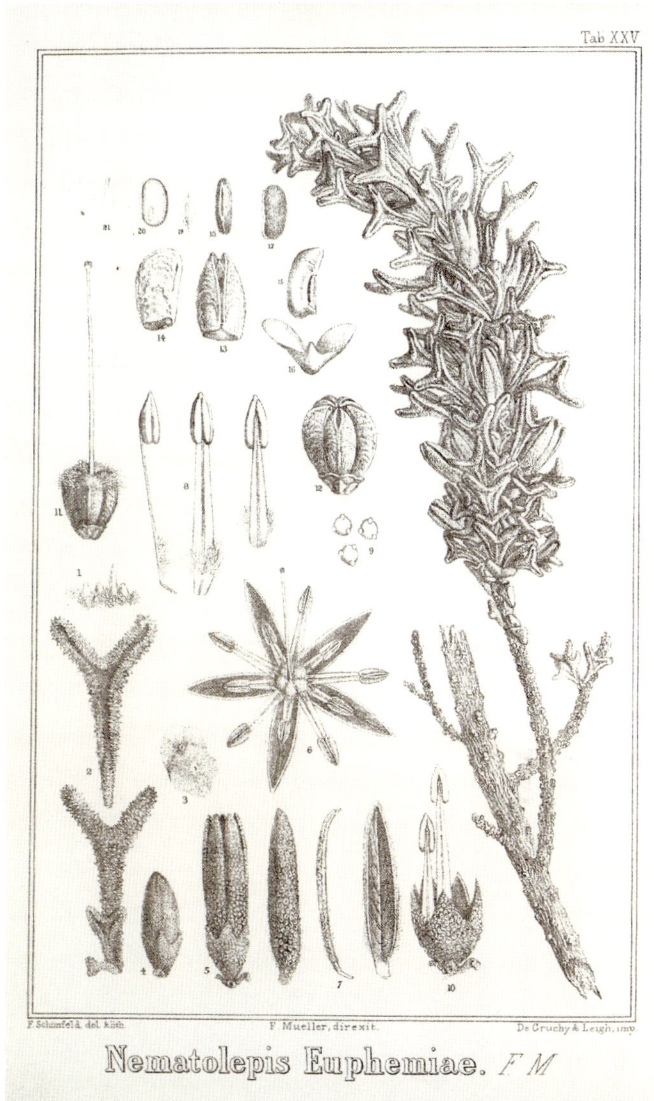

Goodenia macmillani F.Muell. and *Rhadinothamnus euphemiae* (F.Muell.) Paul G. Wilson
(= *Nematolepis euphemiae* F.Muell.)
Uncoloured lithographs drawn and lithographed by Friederich Schoenfeld and published in Ferdinand von Mueller's magnum opus, *Fragmenta Phytogeographiae Australiae*, 1858–1881.
Nematolepis euphemiae was named in honour of the artist Euphemia Henderson, with whom von Mueller had an almost whirlwind romance.
[CANB 18.2 x 11.0]

herbaria. Whilst von Mueller's own collection was extensive and relatively representative, it was far from being as rich as that which had accumulated in England over the previous, almost ninety, years. Moreover, most of the type specimens were in England and it was desirable that he see these and compare Australian material with that of nearby countries.

Von Mueller struggled to do things his own way but in the end he was obliged to stand aside and cooperate from afar. While the struggle seems to have been as much with himself as with authorities or with others aspiring to do the work in England, the Flora would not have happened as it did if von Mueller had had the courage, or whatever required, to travel there.

Eventually the job was given to George Bentham[4] at Kew and von Mueller cooperated. The entire Melbourne Herbarium collection was gradually packed for England, studied by Bentham, and systematically returned. It is possible to identify the specimens which were seen by Bentham, because he consistently inked a 'B' on the label of each.

Bentham wrote of this:[5]

> When indeed it was first contemplated to bring out a general Flora of Australia under Government sanction, Dr Mueller was naturally looked to as the botanist best qualified for undertaking the task of preparing it; and in the hope that it would be entrusted to him, he had devoted his utmost energies to collecting the necessary materials. But there was one indispensable step, the examination of European herbaria where the published types were deposited, which he was unable to take; and it is a signal proof of the generosity of his disposition and the absence of all selfishness, that when it was proposed to him that the preparation of the Flora should be confided to me, on account of the facilities which my position here gave me for the examination of the Australian collections I have mentioned above, he not only gave up his long-cherished projects in my favour, but promised to do all in his power to assist me, a promise which he has fulfilled with the most perfect faith.

Von Mueller settled down to his own publishing and to a romance with Euphemia Henderson.[6] Ferdinand and Euphemia had an almost whirlwind and wonderful romance early in the 1860s which died by the end of 1863. Euphemia was a capable artist and painted wildflowers for Ferdinand. These did not make their way in to science but remain in the collection at the National Herbarium of Victoria, a reminder of the personal life of Ferdinand.

In those early days conditions were austere to say the least and von Mueller had gained more responsibility but no more money. This is shown in the types of paper he used for labels — even used envelopes sufficed. In spite of this he managed to have a little money to pay for some illustrations. More importantly he managed to locate some extremely competent artists resident in Melbourne. The German Club and the Philosophical Society in Melbourne attracted a number of people with particular expertise. In both these places von Mueller found artists, Ludwig Becker having the honour of being the first.

Ludwig Becker (1808–1861) was born in Darmstadt in Germany. He studied classics, natural sciences and painting. Perhaps, during his university days, he became friendly with Johann Jakob Kaup, a prominent German zoologist. Becker executed some zoological illustration for Kaup before moving to Frankfurt to study lithography. Becker then became Court painter for the Archduke Ludwig III of Hesse-Darmstadt. While in that posting, Becker became a specialist miniaturist painter, especially of portraiture. He also became something of a dilettante and accumulated a very valuable fine art collection. He left Germany in the 1840s to travel in other parts of Europe, always with a foot in a scientific camp. Becker came to Van Diemen's Land in 1851. The goldrush on the mainland tempted

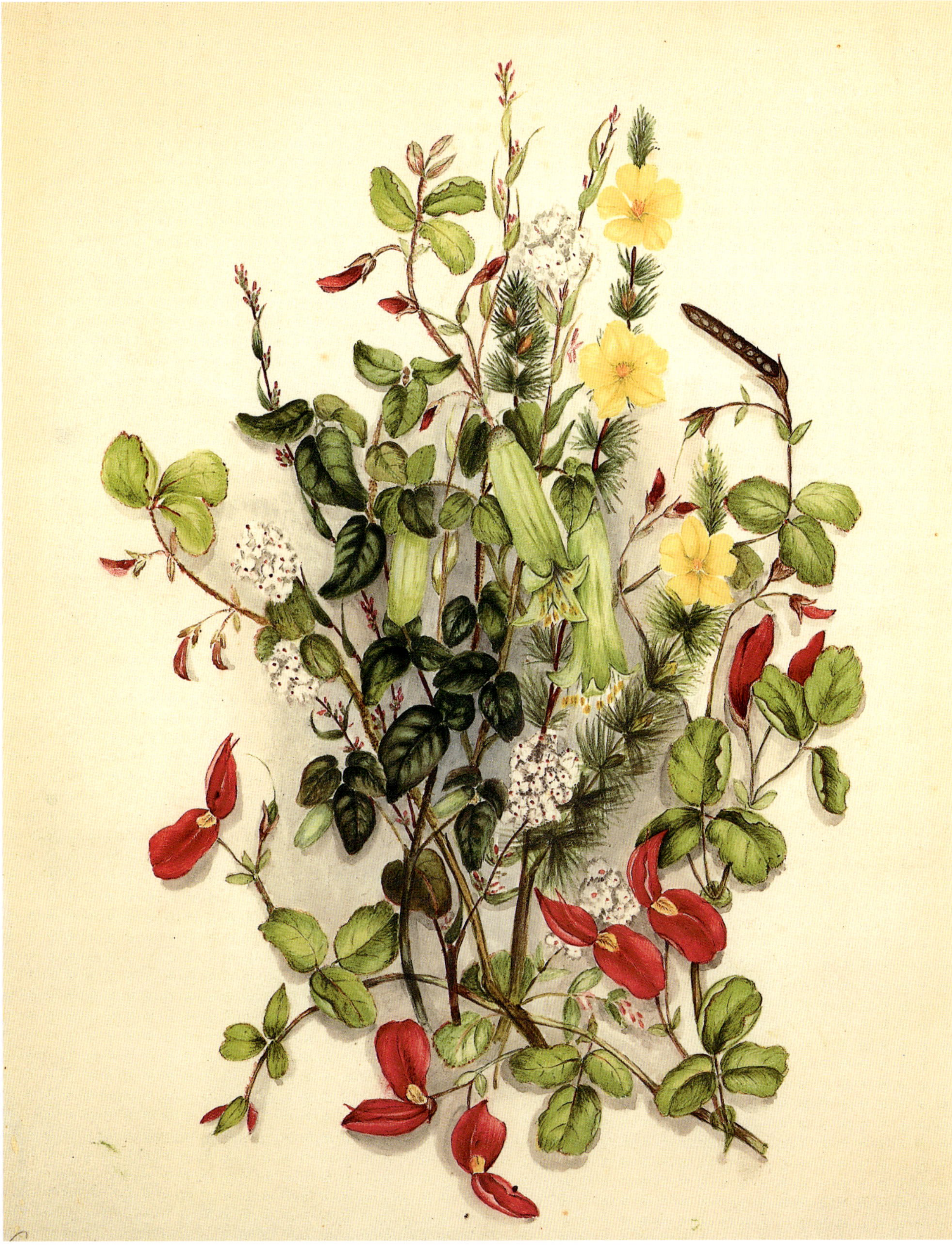

Watercolour painting by Euphemia Henderson. Wildflowers include *Correa, Hibbertia* and *Styphelia*.
[MEL 30.5 x 24.0]

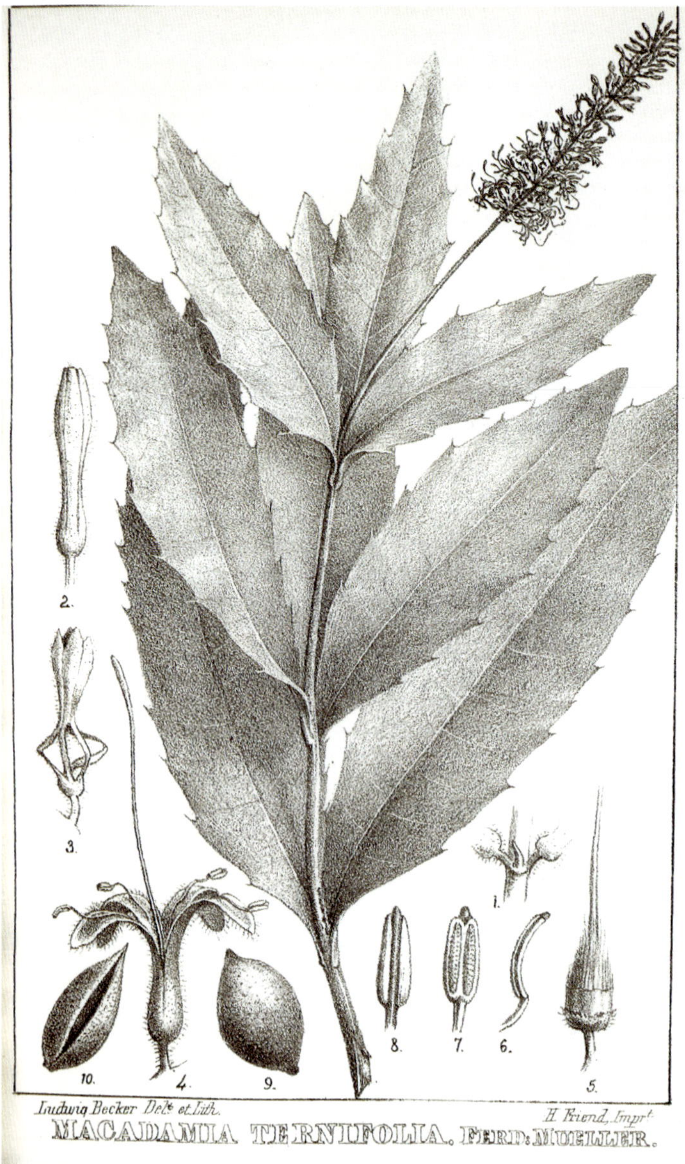

Macadamia ternifolia F.Muell.
Uncoloured lithograph drawn and lithographed by Ludwig Becker and published in the *Transactions of the Philosophical Institute of Victoria* in 1857.
{NLA 19.2 x 11.2}

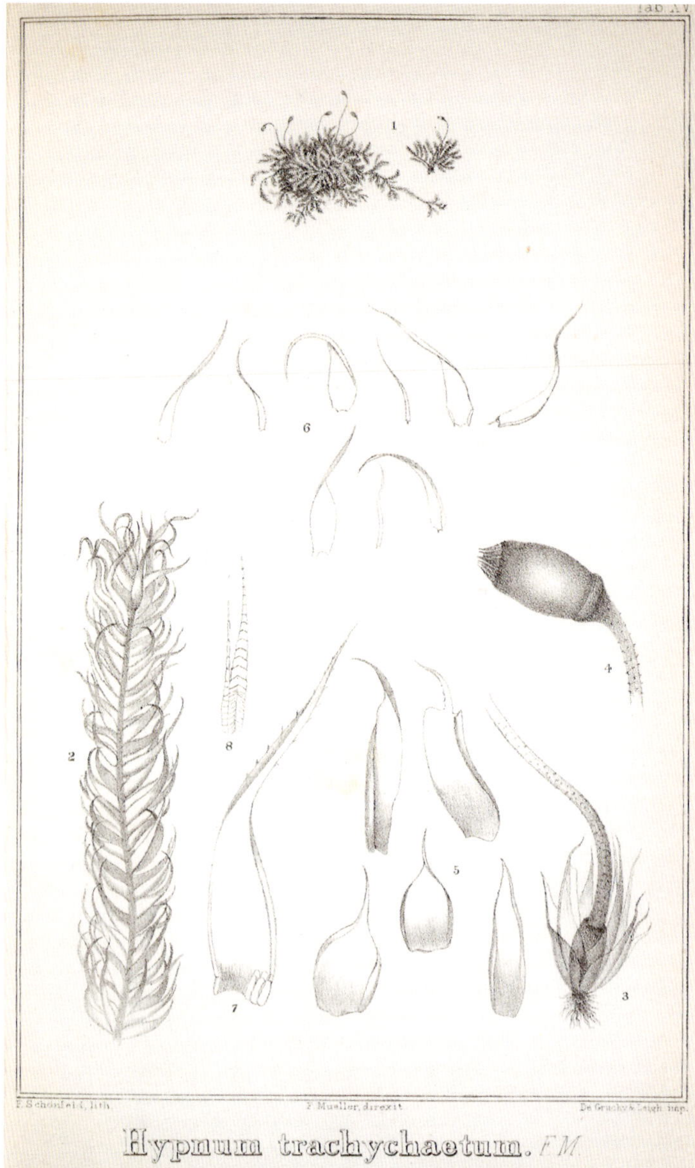

Sematophyllum leucocytus (C.Muell.) Sainsb. (= *Hypnum trachychaetum* F.Muell.)
Uncoloured lithograph drawn and lithographed by Friederich Schoenfeld for publication in Mueller's *Analytical Drawings of Australian Mosses* in 1864.
{CANB 18.6 x 11.4}

him to Bendigo in 1852. In 1854 he moved to Melbourne and became active in the Philosophical Society. Becker earned a living painting miniature portraits, but he was a multi-talented person. He was a natural historian, geologist and singer, as well as an artist and illustrator.

In 1857 von Mueller published *Octoclinis macleayana* in the *Transactions of the Philosophical Institute of Victoria* 2: plate 1 (Sept. 1857). This was drawn and lithographed by Becker and appears to be the first formal scientific illustration of an Australian plant to be published in Australia. There are arguable exceptions. One is a whole plant study of *Richea pandanifolia*.[7]

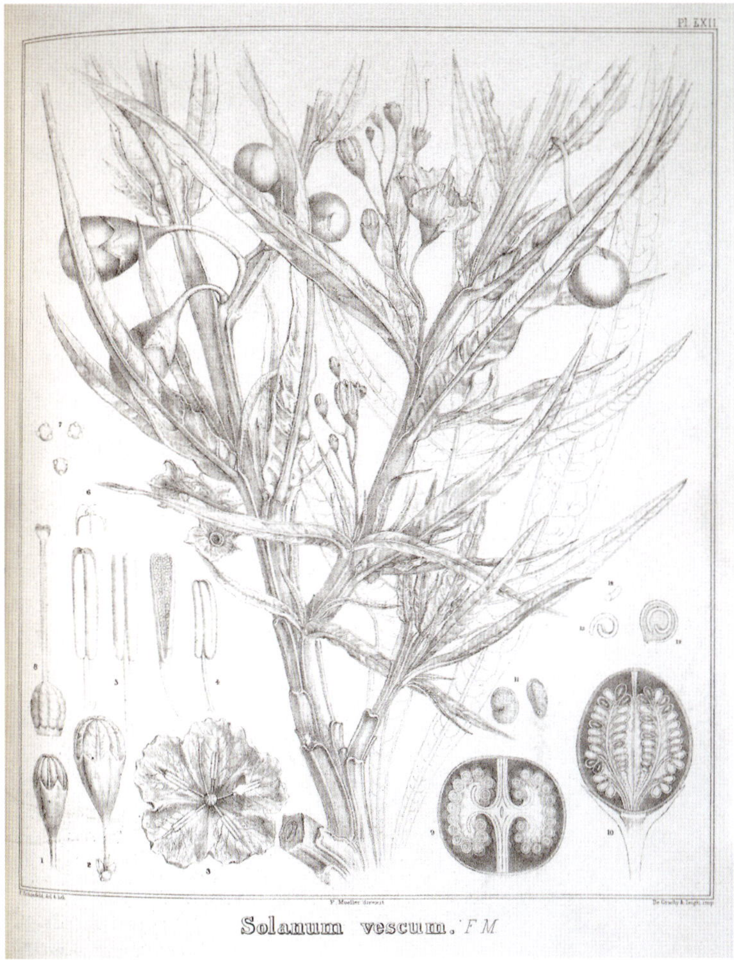

Solanum vescum F.Muell.
Uncoloured lithograph drawn and lithographed by Friederich Schoenfeld for publication in Mueller's *Plants Indigenous to the Colony of Victoria* in 1864–1865.
[NLA 24.3 x 20.0]

Prior to this, drawings or paintings may have been made in Australia, but the actual printing took place in England or Europe.

Macadamia ternifolia appears later in the same journal and is illustrated with two plates. One is drawn and lithographed by Becker and the other (supervised by Becker) is a leaf-surface study, the first example of nature printing of Australian plant material.

In 1858 Volume 1 of Mueller's *Fragmenta Phytographiae Australiae* appeared and included ten plates. These are all lithographs and were executed by either Ludwig Becker or Friederich Schoenfeld (1810–1868), sometimes spelled, Schönfelt. Schoenfeld was born in Switzerland. He came to Australia in 1858 and worked freelance as a painter, lithographer and art teacher in Melbourne. Both Frederick McCoy at the Museum and von Mueller at the Herbarium employed him from time to time to illustrate their respective publications. Meanwhile, Schoenfeld worked for Hamel & Co. Printers and taught art at Melbourne's German clubs.

The Plants Indigenous to the Colony of Victoria (1860–1865) was illustrated with lithographs by Becker and Schoenfeld. Schoenfeld's work is outstanding. Even in the international context, Schoenfeld's work is superior to that of contemporaries, Fitch included. Through careful inspection of Schoenfeld's later plates one can fully appreciate Dunthorne's comments about the achievement of tone in lithography. Schoenfeld was a master.

Schoenfeld's work also provides an example of artistic licence in science. Von Mueller, probably for efficient use of a page, directed Schoenfeld to prepare composite illustrations which embraced buds, flowers and fruits on the one branch or juvenile and adult leaves on the one branch, where in nature they would not have occurred in such close juxtaposition.

There is a tendency to overlook small and less obvious plants, such as mosses and liverworts, both in nature and in works such as this, however von Mueller's *Analytical Drawings of Australian Mosses* (1864) must be included. This was illustrated with twenty lithographs by Schoenfeld and provides delicate drawings of tiny moss plants.

The lack of funding to support more than occasional illustration contributed indirectly to the early death of both Becker and Schoenfeld. On the recommendation of von Mueller, Becker joined the Burke and Wills Expedition to the Gulf of Carpentaria in 1860 as an artist and naturalist. Becker did not receive the support he needed from Burke and was forced to work rather as a lackey. As they proceeded north Becker recorded the natural history, landscape, weather and Aboriginal culture. His physical role in the day-to-day progress of the expedition forced him to work long into the night keeping his records and artwork up to date. He eventually succumbed to exhaustion caused by a combination of starvation, dehydration and scurvy, or beriberi from nardoo poisoning.[8] Schoenfeld, on the other hand, was not as versatile as Becker and was unable to attract sufficient work to support himself. Lack of self-esteem and depression caused him to commit suicide by drowning.

The tragic deaths of both these men were a loss for which Australian botany, in particular, was much the worse. Only Ludwig Becker[9] was honoured with a plant name.

Botanical art was not restricted to men, although the early history of illustration of Australian plants reveals a male domination. Notable exceptions are there and Sarah Drake, working for John Lindley in London, would be high on the list. Flower painting, on the other hand, seems to have been a regular indulgence of women in royalty and other aristocratic families in Europe.

The story in Australia has some parallels and some distinctions. One distinction was a fundamental need to show families back in England just what Australia was like. In the absence of the modern phenomenon, the camera, many people had a box of paints and produced very acceptable images which, of course, included plants and animals. The nineteenth century, therefore, saw the emergence of a number of very competent botanical artists who were rarely painting for science, but many of whom sought scientific endorsement. This group included a number of women whose work has not been lost or even forgotten. Many of them were in communication with Ferdinand von Mueller and many of them also collected plants for him. Von Mueller, in turn, identified their subjects.

Louisa Ann (Twamley) Meredith[10] was among the first of these. She was born in Birmingham, England, and in her youth painted portraits and miniatures. She came to Australia in 1839 and settled in Van Diemen's Land. Louisa was something of a poet and illustrated her own works, of which there are several. To claim that these were scientific would be to stretch the definition. Rather, both the poetry and the artistic presentation usually fell into the naive-romantic genre. Even so her portrayal of plants is accurate and she could have illustrated for science if given the opportunity. Australian books published by her include *My Home in Tasmania* (1852–1853), *Some of my Bush Friends in Tasmania* (1860), *Tasmanian Friends and Foes* (1869), *Our Island Home: A Tasmanian Sketchbook* (1879), and *Bush Friends in Tasmania* (1891). Meredith communicated with von Mueller, at least for identification of some of her subjects.

Meredith hoped to carry a conservationist message through her publications but rather her writings have become an important record of society and its attitudes in Australia during the second half of the nineteenth century.

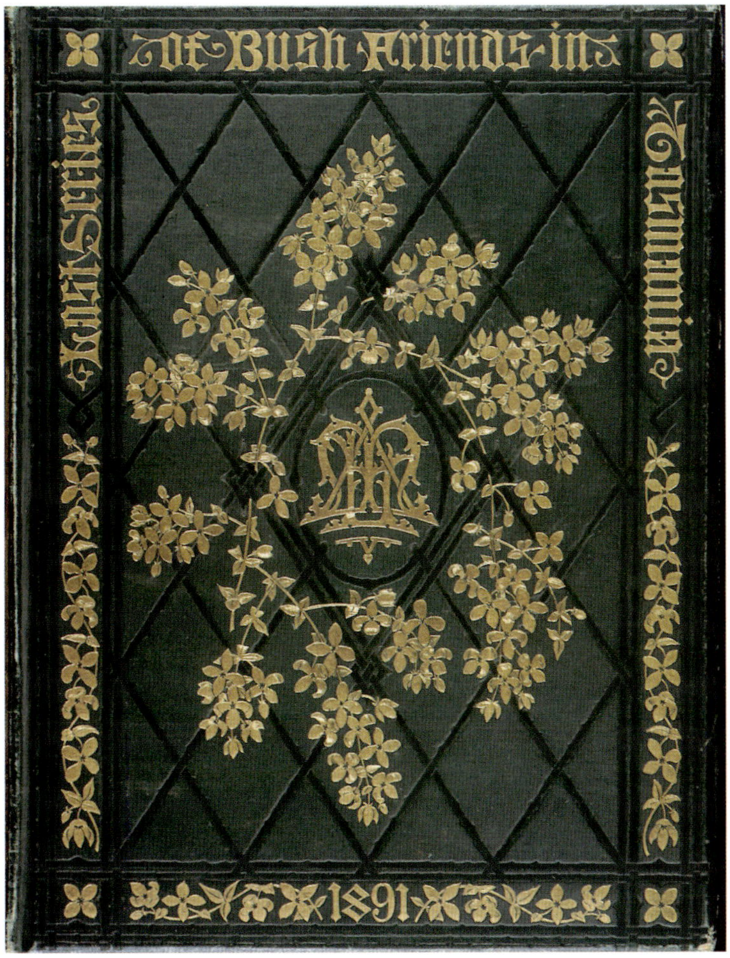

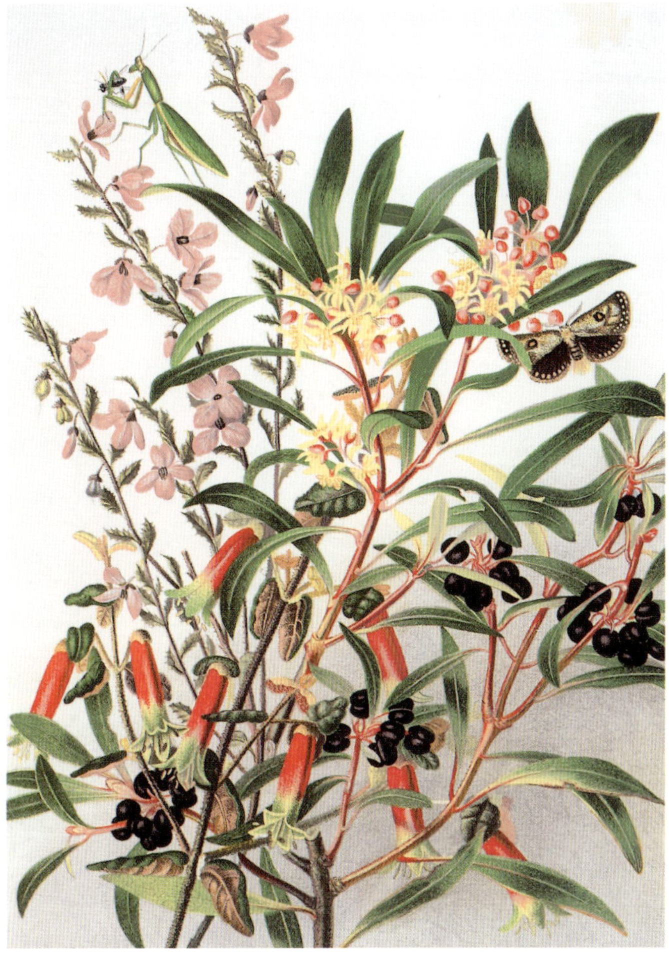

Bush Friends in Tasmania by Louisa Ann Meredith, published in 1891, was the last in a series. The cover features an embossed design incorporating *Tetratheca*. The book is illustrated with paintings by Meredith, reproduced as chromolithographs.
[NLA 38.0 x 27.5] [NLA 33.0 x 22.8]

Fanny Ann Charsley came to Australia in 1856 with members of her family. They lived in Melbourne for ten years before returning to England. During her stay in Melbourne Charsley painted the wildflowers of the surrounding countryside. In doing this she communicated with, and collected for, von Mueller. Von Mueller provided the naming and classification of the species which she painted, and on her return to England Charsley had her collection published in London as *The Wild Flowers around Melbourne* (1867). And so she joined the army of correspondents and collaborators of von Mueller.

Von Mueller named a central Australian everlasting daisy in her honour — *Helipterum charsleyae* (Asteraceae), stating: [11]

> Species nunc geret nomen F.A.Charsley, quae multas plantas pulchritudine eximias circum urbem Melbourne se ipsa selectas eleganter delineavit et pinxit, tunc sua manu lapidi incidit et in opere 'Wild Flowers around Melbourne' edidit.

This book is rare and unfortunately cannot have contributed greatly to helping the local Melburnians identify their native plants.

Helene (Scott) Forde[12] and Harriett (Scott) Morgan, sisters, were born in Sydney and raised on Ash Island in the middle of the Hawkesbury River. They were natural history painters and Helene, in particular, had zoological paintings published to illustrate works for their father, A.W. Scott, in *Australian Lepidoptera and their transformations* (1864, 1890); for J.C. Cox in *Monograph of Australian Land Shells* (1868); and for J.L.G. Krefft in *Snakes of Australia* (1869) and *Mammals of Australia* (1871). The sisters were trained by their father and influenced by both Conrad Martens and Elizabeth Gould. Helene began to prepare several paintings of plants of western New South Wales with a view to publishing a Flora of the Darling River in the vicinity of her husband's property. She abandoned the work following the death of her husband and provided collections for this enterprise to both Reverend William Woolls and von Mueller. Helene was a friend of Louisa Atkinson. No doubt they shared experiences and methods — there are strong resemblances in style.

Caroline Louisa Waring Atkinson (Calvert)[13] was born at Oldbury, near Berrima, and spent most of her life at Kurrajong Heights. Atkinson had an inherent interest in nature and soon made contact with Reverend Woolls and von Mueller, corresponding with, and collecting for both of them. She was also an author of novels, children's books and popular natural history articles, most of which were published in the press of the day. She illustrated her writings with charming paintings, reminiscent of Beatrix Potter's natural history paintings.

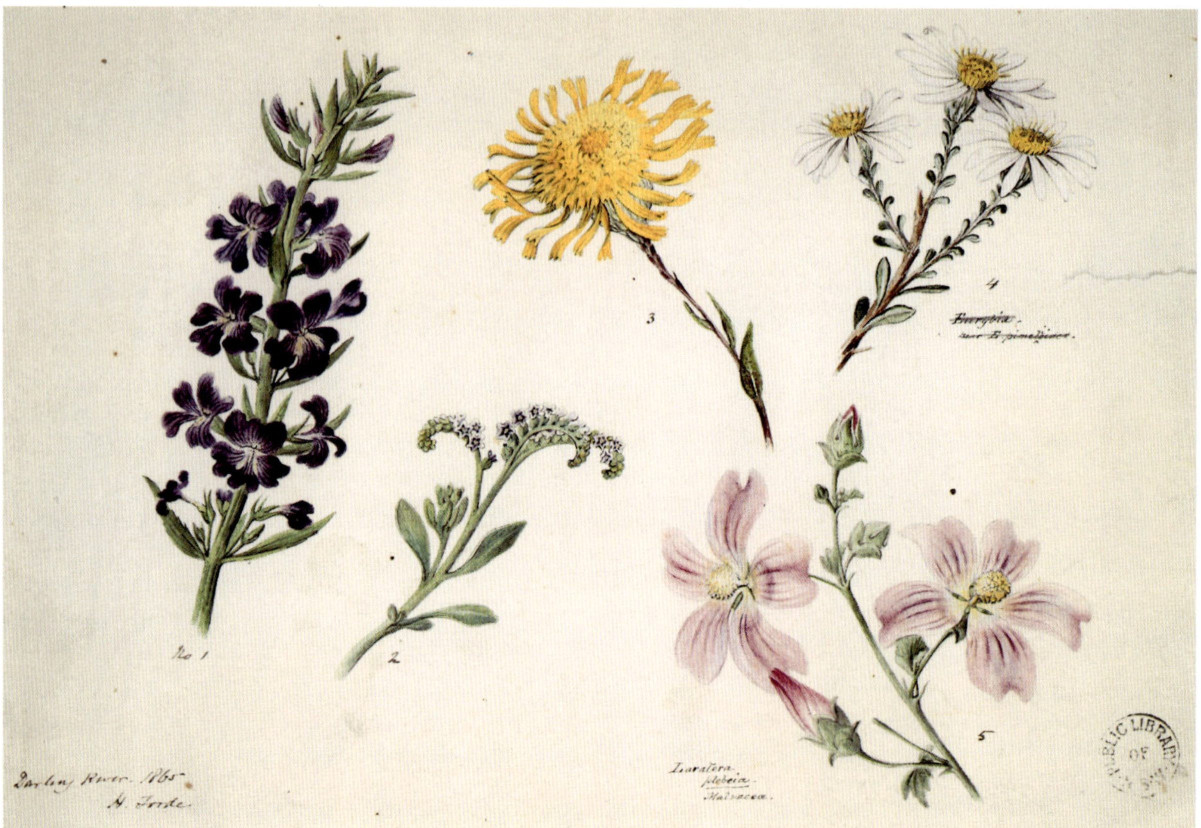

Ajuga australis R.Br. (1), *Heliotropium curassavicum* L. (2), *Podolepis jaceoides* (Sims) Voss (3), *Olearia pimelioides* (DC.) Benth. (4), and *Lavertera plebeia* Sims (5)
Watercolour paintings by Helene Forde for the proposed Flora of the Darling River.
[ML 17.0 x 25.0]

Atkinson married James Snowden Calvert in 1870. Calvert had participated in the Ludwig Leichhardt expedition to Port Essington in 1844–1845 as the botanist. John Gould's collector, John Gilbert, was on this expedition too. In a skirmish with some Aborigines, Calvert was injured and Gilbert was killed. It is a tragedy for the popular communication of natural history that Louisa died in child-birth two years later.

Fanny Elizabeth de Mole came to Adelaide in 1856, suffering from tuberculosis. She and other members of her family were accomplished artists. De Mole painted the wildflowers in the vicinity of the family home at Burnside, with the encouragement and support of her family. She generously aimed to publish her collection to raise funds for the building of a local church in the Adelaide Hills. The paintings appear to have been sent to England for lithography and returned for colouring by hand. Collaboration with von Mueller seems to have been possible, but has not been confirmed. Certainly, de Mole appeared to have sought the aid of a botanist for accurate observation and correct identification. The book, *Wild Flowers of South Australia*, was published in 1861 as a limited edition of a hundred copies. A facsimile was produced in 1981.

Marian Ellis Rowan[14] was also an Australian-born artist. She was born in Melbourne into the pioneering squattocracy. Rowan taught herself to paint and was encouraged by her husband while they were resident in New Zealand. She had a desire to record the Australian plants for the Australian people and tried to interest the government in supporting this venture.

Meanwhile, she made contact with von Mueller. He intended to publish an illustrated Flora and he felt Rowan could be trained to be the artist. Some of her work, in which she attempted analysis of flowers and fruits, demonstrates that she really had no idea of floral structure and how to analyse and illustrate it. Von Mueller created the chance for her to learn the science of botanical illustration in Germany. However, Rowan did not follow this through and ended up spending considerable time in America illustrating popular American Floras for Alice Lounsberry — *A Guide to the Trees* (1900) and *Southern Wildflowers and Trees* (1901). The closest Rowan attained to scientific illustration in Australia was to have sixteen of her plates used by Frederick Bailey to illustrate his *Comprehensive Catalogue of Queensland Plants* (1909).

Rowan's work won her significant art awards in Australia and other parts of the world. Certainly much of her work is excellent, however it is patchy and she made copies of many of her paintings to sell. No doubt each copy reduced the freshness, accuracy and detail of the original. Perhaps, too, her eyesight was letting her down in later years. Nevertheless she is one of the best known botanical artists of her time in this country. It is a great pity that she did not pursue the opportunity offered her by von Mueller.

Rowan did some publishing of her own. *A Flower Hunter in Queensland* (1898) and *Bill Baillie: his life and adventures* (1908) record different aspects of her field and painting experiences.

Marianne North was English-born and did not settle in Australia. She undertook several world trips and painted whatever took her fancy along the way. She visited Australia in 1880 and made contact with Ellis Rowan. During that time it is thought that North taught Rowan the art of painting with oils. An extensive collection of North's paintings were

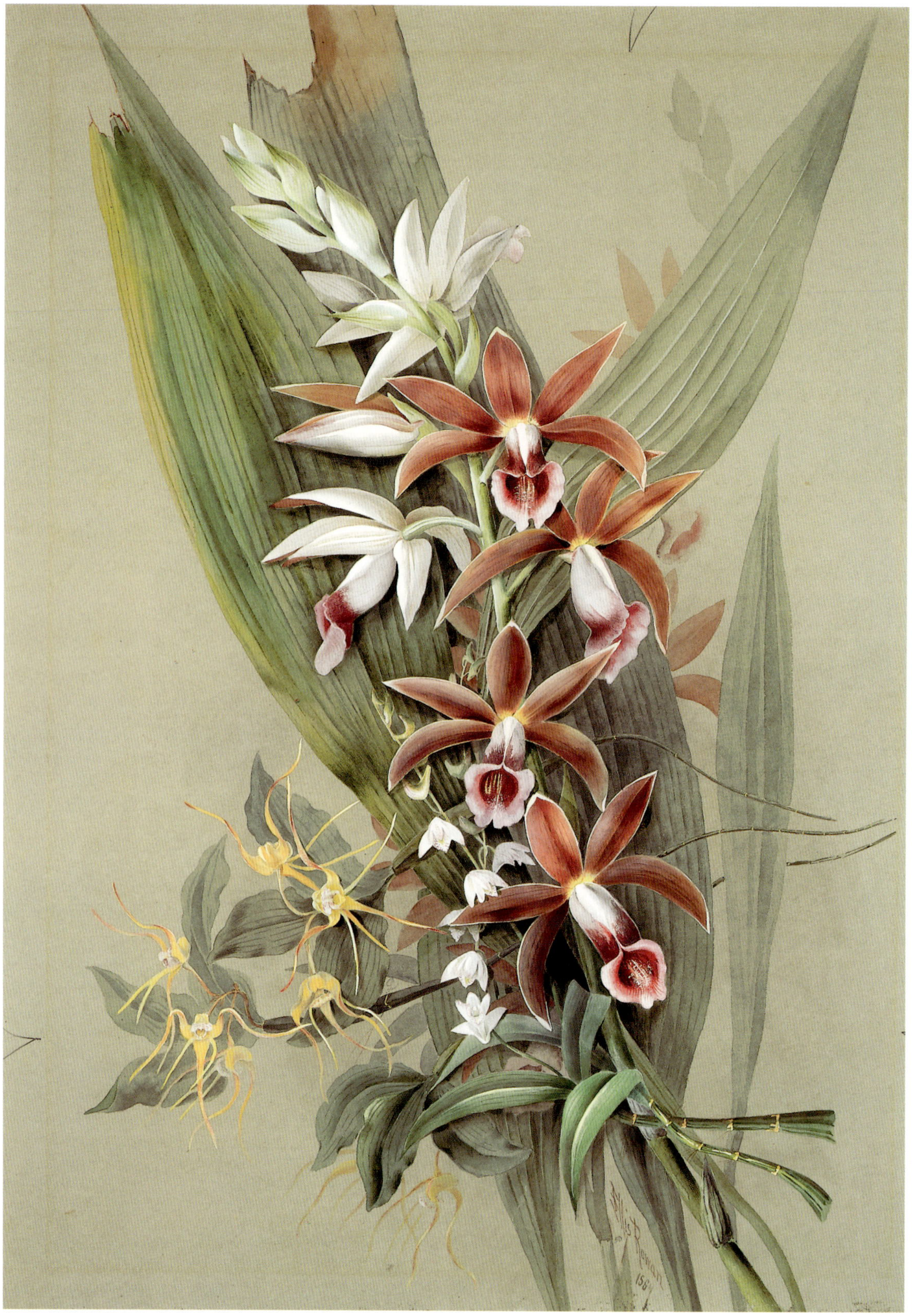

Phaius tankervilleae (L.Hér.) Blume (= *Phaius grandifolius* Lour.), *Dendrobium cacatua* M.A. Clements & D.L. Jones and *Dendrobium x delicatum* (Bailey) Bailey
Watercolour and gouache painting by Ellis Rowan, 1867.
[NLA 54.7 x 38.0]

donated to Kew together with the resources to build the North Gallery to display them. There are about eight hundred North paintings and the collection includes Australian subjects. North used oil as her medium and consequently the work is heavier than that of the watercolour artists. Oil allows the collection to be permanently on display — watercolour pigments are more fragile and soon fade when exposed to bright light.

Another feature of North's works is that the subjects are painted in the setting of their environment and so, being partly landscapes, the works communicate more of the distant and exotic parts of the world than if only the plant were depicted.

Rosa Catherine Fiveash was born in Adelaide. She was attracted to native plants and specialised in botanical art. This led to a commission to paint the plates for John Ednie Brown's *Forest Flora of South Australia* (1882–1890). Fiveash also painted for Richard Sanders Rogers, an orchid enthusiast. *An introduction to study of South Australian Orchids* (1909, 1911) is one example of Rogers' work in which Fiveash's paintings were published.

Fiveash was trained in art at the Adelaide School of Design and School of Painting, but developed a love of painting flowers and essentially trained herself to become a botanical artist. She trained future artists, such as Alison Ashby.[15] Fiveash's association with Richard Rogers led to her becoming a prominent illustrator of orchids for many decades. In later life her eyesight failed and perhaps the quality of some of her later work suffered because of this. While it is probable that she collaborated with von Mueller, this has not been proved.

Other women were painting native flowers too including Elizabeth Macarthur, Elizabeth Darling, Mary Hindmarsh, Marianne Collinson Campbell, Margaret Forrest and Annie Walker. It is interesting that some of these women artists were also authors, such as Louisa Meridith, Louisa Atkinson and Ellis Rowan.

For several of these artists von Mueller provided the scientific identity of their subjects and gratefully received voucher specimens of their paintings for incorporation in the Melbourne collection. Indeed von Mueller developed an amazing network of collectors. He actively sought participation from the remotest districts to obtain the widest possible representation of the flora. In 1883, for example, von Mueller advertised in the *West Australian* for collectors. No doubt he did this in the other Australian colonies too. The entry read,[16] 'urge ... to bring ... specimens of *all sorts* of plants to be dried at the stations and forwarded to me by post ... The minutest annuals should not be overlooked.' Many women, other than artists, became deeply involved in his network.

Konkordie Amalie Dietrich[17] was born in Siebenhelm, Germany. Amalie married Wilhelm Dietrich in 1847. Wilhelm was a pharmacist but preferred botany and earned his living as a botanical and natural history collector. Amalie Dietrich, already enchanted by plants, joined the collecting enterprise and thereby no doubt learned a great deal of botany. The marriage did not endure, but Dietrich's botanical pursuits did. She managed to gain employment with the Godeffroy Museum of Natural History, Hamburg, as a botanical and natural history collector in Australia. She arrived in Brisbane in 1863 on the *La Rochelle* and began an amazing expedition collecting alone in the wilds of Queensland.

Dietrich made contact with Ferdinand von Mueller shortly before departing Australia for the Pacific islands in 1871. She did not set out to collect for von Mueller and so did not strictly belong to his network. However, a duplicate set of many of her collections was ultimately sent to him.

Jessie Louisa Hussey[18] of Port Elliot, South Australia, joined the von Mueller collecting network after reading one of his advertisements. Von Mueller encouraged her to take an interest in the seaweeds of the coast near her home and Hussey became one of the most important Australian algal collectors in the second half of last century. Hussey did not restrict her collecting efforts to seaweeds but also made some important contributions to von Mueller's flowering plant collection.

Other South Australian women collectors included Annie Richards,[19] who collected from Streaky Bay and other localities in the state, and von Mueller's sister, Clara Christine Marie (Mueller) Wehl[20] and his niece, Louisa Therese Wehl, who collected from Lake Bonney and Mt Gambier respectively. In Western Australia Sarah Theresa Brooks[21] of Balbinia Station, Israelite Bay, joined the network.

Von Mueller's enormous postal collection network of course also included many men. William Vincent Fitzgerald, early in his career in Tasmania, Charles Walter, an itinerant collector in Victoria, and William Woolls, of New South Wales, were among their number.

The Reverend William Woolls[22] was born in Winchester, England, and trained in Göttingen, Germany. He came to Australia in 1827 and taught in several Sydney schools. He was a very knowledgeable amateur botanist who seemed to attract like-minded people, rather more so than had Charles Moore. His network included some local artists and also, of course, von Mueller. Woolls had several works published on the flora of New South Wales and established himself as something of a local authority. One of his greatest contributions to Australian botany was to teach the subject to Joseph Maiden.

Most of the major early collectors have been expounded in this book, however the number of collectors increased enormously over the years. Joseph Hooker listed some in his introductory text to the *Flora Tasmaniae*, and George Bentham listed others in the preface to his *Flora Australiensis*. Ferdinand von Mueller became a giant among these collectors but the growth of von Mueller's great herbarium was significantly augmented with contributions by numerous lesser known, willing, collaborators.

While this activity was going on in Australia, George Bentham worked on inexorably at Kew and finished the *Flora* in 1878. This massive job was methodically approached and resulted in a Flora which was to serve Australia for a whole century.

Flora Australiensis (1863–1878) came out in seven volumes, the sixth Australian Flora, but still unillustrated. Illustrated or not, it was the most thorough and most comprehensive to date, including collections from all regions of the continent. It demonstrated a knowledge of over eight thousand five hundred species occurring in Australia and is renowned for the wisdom of taxonomic interpretation undertaken by Bentham. The role of both Robert Brown and Ferdinand von Mueller must be acknowledged as underpinning and augmenting Bentham's wisdom.

If the enterprise had been undertaken in Australia, there is no doubt that von Mueller would have prepared a similarly competent treatment. It also seems certain that von Mueller would have attempted to organise at least some illustration to enrich it.

A few short years after *Flora Australiensis* was completed von Mueller, in his usual enthusiastic way, compiled from Bentham's *Flora* and from his own updates a *Systematic Census of Australian Plants* (1882). Von Mueller dedicated his *Census* to George Bentham, Joseph Hooker and Alphonse de Candolle, each of whom had made their own particular contribution to knowledge of the Australian flora. Von Mueller's words were: 'These pages are offered as an appreciative tribute from a young colony in antipodal remoteness.' Bentham acknowledged his copy very ungraciously, saying:[23]

> ... it grieves me to think that you should have devoted so much of your valuable time to a work which, botanically speaking, is not only absolutely useless but worse than useless ... Let me entreat you to give up the vain endeavour to attach the initials 'F.v.M.' to so many specific names, good or bad, as possible, and to devote your energies, your great abilities, and the splendid materials at your disposal, to the completion of such classical works as your Eucalyptus and similar monographs.

Well! Von Mueller gave up nothing. He prepared supplements to his *Census* and in 1889 published a *Second Systematic Census of Australian Plants* and removed Bentham from the dedication. Moreover, he also completed the *Eucalyptus* monograph, together with other monographs, before his enormously productive life came to an end in 1896.

Bentham did not visit Australia and cannot have understood the vastness of the country, its political divisions, its climate and its phytogeography. Von Mueller had prepared a very useful adjunct to *Flora Australiensis*, making the latter much more functional for each of the state-based Government Botanists. Von Mueller had also greatly aided the compilation of state-based Floras. One is led to wonder what Bentham would make of the huge databases being compiled these days. One can be sure von Mueller would have revelled in them.

Von Mueller prepared several illustrated monographs. These included the *Eucalyptographia* (1879–1884), *Description and illustration of the Myoporinus plants of Australia* (1886–1887), *Iconography of Australian species of Acacia* (1887–1888), *Iconography of Australian Salsolaceous plants* (1889–1891) and *Iconography of Candolleaceous plants* (1892).

All were generously illustrated, as the titles imply, with uncoloured lithographs. *Eucalyptographia* was illustrated by R. (?Richard) Austen, Emil Todt,[24] and L. (?Louis) Rummel. In this work Austen did his own lithography and Todt's and Rummel's drawings were lithographed by C. Troedel & Co., the particular artisan not being acknowledged. Austen also drew eucalypt species for von Mueller for publication in *Report on the Forest Resources of Western Australia* (1879). This report was published in London and the plates were lithographed there by John Nugent Fitch, a nephew of the famous Walter Fitch. John carried out lithography for *Curtis's Botanical Magazine* for some years, starting in 1878.

Emil Todt was born in Berlin and trained as a sculptor. He came to Victoria in 1852 and continued sculpting, though little of his work has survived. In naming *Eucalyptus todtiana* in his honour von Mueller wrote of Todt:[25]

> ... this species bears the name of Mr Emil Todt, whose artistic talent became devoted only to illustrating plants at a venerable age, when most of those engaged in such pursuits, have ceased to work professionally. Therefore all the more praise is due to this gentleman for the youthful ardour, which he still brought to bear on the extensive furtherance of the present publication.

Von Mueller did not usually acknowledge his artists in his publications let alone praise them, so it is assumed that Todt really deserved the accolade. Rummel worked at drawing microscopic and anatomical features for *Eucalyptographia*.

For *Myoporinus plants* and *Acacia* Robert Graff[26] did the drawings and C. Troedel & Co. did the lithography; and for *Salsolaceous plants* and *Candolleaceous plants* Graff did both the drawing and lithography.

Robert Graff was described by von Mueller as having 'patient perseverance as well as artistic accomplishment'.[27] He certainly was von Mueller's most prolific artist. Moreover, von Mueller also found that Graff was a competent woodcut engraver and it is assumed, but not confirmed, that Graff executed a number of woodcuts for him. Only one woodcut has been identified as Graff's.[28]

Von Mueller used woodcuts to illustrate a number of his more 'popular' works, such as *Introduction to Botanic Teachings* (1877); *The Native Plants of Victoria* (1879) which was 'done with as much celerity as the means for the

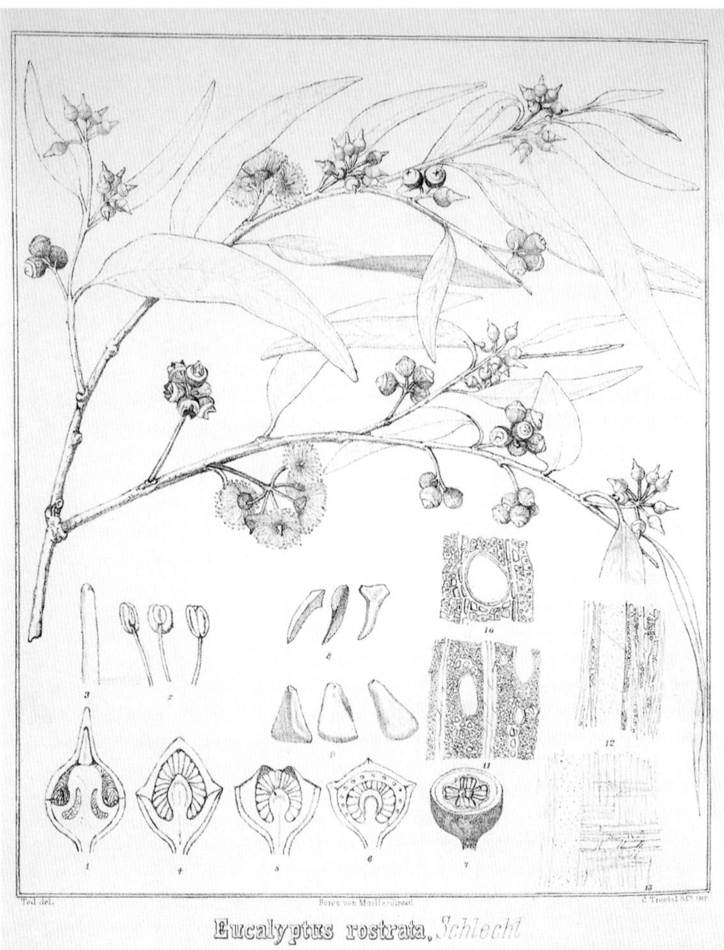

Eucalyptus camaldulensis Dehnh.
(= *Eucalyptus rostrata* Schlecht.)
Uncoloured lithograph drawn and lithographed by Emil Todt and published in Ferdinand von Mueller's *Eucalyptographia* between 1879 and 1884.
[CANB 24.3 x 19.7]

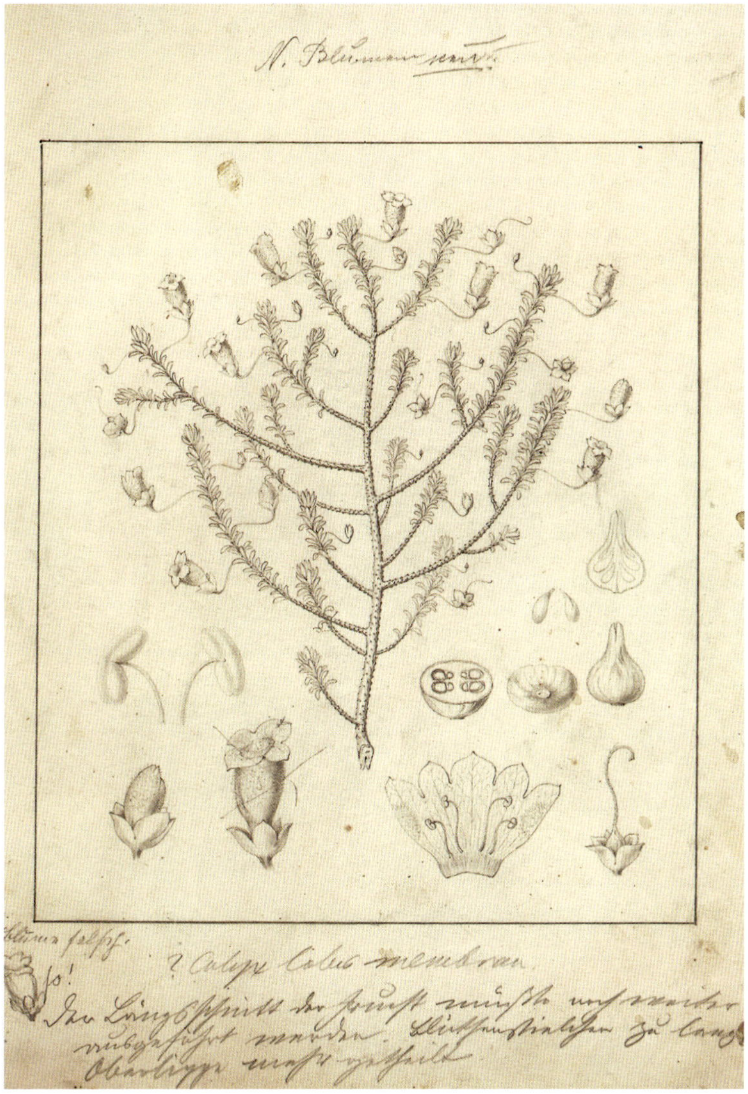

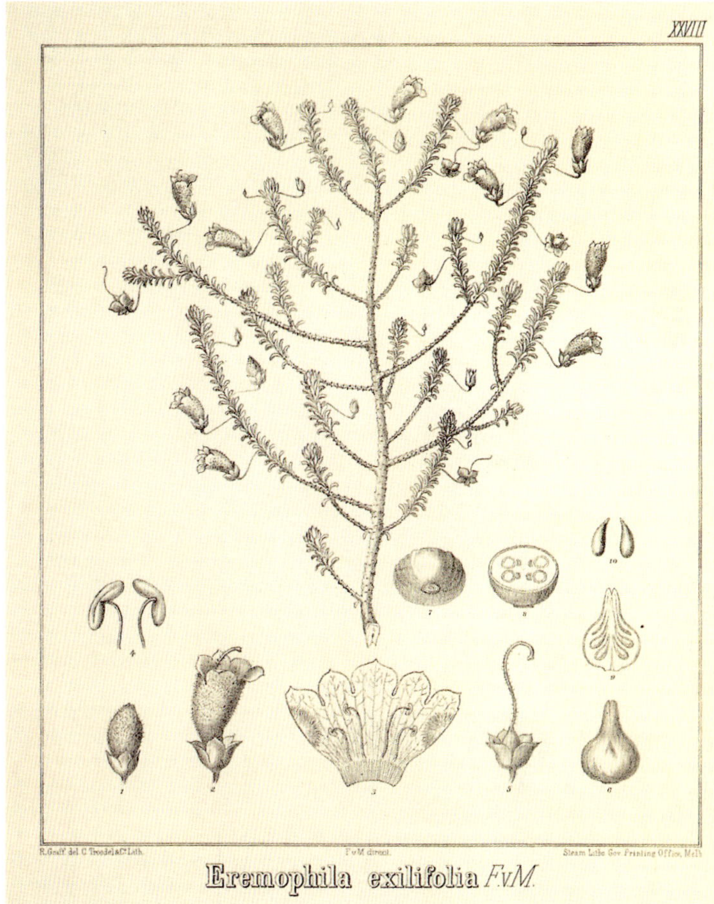

Eremophila exilifolia F.Muell.
Pencil sketch by Robert Graff with corrections by Ferdinand von Mueller (left). The corrected drawing was lithographed by an unknown lithographer working for C. Troedel & Co., and published in von Mueller's *Description and illustration of the Myoporinus plants of Australia* in 1886–1887.
[MEL 25.0 x 20.5]

preparation of the needful xylographic illustrations will admit of';[29] and *Key to The System of Victorian Plants* (1885).[30] H.B. Smith is the only engraver to have signed any of the woodcut engravings in these works. Close scrutiny of these indicates that the engravers used the published illustrations of Becker, Schoenfeld and other artists as the basis for many of the woodcuts. Moreover, von Mueller used them in more than one publication. Speed, economy and usefulness were all issues for von Mueller. In 1879 he wrote:[31]

> Perhaps Providence may spare me life and strength to carry this plan yet into effect for any such future editions of this work as may still appear, especially as a likely demand for this book in the surrounding colonies for identifying there also at least the commoner plants may perhaps early absorb the present edition, and as by providing now once for ever the xylographic blocks any subsequent issues can be rendered still so very much less expensive.

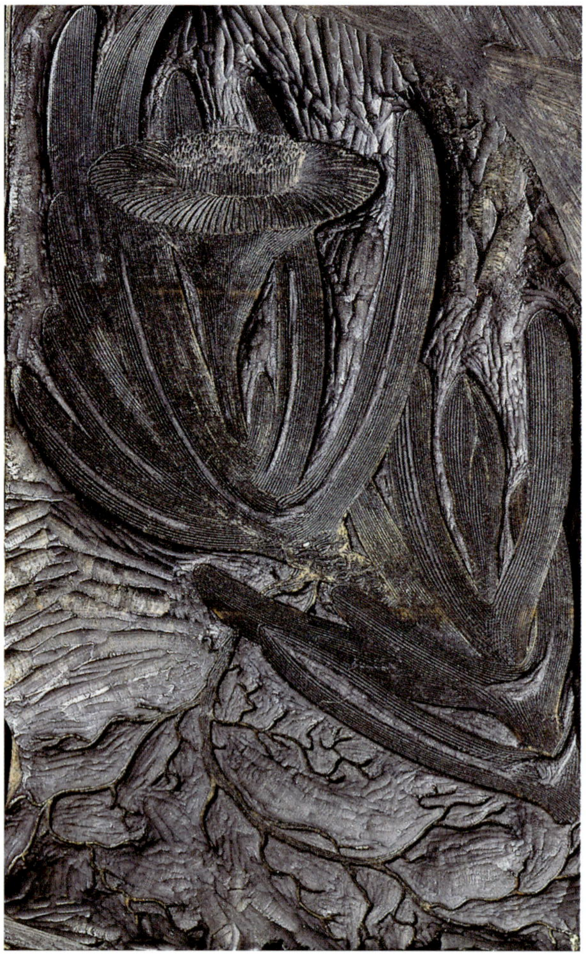

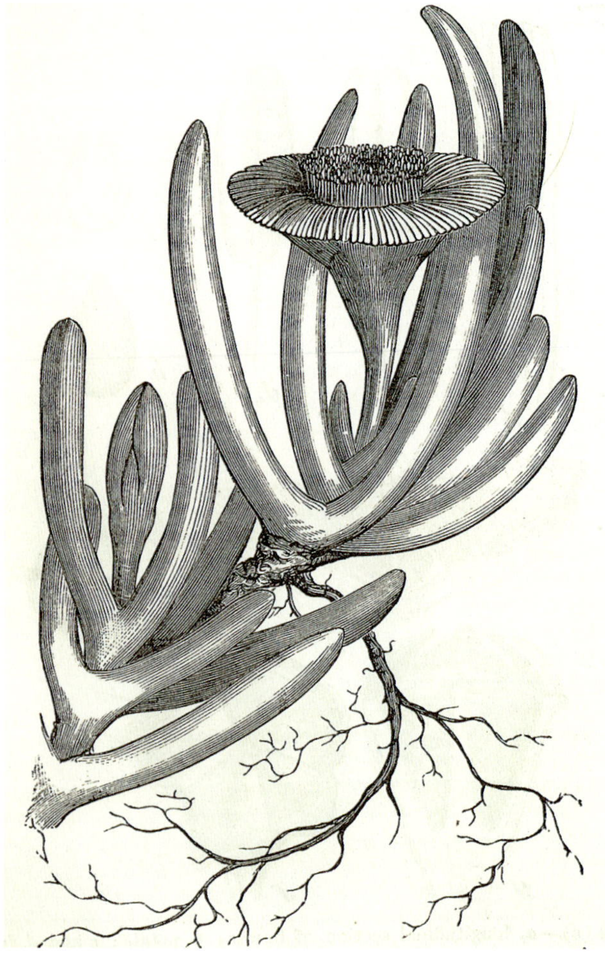

Disphyma crassifolium (L.) L.Bolus (= *Mesembryanthemum australe* Ait.)
Wood block engraved and inked for printing (left). The print made from this block was published in Ferdinand von Mueller's *The Native Plants of Victoria* in 1879.
[MEL 12.5 x 7.6]

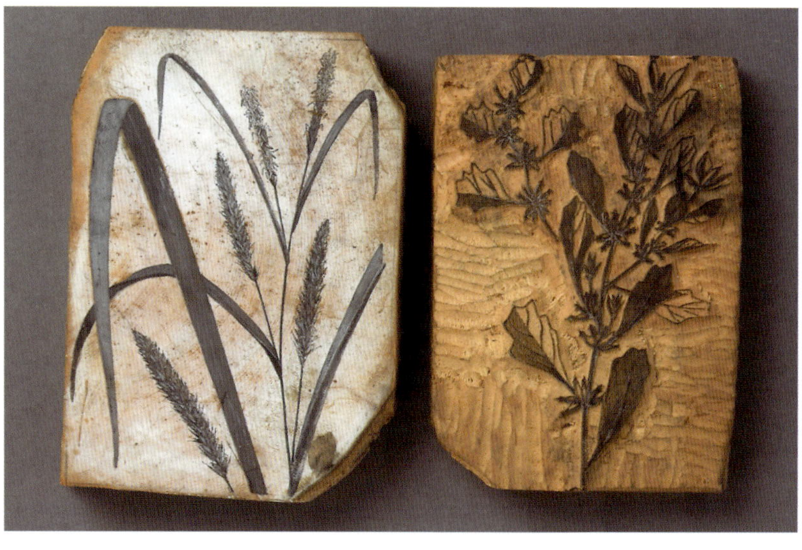

Carex alsophila F.Muell. and *Atriplex leptocarpum* F.Muell.
Wood blocks in the early stages of engraving prior to printing. *Carex* (left) shows the original drawing and *Atriplex* (right) shows the unfinished carving prepared for proofing.
[MEL 12.6 x 8.9, 11.7 x 8.0]

Von Mueller also released a small collection of voucher specimens to be housed in Mechanics Institutes in Victoria as references for people attempting to use his educational books to identify plants.

A number of other interesting items were produced by scientists and artists working concurrently with von Mueller in Australia.

Robert Desmond Fitzgerald[32] was born in Ireland, trained as a civil engineer and came to Australia in 1856. He developed an interest in Australian orchids and ferns. This he coupled with his ability as a botanical artist to produce a two-volume monograph entitled *Australian Orchids* (1875–1894) which he dedicated to Charles Darwin. The work appeared in parts, the cover engraving of *Sarcochilus divitiflorus* staying the same. The plates are hand-coloured lithographs based on Fitzgerald's paintings, the lithographer being Arthur James Stopps. Fitzgerald drew the very early lithographs on stone and all were coloured by hand.

Fitzgerald worked in the Lands Department in Sydney as a draftsman until he took up the post of Deputy Surveyor General in 1873. No doubt he and Arthur Stopps (also on the staff) worked together on many official projects. All the while Fitzgerald indulged in his hobby of natural history and drew and painted local flora.

Arthur Stopps came to Australia early in the 1850s to try his luck on the Victorian goldfields. He was soon, however, gainfully employed working in his profession as a watercolourist and lithographer and he did lithography for S.T. Gill, whose work depicted life on the goldfields. In 1864 Stopps moved to Sydney to work as the lithographer with the NSW Lands Department. In this

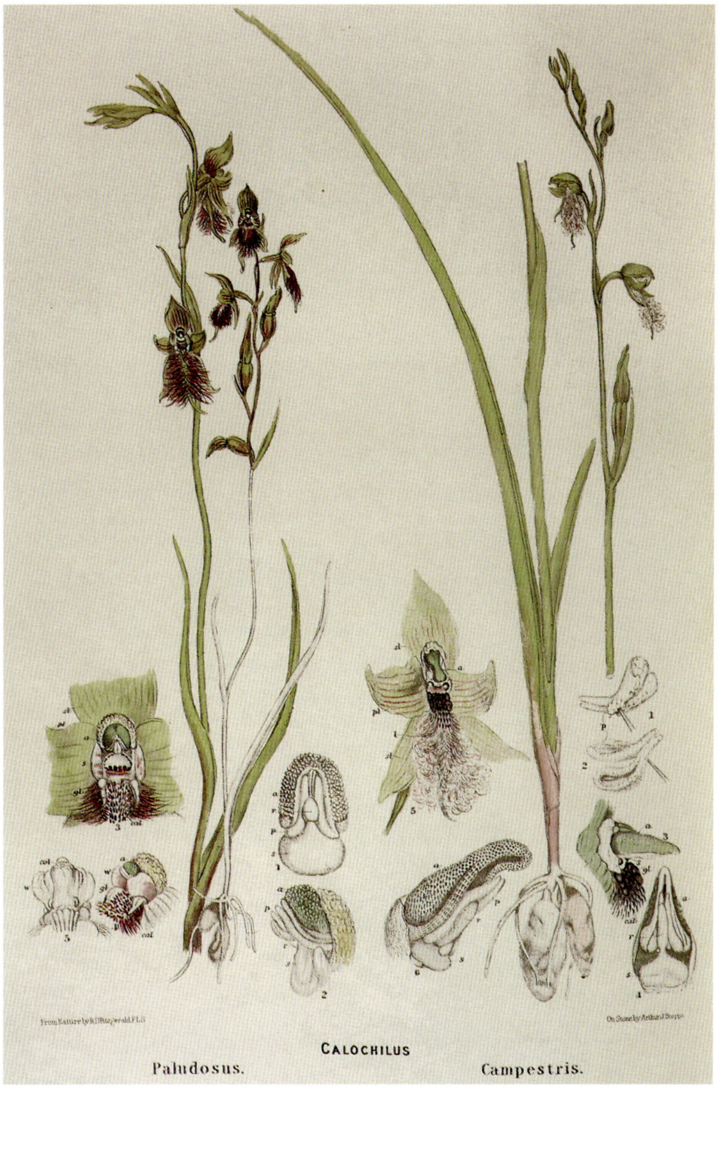

Calochilus paludosus R.Br. and *Calochilus robertsonii* Benth. (= *Calochilus campestris* R.Br.) Hand-coloured lithograph drawn by Robert Fitzgerald and lithographed by Arthur Stopps. Published in Fitzgerald's *Australian Orchids* in 1878.
[CANB 50.0 x 22.0]

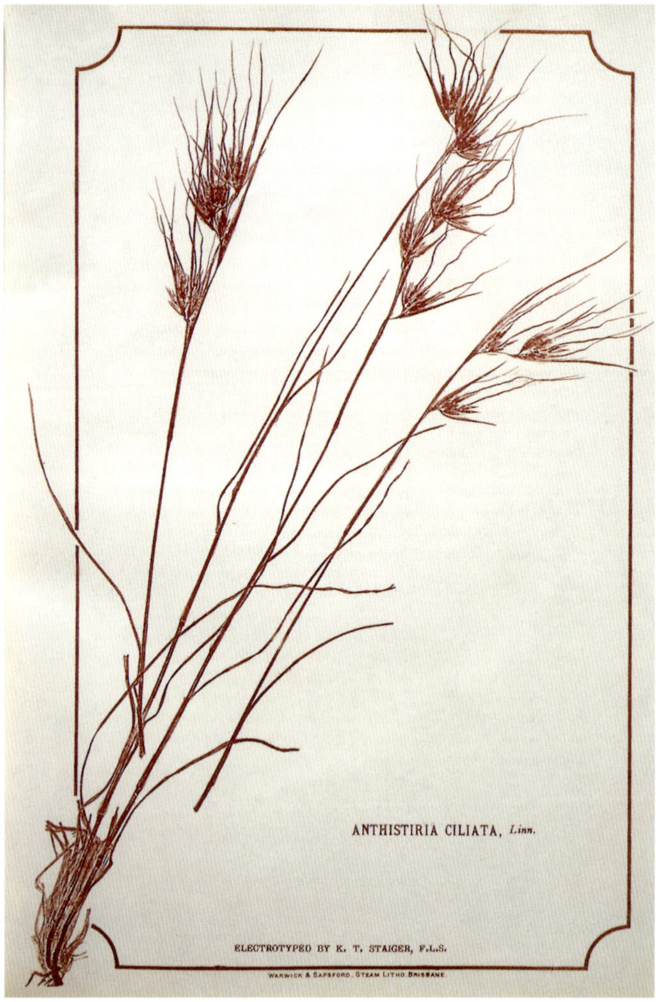

Themeda triandra Forsskal. (= *Anthistiria ciliata* L.f. and *Themeda australis* (R.Br.) Stapf.)
Nature print prepared by electrotype by Karl Staiger and published in *An Illustrated Monograph of the Grasses of Queensland* in 1878.
[CANB 37.0 x 26.5]

Frederick Manson Bailey[33] (1827-1915) produced some particularly interesting items. In 1878, together with Karl Theodore Staiger,[34] he produced *An Illustrated Monograph of the Grasses of Queensland*, largely for farmers and graziers. Each plate is a nature print from an herbarium specimen. The plates and the text are monochrome, being either brown or magenta. For printing purposes the plates were prepared by an electrotype process, presumably for a larger print run than usual. The limitation of the nature print is that much detail is obscured and the result inadequate for close botanical analysis. However, the one-to-one, gross detail is quite adequate for communicating identity to farmers and other people who are unused to formal botanical description.

Nature printing does, however, work reasonably well for fern fronds which are relatively flat and have strong veins which are frequently diagnostic. In 1892 Bailey produced *Lithograms of the ferns of Queensland*. This consists entirely of a compendium of nature prints of fern fronds accompanied by an index to their scientific and common ('trivial') names. The images in this case were transferred to stone for printing by inking the specimens and pressing them onto the stone. Bailey explained that the book was aimed at:[35]

> ... lovers and cultivators of our indigenous fern, who, while having a desire for their names, find difficulty in determining them from published descriptions, be such ever so plain and devoid of technicalities.

In 1887 Bailey teamed with Patrick Robertson Gordon, Chief Inspector of Stock, to prepare *Plants reputed Poisonous and Injurious to Stock*. The plant description is brief and a reference to a more expansive botanical description is given together with a drawing by Margaret A. Hope. Of the drawings, Bailey and Gordon wrote:[36]

> The illustrations will greatly assist in identifying the plants ... [though] ... in most cases the drawings were made from herbarium specimens, and some of

> them only fragmentary; so that really much credit is due to the artist ... for the work ... This little pamphlet has been prepared for the stockowners and farmers, few of whom have leisure for botanical pursuits, a general representation of a portion of the plant, just sufficient for its identification, has been given.

The drawings are reproduced as lithographs and have been printed as if they were pencil sketches, with very pleasing effect.

Frederick Bailey came to Australia in 1839 with his father, John, who had been appointed Government Botanist in Adelaide. No doubt John stimulated Frederick's interest in plants. Indeed John Bailey founded a botanical family, a number of his descendants making important contributions.

Frederick had nursery and seed-store experience before taking a post as a botanical pathologist with the Queensland Government. In 1881 he was appointed as Queensland's Colonial Botanist. His practical and applied background and his sympathy with the layman show through in his many publications. Frederick went on to become a very important figure in Australian botany and certainly became the founder of the botany of Queensland's plants.

John Ednie Brown published *The Forest Flora of South Australia* in parts from 1882 to 1890. Brown was Scottish-born and practised in forest management. He came to South Australia in 1879 to take the post of Conservator of Forests. Brown was elected a Fellow of the Linnean Society in the same year. He held a similar post in New South Wales from 1890 and in Western Australia from 1895. During Brown's employ in South Australia he prepared the *Forest Flora*. It is elephant-sized and illustrated with forty-five plates which are high-quality chromolithographs, originally drawn on stone from paintings and photographs. Most of the plates were painted by Rosa Fiveash. Camilla Hammond and Miss Stuckey both provided some artwork for five plates in Part 1. H. Barrett was the lithographic artist who transferred the paintings onto stone and crafted the colour printing.

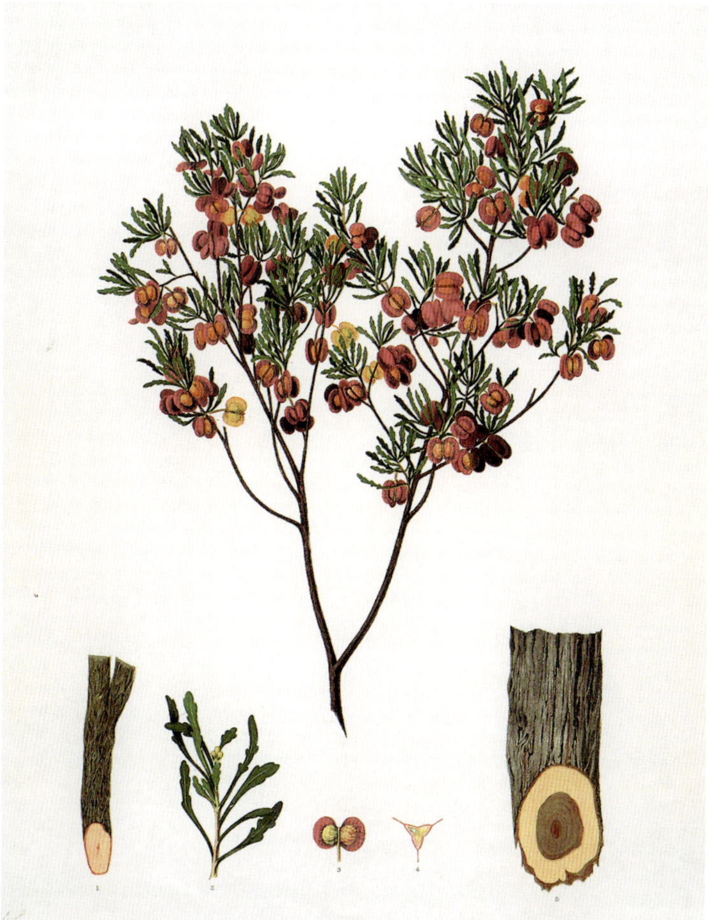

Dodonaea lobulata F.Muell.
Chromolithograph painted by Rosa Fiveash and lithographed by H. Barrett. Published in John Brown's *The Forest Flora of South Australia* between 1882 and 1890.
[NLA 44.5 x 34.0]

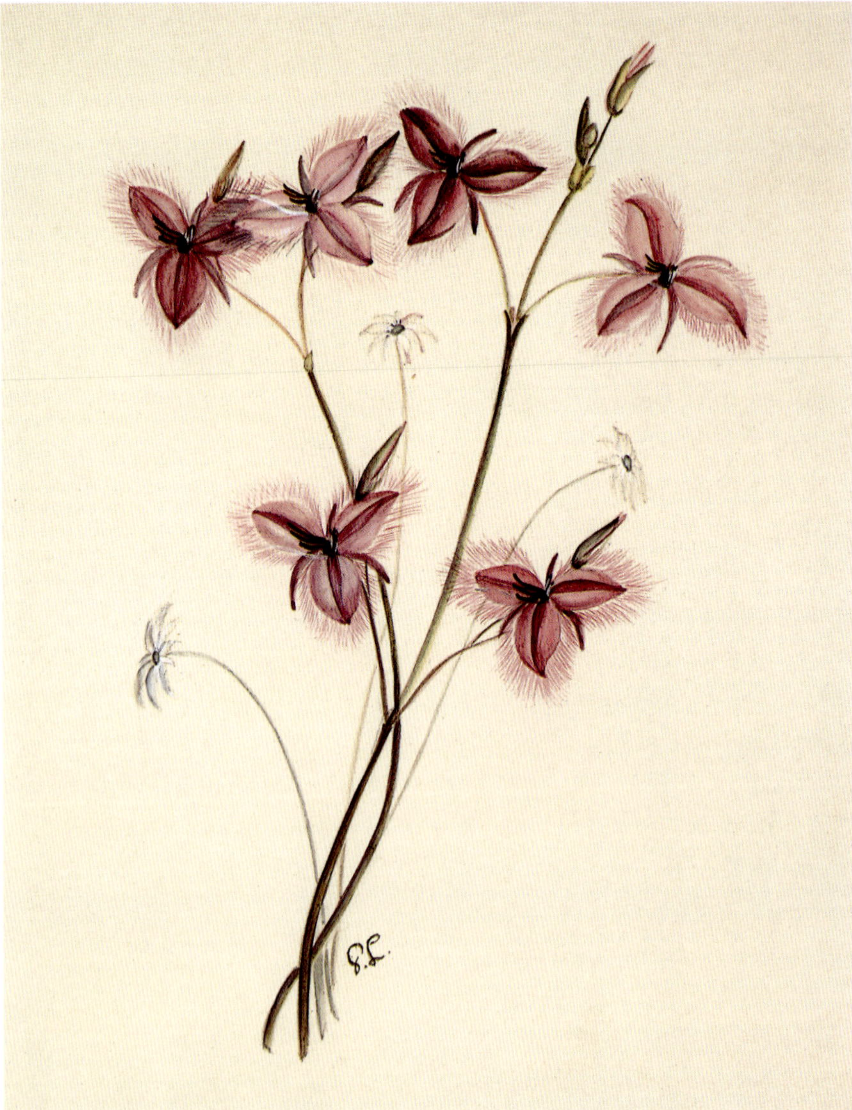

Thysanotus juncifolius (Salisb.) J.H. Willis and *Actinotus minor* (Smith) DC.
Watercolour painting by Gertrude Lovegrove.
[ML 26.2 x 17.6]

The process is described in Part 2 as:

> ... requiring the greatest skill and care, for not only has the lithographic artist to pick out the various shades of colour and place them on different stones, so that when one is printed after the other it shall come in exactly the right spot on the sheet, but he has also to modify his colours, so that in drying they shall develop the proper tint.

Barrett also drew the cover illustration. He worked as a chromolithographer for the South Australian Government Printer. H.F. Leader and E. Spiller,[37] Government Printer(s), Adelaide, are named as the lithographers. This is an early, very successful work depicting Australian plants, produced by chromolithography.

Little seems to be known of Gertrude Lovegrove[38] except that she painted in Sydney during the second half of last century. Lovegrove was chosen by William Bauerlen to illustrate a popular Flora of New South Wales, *The Wildflowers of New South Wales*. Bauerlen has been described as a 'circumspect and zealous' collector.[39] He was on the books of von Mueller in Melbourne and Joseph Maiden in Sydney and is known to have made some valuable collections. His *Wildflowers*, however, was not completed. One fascicle was published in 1891. This is an extremely rare publication, so it did little to edify the public about plants. The paintings are mostly in the 'bunches of flowers' style, and are competent but not closely detailed.

By the end of the 1870s the emphasis on taxonomy of Australian plants had well and truly shifted to Australia. In the late 1800s George Bentham and the Australian Government Botanists were active in compiling working tools in the form of synoptic Floras and Keys. The stimulus provided by Bentham and the steady publication of the *Flora Australiensis* was fundamental to much of this activity, providing the reference for the relevant regional material.

In Tasmania William Webb Spicer produced *A Handbook of the Plants of Tasmania* in 1878, and in 1880 Ralph Tate produced a South Australian Census followed in 1890 by *A Handbook of the Flora of Extratropical South Australia*, which was a synoptic Key to South Australian plants. Charles Moore, together with Ernst Betche, published a *Handbook of the Flora of New South Wales* in 1893. Of these only Spicer's book is illustrated.

Bailey compiled a *Synopsis of the Queensland Flora* in 1883, which is unillustrated, followed by three supplements, two of which are illustrated. In the *First Supplement* (1886) Charles Knight of New Zealand drew four lithographs of lichens. They were printed by the Government Lithographic Office in Brisbane. *The Third Supplement* (1890) has several plates illustrating lycopods, ferns, mosses and liverworts and was printed by the Government Printer, James C. Beal. These are nature prints which combine crudity and delicacy. The crudity of nature printing is eliminated in hand-prepared illustration, while the delicacy of features such as venation is virtually impossible to portray in hand-prepared illustration other than by enlargement or magnification.

Attention to Australian plants did not cease in overseas institutions. Individual taxonomists continued to publish in journals or in one-off monographs, and the grand periodicals, such as *Curtis's Botanical Magazine*, continued to include them. Heinrich Gustav Adolf Engler and Karl Anton Eugen Prantl began a world family Flora, *Die Natürlichen Pflanzenfamilien* (1887–1915), destined to be a very important reference text. This was written by numerous authors and was selectively illustrated. The figures were gathered from a number of sources, frequently being borrowed and acknowledged as such. Some were original and generally the artists were unacknowledged. Where appropriate, Australian plants were illustrated.

The *Pflanzenfamilien* illustrations, while being generally of a very high standard, tend to be stylised and almost diagrammatic. In spite of that, they serve to demonstrate the different techniques and styles employed by numerous artists at the turn of the century. The compilation thus serves to provide a standard to which the contemporary Australian artists can be compared.

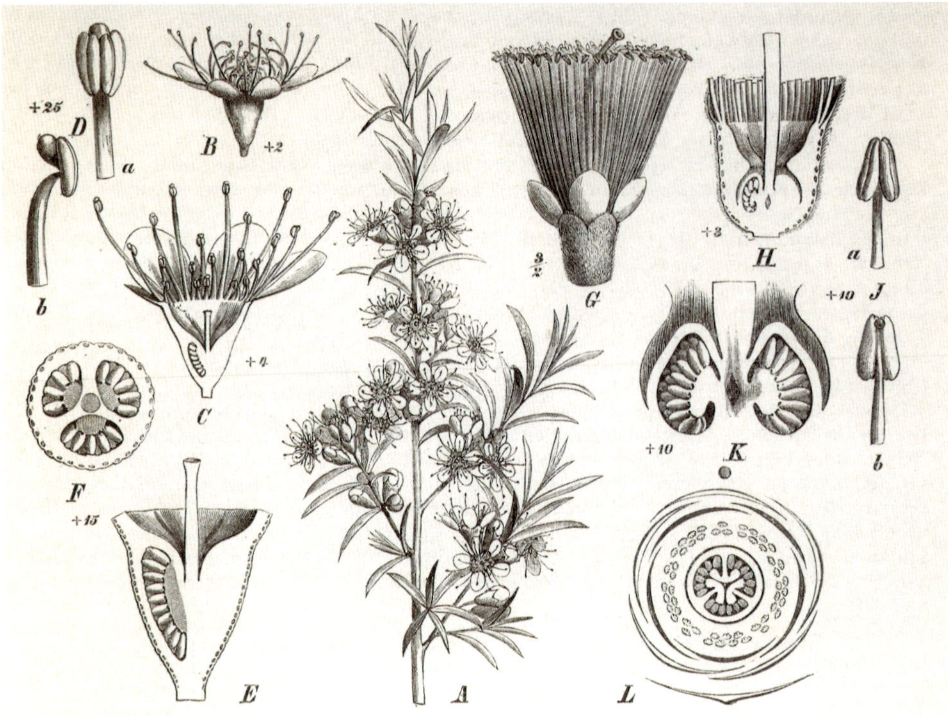

Kunzea ericoides (A.Rich.) Joy Thompson (= *Kunzea peduncularis* F.Muell.) and
Callistemon citrinus (Curtis) Skeels (= *Callistemon lanceolatus* (Smith) Sweet)
Metal engraving published in *Die Natürlichen Pflanzenfamilien* in 1895.
The artist and engraver are not acknowledged.
[CANB 9.5 x 12.5]

Ferdinand von Mueller died in 1896. For half a century he had participated in and overseen, the establishment of Australian systematic botany in a way that no predecessor or successor did, or could hope to do. He was a unique character. Australia is fortunate that he made his best efforts to have plants illustrated with the facilities which his resources and manpower permitted.

Ferdinand von Mueller was honoured in so many ways that it would take a page to list them all. He was a proud man and usually documented his degrees, honours and awards on the title page of his publications. He was elected a Fellow of the Linnean Society in 1859 and of the Royal Society in 1861. In 1869 the King of Württemberg bestowed the title 'von' on Mueller and two years later raised him to the hereditary title of Freiherr or Baron. In 1873, much to his consternation, he lost the post of Director of the Melbourne Botanic Gardens and he was only partly consoled by receiving a knighthood in 1879. Von Mueller was honoured with *Sirmuellera* (Proteaceae) by C.E.O. Kuntze in 1891, *Muellerina* (Loranthaceae) by P.E.L. van Tieghem in 1895, *Austromuellera* (Proteaceae) by C.T. White in 1930 and *Muelleranthus* (Fabaceae) by J. Hutchinson in 1964, all of which are flowering plants; a moss and an alga; and about a hundred species names. The house journal of the National Herbarium of Victoria, *Muelleria*, is also named in his honour: 'no more worthy memorial to such a distinguished man of science could be conceived than a journal, bearing his name and emanating from the very institution he had founded.'

9. Joseph Maiden and the Floras

References are always given to Bentham's Flora Australiensis, *which still is, and will long remain, the standard work on our flora. The greater one's experience with it, the more sincere is one's admiration of it.*

Joseph Maiden[1]

The deaths of Charles Moore in New South Wales and Ferdinand von Mueller in Victoria wrought significant changes in taxonomic botany in Australia, especially in the south-east. Moreover, Australia was entering a period of development in which a knowledge of the flora was really necessary, especially for those working in primary industry — farming, grazing and forestry.

In New South Wales, Moore had not made a particularly significant contribution to taxonomy and had failed to establish a research herbarium, the fundamental resource for the science of taxonomy. In Victoria von Mueller, by contrast, made a giant contribution both to taxonomy and to the establishment of a very important research herbarium. Upon their deaths, however, the rate of progress made in their respective states, as the colonies became at the turn of the century, was reversed. Where von Mueller was not replaced with a dynamic personality or greater financial resources, Moore was succeeded by the energetic Joseph Maiden. In Queensland, Frederick Bailey continued to make his significant contribution.

Joseph Henry Maiden[2] (1859–1925) was born in London and educated at the University of London. He came to Australia for health reasons in 1880 and soon became involved in the establishment of the Technological Museum of Sydney. The following year he was appointed

Curator. Maiden's growing interest in the local flora led to his taking instruction in botany from William Woolls. Clearly Maiden was a good student and his career as a botanist soon went from strength to strength. By 1890 Maiden was appointed Consulting Botanist to the Department of Agriculture and so he began his botanical career in an institution demanding very applied outcomes.

In 1896 Maiden was appointed Government Botanist and Director of the Sydney Botanic Gardens. Maiden quickly appreciated the value of what von Mueller had established for Victoria and he set about establishing a research herbarium and library for New South Wales. He put Ernst Betche in charge of the herbarium in 1901 and began to establish himself as a specialist botanist in the two large Australian genera, *Eucalyptus* and *Acacia*, publishing a very important monograph on the former.[3] His studies and field investigation were not, of course, limited to those genera. In fact, Maiden's interests were diverse. He also published on applied botany and on history. His interest in the history of Australian botany and botanists has proved invaluable for later publications on those subjects.

The era of exploration and discovery was largely over and it was time for consolidation and development. The twentieth century saw Federation and the birth of a nation in 1901. Initially this did not have much apparent impact on science and botany yet it served to reinforce the role of state governments, and their Government Botanists, in documenting and managing their plants and animals. Government Botanists had a responsibility to serve economic botany and the needs of those working in primary industry. Universities had broader needs to serve such as medicine, pharmacy, science and education as well as agriculture and forestry.

These applied needs meant that Floras had a practical objective. This in turn resulted in illustrations with a more diagnostic focus instead of full representation and interpretation. The luxury of colour remained very restricted and, when used, the reproduction was generally very poor, largely due to the methods of colour printing.

Updated editions of extant Floras, or new Floras, and Censuses were produced for some states. Frederick Bailey prepared *The Queensland Flora* (1899–1902). This was based on Bentham's *Flora Australiensis* and included six volumes. In 1903 Leonard Rodway produced *The Tasmanian Flora*, and in 1909 John Black produced *The Naturalised Flora of South Australia*.

All three authors arranged for at least some illustration in these works. Bailey's *Queensland Flora* was illustrated with a limited number of plates of rare species, mostly drawn by F.C. Wills and lithographed by F. Elliott. Wills did some of his own lithography in later volumes and Elliott drew one plate, but may have done more which were unacknowledged, especially in the sixth volume. An interesting exception is *Geijera helmsiae* (Rutaceae) which was drawn by Mrs R. Helms[4] and named in her honour by Bailey. Mrs Helms collected the type specimen at Childers. The plates in Volume 2 were drawn by Wills, and reproduced by a halftone process by the Government Lithographic Office in Brisbane. The other volumes were printed by a photolithographic process.

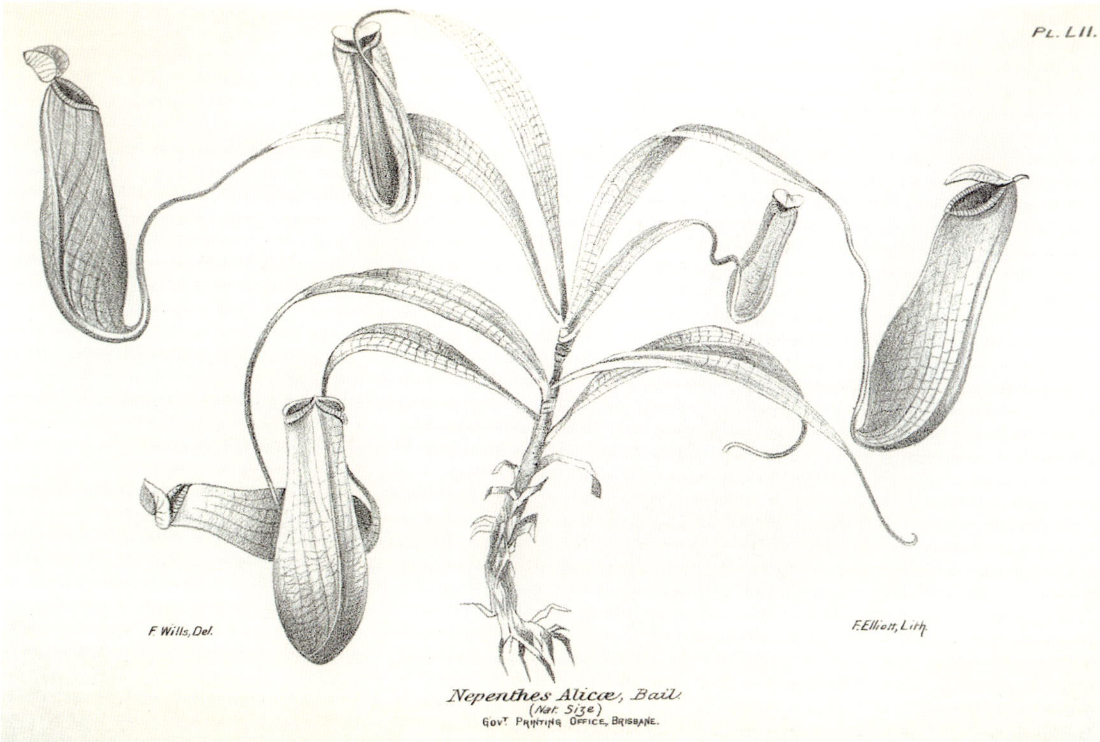

Nepenthes mirabilis (Lour.) Druce (= *Nepenthes alicae* Bailey)
Uncoloured lithograph drawn by F. Wills, lithographed by F. Elliott, and published in Frederick Bailey's *The Queensland Flora*, 1899–1902.
[NLA 14.8 x 23.8]

Leonard Rodway[5] illustrated his *Tasmanian Flora*, choosing to draw some typical species:[6]

> The Plates ... are faithful copies of the author's drawings. Had the departmental artists been permitted to idealise them some departure from the faithful copy of nature must have resulted, besides which increased cost would have necessitated a considerable decrease in number, which may have been the reverse of desirable.

They were reproduced by a photolithographic process. Leonard's drawings are not up to the standard of those by Wills and Elliott, but must be regarded as adequate.

Leonard Rodway was an enthusiastic amateur botanist, who earned sufficient respect to be appointed Honorary Botanist to the Tasmanian Government in 1896.

The *Naturalised Flora of South Australia* (1909) is illustrated with numerous vignette drawings all by author John Black. He desired that, 'by means of the figures and the glossary of scientific terms, even those previously unacquainted with botany may be able to follow the descriptions'.[7] These are pen drawings and are printed using process engraving. The printing industry had reached an important stage of development, which enabled scientists and their artists to be freed from the reproduction process. From the early 1900s line drawing was to become the norm for formal scientific illustration.

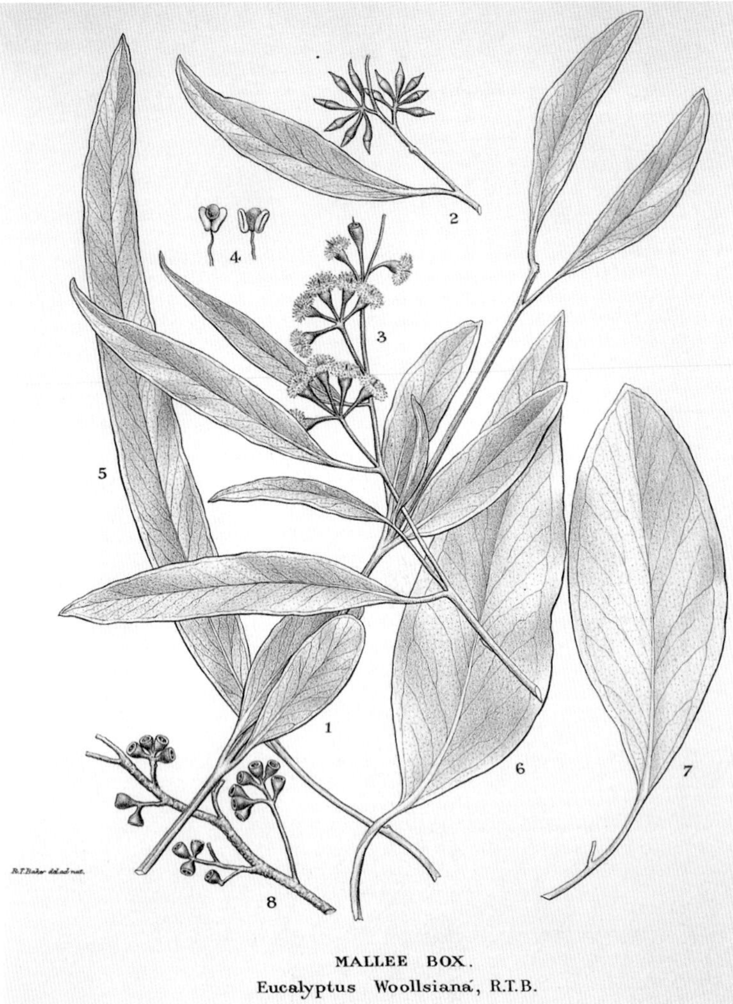

Eucalyptus microcarpa (Maiden) Maiden (= *Eucalyptus woollsiana* R. Baker) Uncoloured lithograph of a painting by Richard Baker, lithographed by an anonymous engraver from the Government Printing Office, and published in Baker's and Smith's *A Research on The Eucalypts and their Essential Oils* in 1902. In the second edition in 1920 the plate was an uncoloured screen print.

John McConnell Black[8] (1855–1951), another self-trained botanist, began to publish on botanical subjects in 1909, when he was fifty-four. In spite of his enormous contribution to the knowledge of South Australian flora, he never held a professional position as a botanist. Indeed, most of his work was conducted after he retired.

Meanwhile, Joseph Maiden's successor as Curator at the Sydney Technological Museum, Richard Baker,[9] played a role in producing technical literature. Baker teamed with Henry George Smith to produce publications on Australian softwoods and hardwoods. Baker and Smith made good use of photography, both black and white and colour, to illustrate their texts. One of these publications, *A Research on The Eucalypts and their Essential Oils*, came out in two editions. The first edition (1902) was illustrated with lithographs of Baker's drawings and of others unsigned. The lithographer, an employee of the Government Printing Office, is not acknowledged. In addition, a correlation between essential oils and leaf anatomy is illustrated with nine monocolour photos of leaves, to demonstrate the venation.

The second edition (1920) is more remarkable for the range of illustration methods and associated printing methods. The monocolour leaf photos are inked in a paler green and are not nearly as effective in demonstrating the venation as the darker ink used in the first edition. There are more black and white whole plant photos, with halftone colour plates of bark and colour photomicrographs of sections of leaves. The more classical botanical illustrations of fruits are cut into the text by process engraving. The black and white drawings of the first edition are included, with the addition of several paintings printed in halftone colour. Regrettably this is another example of artists remaining unacknowledged. However, some of the paintings are signed, revealing Baker, W. Marshall and C. Toms as the artists. The drawings of fruits are not acknowledged. The halftone reproduction of both the black and white and colour plates demonstrates a quality that is noticeably reduced compared with the first edition.

Richard Thomas Baker came to Australia in 1879 and taught art and science at Newington College for several years before gaining a post at the Sydney Technological Museum. The artwork in *Eucalypts & their Essential Oils* demonstrates that he was a very competent artist. Another work by Baker exposes his love of art. In 1915 he published *The Australian Flora in Applied Art. Part 1. The Waratah*. This is a marvellous exposé of the way in which the Waratah had been embraced into Australian design for a range of practical uses.

C. Toms is revealed as an artist in the employ of the Technological Museum and lecturer in decorative art at Sydney Technical College. Both Toms and W. Marshall illustrated *Eucalyptus* species very competently for Baker and Smith.

Collection and incorporation of material into Australian herbaria was, by now, taking place on a wide scale by official collectors (associated with the herbaria and their staff), privately by visiting European scientists and professional collectors, and piecemeal by enthusiastic amateurs and people seeking identifications.

A particularly interesting collector was Max Koch.[10] He was German-born and came to Australia in 1878 on a merchant ship, disembarking in Adelaide. Working in the rural industry in South Australia for many years, he developed an interest in plants and began collecting to supplement his income. Mt Lyndhurst Station near Maree, where he worked, was a major collecting site in the late 1890s and early 1900s. In 1904 Koch settled in the south-west of Western Australia and continued collecting both herbarium specimens and seed for sale. Koch is not directly linked with publications of his own, but he certainly contributed to the bank of material from which later drawings were derived.

While all of this activity was going on, Friedrich Ludwig Emil Diels,[11] together with Ernest Pritzel, visited Australia and carried out extensive field work, especially in Western Australia. Diels was born in Hamburg in 1874 and studied in Berlin under Adolf Engler. Diels' interest in phytogeography and phytobiology, together with funding from the Humboldt Siftung, led him to his extensive field trip in the southern hemisphere between 1900 and 1902. The prime aim was to study the vegetation of Western Australia and his collection amounted to 4700 specimens.

By now field work in Australia had ceased to be as hazardous as it had been even as little as thirty years previously. Diels wrote of his trip as though he experienced none of the hardships endured by earlier scientists and explorers. He stated that recent progress gave:[12]

> ... advantages over our forerunners which could hardly be overestimated. The rapid construction of railways since 1890 permitted a freedom of travel which no-one had experienced before ... We were thus able in the short period of 14 months to see so much and to collect such large numbers of plants that we could have worked almost independently of any earlier collections.

Diels compiled the results of his study in *Fragmenta Phytographiae Australiae Occidentalis* which was published in the *Botanische Jahrbücher für Systematik, Pflanzengeschichte und Pflanzengeographie* (1904–1905), a German periodical. The treatment is published in Latin and German and is aimed at scientists. It is sparsely illustrated with drawings of a high standard by Miss G. Bartusch. Diels' *Fragmenta* served Western Australian botanists for over half a century as a major reference.

Karel Domin,[13] a Czechoslovakian botanist, carried out field work between 1909 and 1910 in New South Wales and Queensland. On return to Prague he studied his own collections together with material housed at Kew, England, and Göttingen, Germany. In 1915 he published his *Beiträge zur Flora und Pflanzengeographie Australiens* in Volumes 85 and 89 of the *Bibliotheca Botanica*. In Volume 85, Domin teamed with Obenberger to prepare drawings for lithography by G. Reisacher. Later lithographs were executed by E.A. Funke. A few coloured plates were painted by Obenberger alone, lithographed by Funke and reproduced by chromolithography. In Volume 89, Ella Domin did most of the drawings with a couple being drawn by Karel.

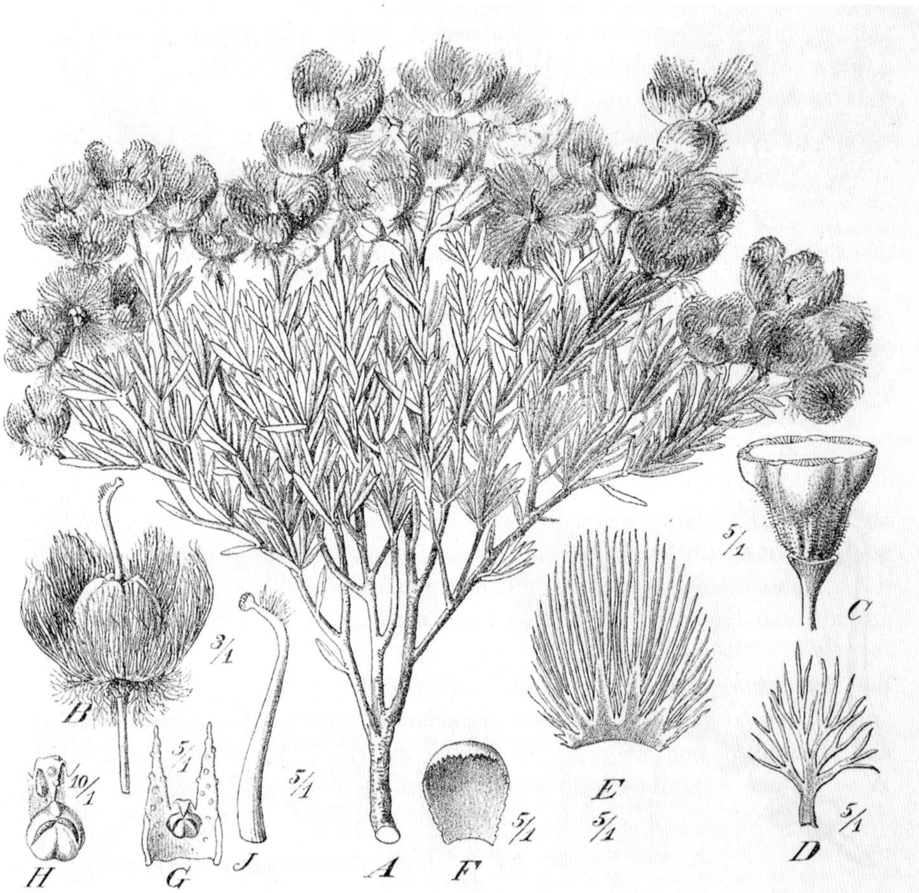

Verticordia pritzelii Diels
A drawing by G. Bartusch and printed in L. Diels and E. Pritzel, *Fragmenta Phytographiae Australiae Occidentalis,* in 1904 by process engraving.
[NLA 11.5 x 11.5]

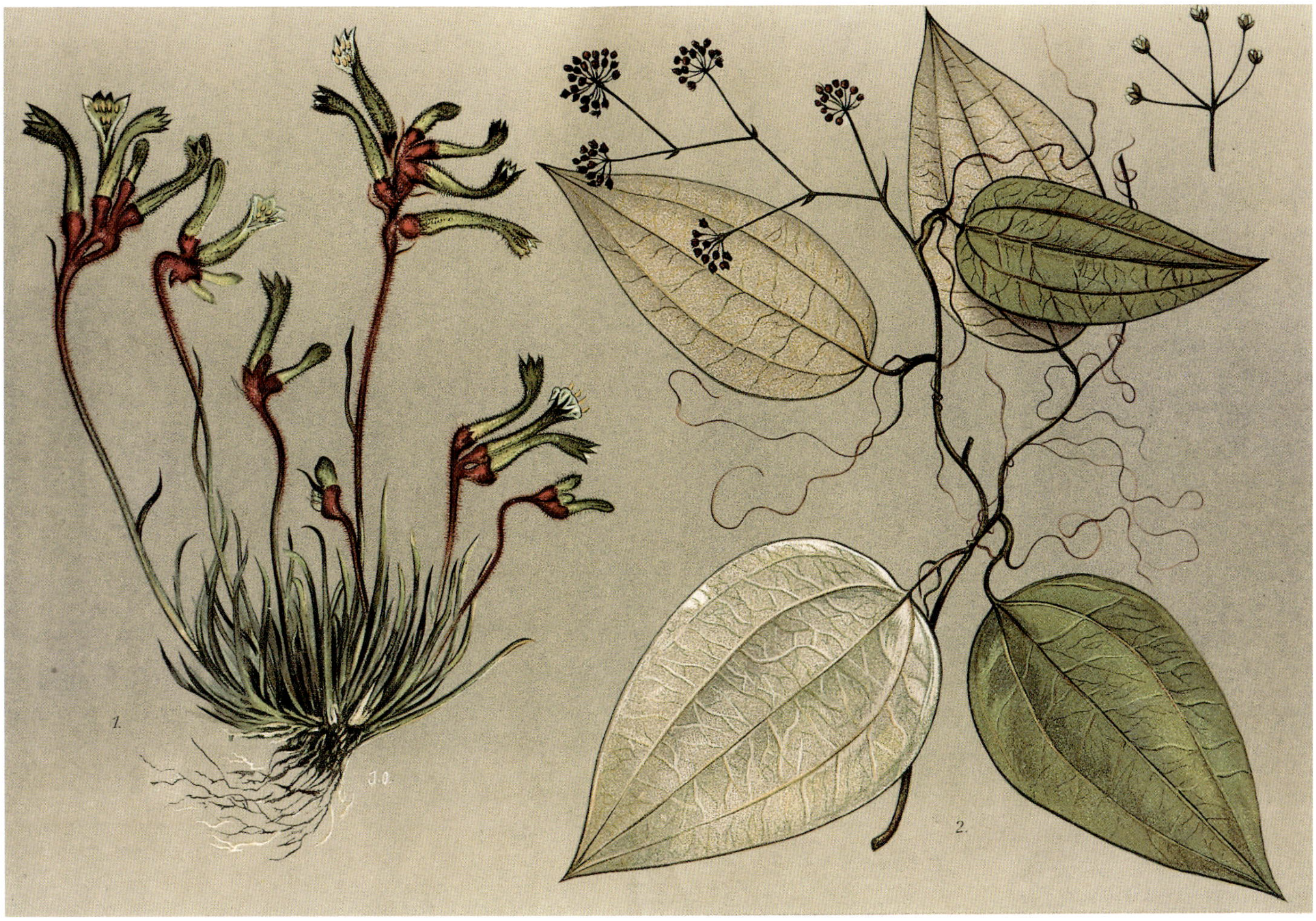

Anigozanthos gabrielae Domin and *Smilax glycophylla* Smith
Painted by J. Obenberger, lithographed by E.A. Funke, and published in Karl Domin's *Beiträge zur Flora und Pflanzengeographie Australiens* in 1915 as a chromolithograph.
[CANB 18.2 x 26.0]

With some notable exceptions, the standard of illustration was at an all time low during the early- and middle-twentieth century. No doubt there were many factors contributing to this, such as the prevailing economic and political climate. Economic collapses in the 1890s and 1930s and two world wars contributed to the first half of the twentieth century being very depressed.

Joseph Maiden experienced this very early in his publishing career. As Government Botanist and Director of the Botanic Garden, he began a beautifully illustrated series, *The flowering plants and ferns of New South Wales*, with especial reference to their economic value (1895–1898). Walter Scott Campbell was co-author. The artists were Edward William Minchin and Henry John Alcock Baron. The plates were reproduced as high quality chromolithographs, however they bear no signature and it is impossible to determine which artist painted which plates. Maiden was optimistic that this work would document all of the 'principal flowering plants and ferns of the Colony', saying 'the price of the work is very reasonable, and it is hoped that the public will support it sufficiently to justify its

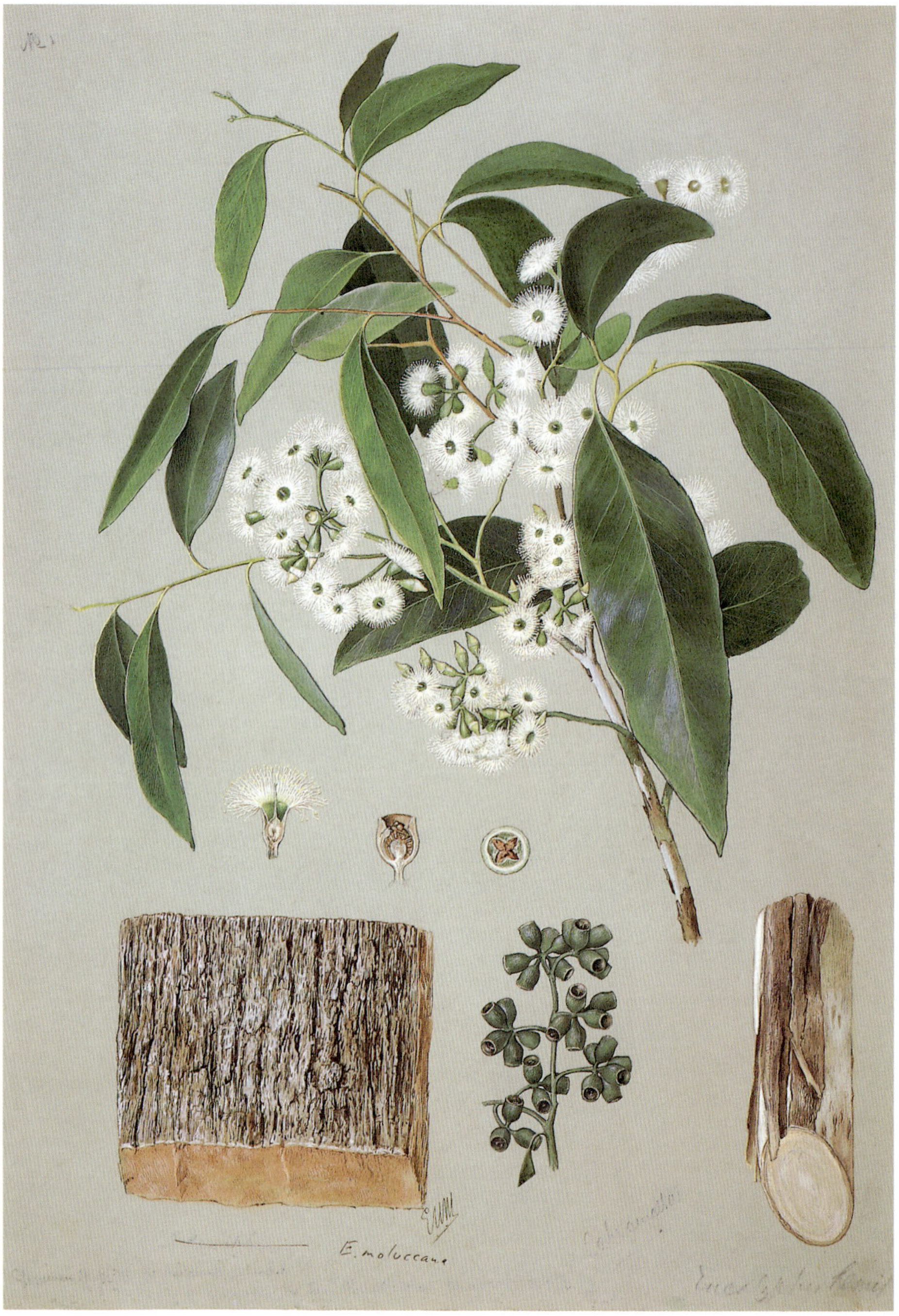

Eucalyptus moluccana Roxb.
Original painting by Edward Minchin, one of Australia's neglected but talented artists, showing signs of its age.
[NSW 38.3 x 27.6]

continuance'.[14] Twenty-eight species were published before the government put a stop to it, accusing Maiden of 'ruining the country by such extravagance'.

Both Minchin and Baron were lithographic artists with the Government Printer during the 1890s. Edward Minchin was born in Perth and moved to Melbourne when young. He worked at various jobs, including going to sea for a few years, before attending National Art Gallery classes

Polyosma cunninghamii Bennett
Lithograph drawn and lithographed by Margaret Flockton, and published in Joseph Maiden's *The Forest Flora of New South Wales* in 1917. Most of Margaret Flockton's work was published as lithographs and most she lithographed herself.
[NLA 24.7 x 20.5]

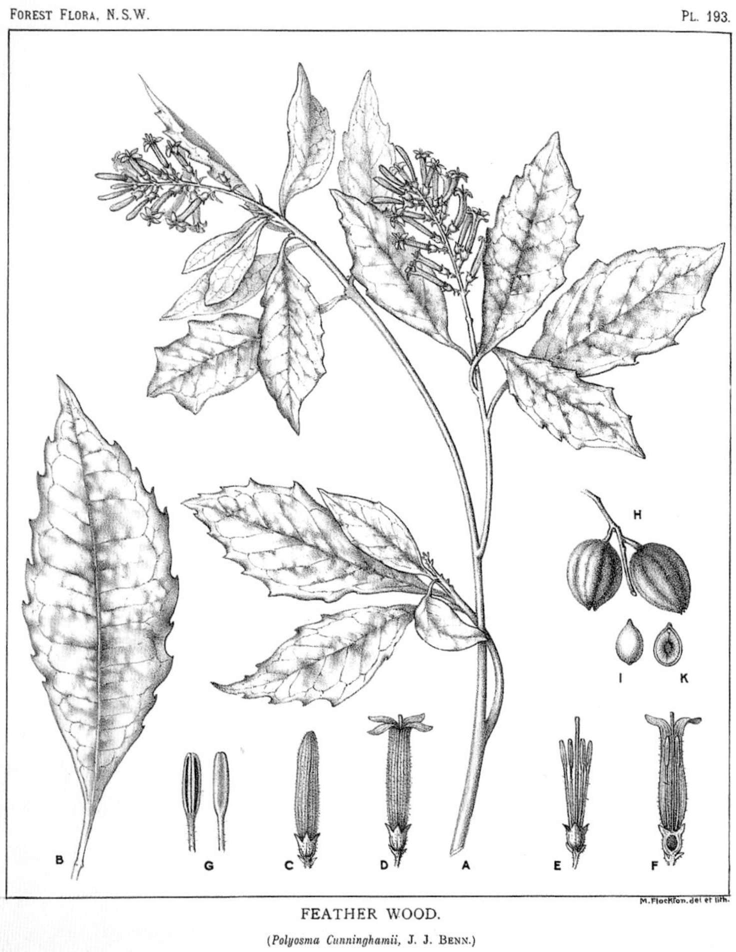

under Clarke and Folinsky. Minchin gained a position as a lithographic artist with the Melbourne Lands Department in 1874. When he moved to New South Wales he worked as a draftsman in the Survey Office from 1879. He transferred to the Government Printer as a lithographic artist in 1894. While there he did work for several scientists, including Maiden. His original paintings and botanical illustrations are very fine examples of draftsmanship.

Maiden succeeded in publishing *The Forest Flora of New South Wales* (1904–1925) in eight volumes. This was almost identical in format to *The flowering plants and ferns*. The plates were drawn and lithographed by Margaret Flockton and printed with a dark green ink. Consequently they are less pretentious and no doubt less expensive. The printer reverted to black ink in later volumes. Plates from 1920–1925 (in the last two volumes) were lithographed by Ethel Anna King.

During these years Maiden also published his magnum opus, a monograph on *Eucalyptus*, *A Critical Revision of the Genus Eucalyptus* (1909–1933), in eight volumes. Part of volume 7 and all of volume 8 were published posthumously, being edited by Richard Hind Cambage and William Faris Blakely.[15] This publication was also illustrated by Margaret Flockton, with lithographs mostly by her. Of special interest is a 'facsimile' printing of Redouté's engraving of *Eucalyptus obliqua* (original printed in 1788) in the first volume. This begs the definition of facsimile, because it is more correctly 'drawn after ....' by Matilda Smith of the Royal Botanic Gardens, Kew, as were a few other plates. The last volume is illustrated with black and white and colour plates of paintings of seedlings, all executed by Margaret Flockton over her entire working career with Maiden. All of these are reproduced by halftone process.

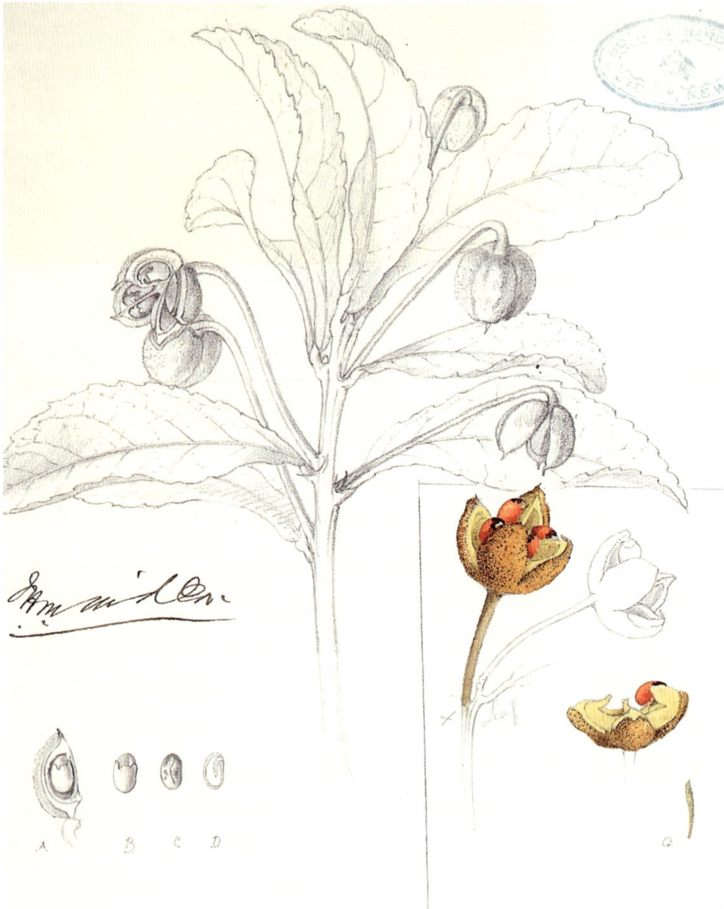

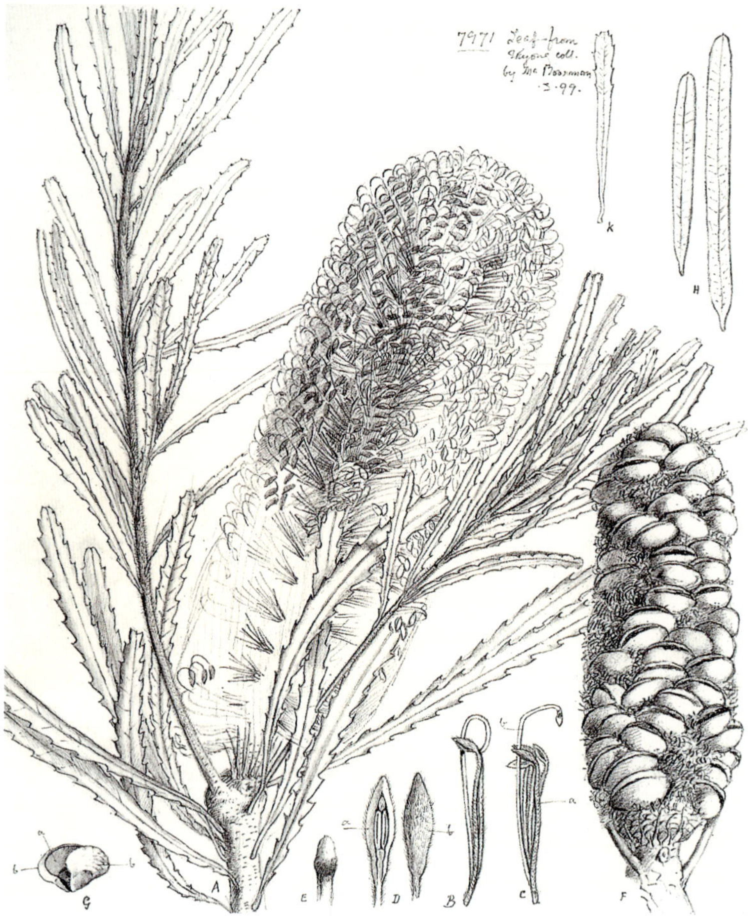

Sloanea australis (Benth.) F. Muell. and *Banksia collina* R.Br.
Original drawings by Margaret Flockton showing her obvious talent. Flockton concentrated on lithography for the major publications of Joseph Maiden who required her particular skills as a draftsman and lithographer. Unfortunately very little of Flockton's coloured work exists. *Banksia collina* (right) was lithographed for publication in *The Forest Flora of New South Wales*.
[NSW 20.8 x 24.9] [NSW 20.8 x 24.9]

Margaret Lilian Flockton[16] proved to be an extremely competent and productive artist and became a valuable member of Maiden's staff. Flockton came to Australia around 1880. She joined the staff at the Sydney Botanic Gardens in 1901 and prepared illustrations for many of Maiden's publications for a number of years, thus being the first fully employed illustrator for an Australian herbarium. Flockton's artistic skills were very high, and she is thought to have assisted Maiden with word descriptions too, proving her great value to him, especially when his health was failing. In 1978 Norman Hall said that:[17]

> ... she was a member of the Royal Art Society and had a high reputation as a lithographic artist at a stage when she was Australia's only woman exponent of the art.

In later years she was assisted with lithography by Ethel King. Flockton ceased employment with the Sydney Botanic Gardens in 1927.

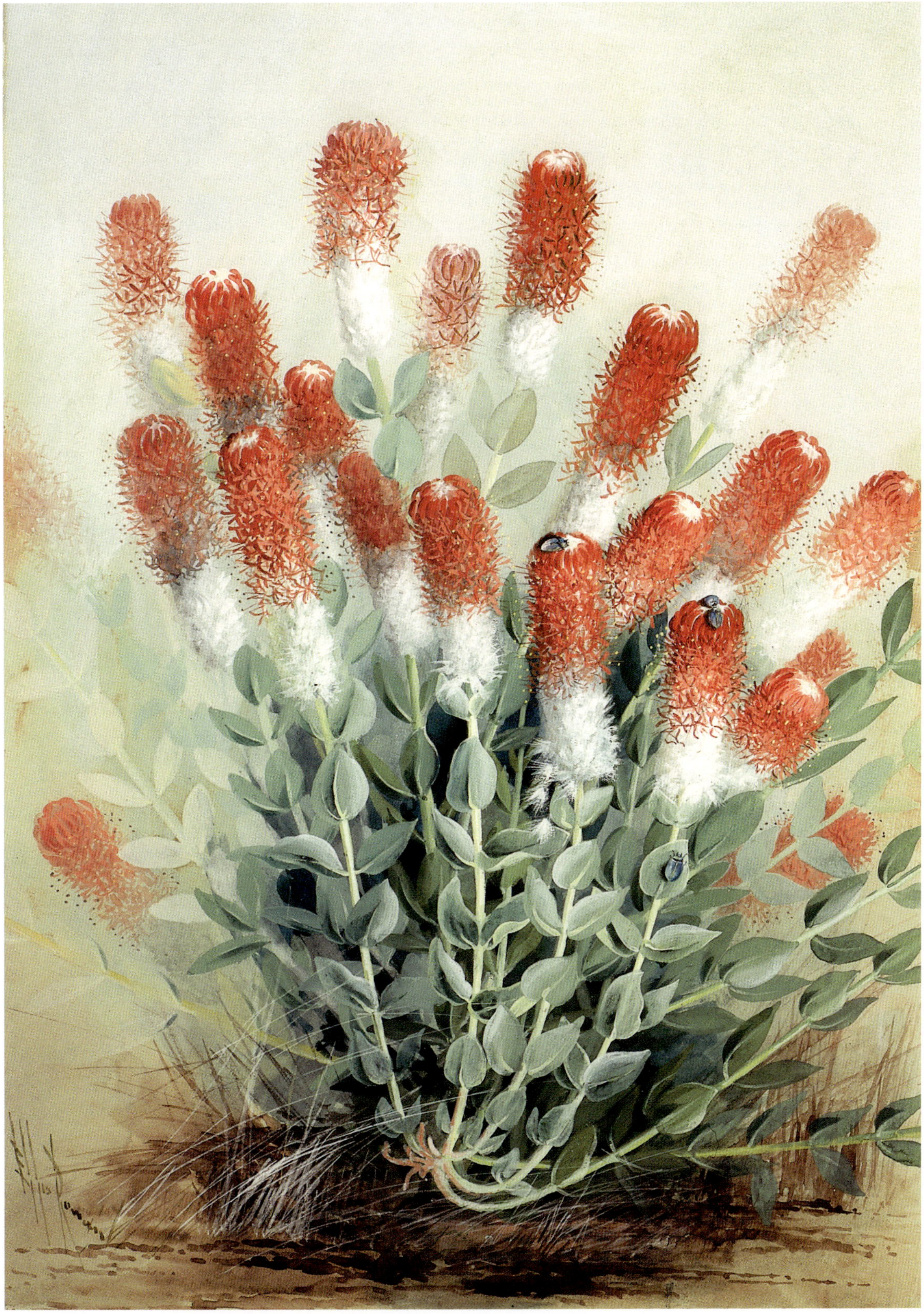

Pimelea decora Domin (= *Pimelea haemostachya* F.Muell.)
Original painting by Ellis Rowan. A portion of a very similar painting was published in Frederick Bailey's *Comprehensive Catalogue of Queensland Plants* in 1909.
[QM]

Bailey's *Comprehensive Catalogue of Queensland Plants* (1909) is illustrated by several means, making it a very interesting publication. Most of the plates are very poor pen drawings by Cyril Tenison White, reproduced by process engraving.[18] The introductory text is illustrated with photographs which are reproduced in halftone. One of these is of a sketch of *Bauhinia hawkesiana* F.M. Bailey, the sketch being executed by Mrs R.W. Garraway.[19] Mrs Garraway collected the type specimen and requested that the plant be named in honour of Claude Somerville Hawkes. In addition there are sixteen colour plates from paintings by Ellis Rowan. These are reproduced in halftone, the colours frequently being out of register.

In spite of World War I Maiden, together with Ernst Betche, published *A Census of New South Wales Plants* (1916). This is unillustrated.

Alfred James Ewart[20] and Olive Blanche Davies[21] published *The Flora of the Northern Territory* in 1917. This is a wonderful example of collaborative publishing, involving five scientists and five artists. On the title page the plates are attributed to Ethel Irene McLennan, Isabel Clifton Cookson, Mary Ellinor Lucy Archer and Margaret Lilian Flockton. Examination of the plates reveals that not all of them are signed and individually attributable. Moreover a number of the more competent illustrations were actually executed by Olive Davies. The plates of McLennan, Cookson, Archer and Davies were reproduced from pen drawings by process engraving. The plates by Flockton were reproduced in black and white halftone from photos of paintings. A map of the Northern Territory was drafted by A.L. Rossitter.

Alfred Ewart was educated in Liverpool and worked in Germany, Java and England before taking up the Chair of Botany at the University of Melbourne in 1905. For the next fourteen years he worked half-time at the Herbarium as Government Botanist and half-time at the University. This cannot have made for great efficiency, nor was he able to provide a service comparable to that of von Mueller. However, during his working career he prepared some very useful handbooks and two Floras representing major components of the Australian flora.

Olive Davies was born in Melbourne into a wealthy and influential family. She was still at school in the 1890s when the banks collapsed and the bottom fell out of the economy in Victoria. Her family was among those bankrupted overnight. She was, however, able to study at the University of Melbourne and found work in the Botany Department. In this context she became co-author of the *The Flora of the Northern Territory*. Davies married Arthur Lyle Rossitter, a school teacher, and followed his career around country Victoria. She does not seem to have returned to her own career as a scientist.

A minor flurry of popular Floras was published in the years before World War I. In 1913 Florence Sulman published *A Popular Guide to The Wild Flowers of New South Wales*. This was illustrated by Eirene Mort and Dorothy M.Watkins. The plates are reproduced by process engraving from pen drawings. Mort in particular has excellent line and composition, reflecting her graphic art training.

Eirene Mort was born in Sydney. She studied art under Dattilo Rubbo, an Italian artist, and A.H. Fullwood for a year before leaving Sydney in 1898 to study in London. Mort returned in 1905 and opened her own studio. She is described as an illustrator, designer, etcher and embroiderer by Jennifer Isaacs.[22] In Australia Mort worked as a professional designer of furniture, furnishings and bric-a-brac. In later years she specialised in etching,

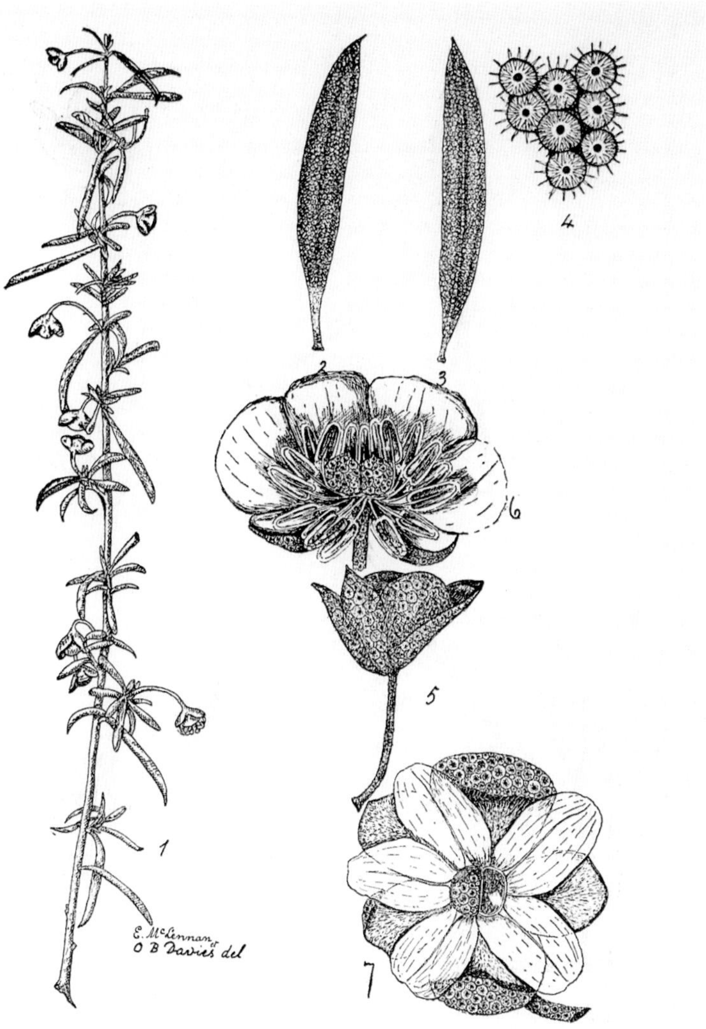

Hibbertia lepidota R.Br. (= *Rossittia scabra* Ewart & O.Davies)
Drawn by Ethel McLennan and Olive Davies and published in Alfred Ewart and Olive Davies, *The Flora of the Northern Territory*, in 1917.
[NLA 24.7 x 20.5]

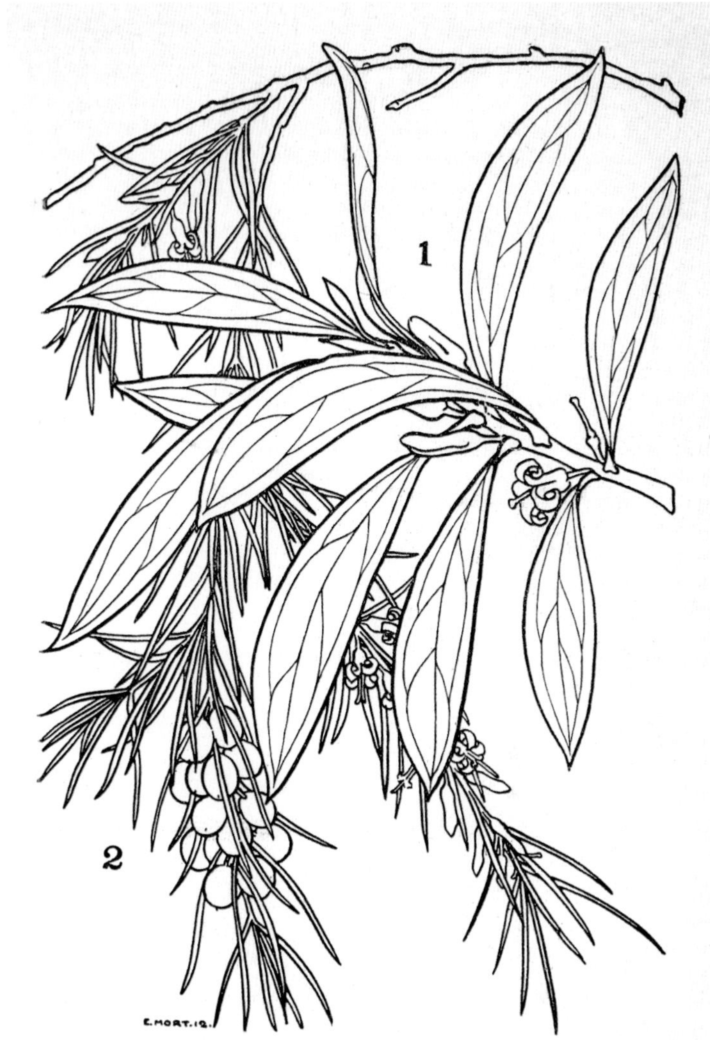

Persoonia levis (Cav.) Domin (= *Persoonia salicina* Pers.) and *Persoonia pinifolia* R.Br.
Drawn by Eirene Mort and published in Florence Sulman's *A Popular Guide to The Wild Flowers of New South Wales* in 1913 by process engraving. Mort's talent as a graphic designer is evident in this plate.
[NLA 14.0 x 9.0]

and woodcut and linocut engraving. This activity led to her becoming a founding member of the Australian Painter-Etchers' Society. Mort did not venture further into botanical illustration, but she used Australian flora and fauna in many of her designs.

The artwork of Dorothy Watkins was reproduced from the *Australian Kindergarten Magazine*.

In 1914 Edward Edgar Pescott published the *Native Flowers of Victoria*. This book is aimed at those considering introducing native flora into their gardens and is well illustrated with photos of possible subjects. The book also has four colour plates painted by Herbert Peabody Dickins. The paintings are more reminiscent of the work of Ellis Rowan than that of the formal botanical illustrators. The presentation of the plates is unusual. They are tipped onto a brown card, which is bound into the book.

In a more academic vein, John McConnell Black prepared the *Flora of South Australia* published between 1922 and 1929. This is illustrated with numerous vignette pen drawings executed by Black. A few of the plates are signed by him with his initials MB. The Orchidaceae are treated in this work by Richard Rogers and the orchid illustrations done by Rosa Fiveash. These plates are reproduced by black and white halftone, from paintings. The Fiveash plates are more in keeping with classical botanical illustration than vignette-type illustrations allow.

In 1930 Alfred Ewart published his second Flora, the *Flora of Victoria*. This was illustrated with numerous vignette pen drawings reproduced by process engraving. There is a range of standard of draftsmanship, but the drawings serve to illustrate diagnostic features. A frontispiece of orchids is reproduced from a painting by Mavis Arnold. This is a halftone chromolithograph. Drawings in the text are not signed but Ewart acknowledged a bank of illustrators:[23]

> The illustrations were prepared under the general supervision of Dr McLennan, mainly by Miss Heward, M.Sc., but a number were drawn by Miss Dixon, M.Sc., Mrs Thomson, B.Sc., Miss McColl, B.Sc., L.A. Thomas, B.Sc., and T.T. Colquhoun, M.Sc., Mrs. Arnold, B.Sc., Miss Baron, M.A., and Miss Cheney, M.Sc. The illustrations are selected so as to dovetail with those already given in the *Weeds of Victoria*, 1909, and in the *Handbook of Forest Trees*, 1925.

Ewart's artists were usually students or junior staff members of the Botany Department at Melbourne University, where he held the Chair of Botany.

The work of William Faris Blakely followed on from Maiden's. The two men had worked together at the Sydney Botanic Gardens since 1900, Blakely having a sequence of positions, culminating in that of Botanist. Maiden and Blakely collaborated closely in the production of *A Critical Revision of the Genus Eucalyptus*. This experience, together with Blakely's inherent capacity as a systematist, led to Blakely's construction of *A Key to the Eucalypts* (1934). The Key is described by Blakely as a 'companion to J.H. Maiden's *A Critical Revision of the Genus Eucalyptus*'. It provides succinct descriptions of 500 species together with comprehensive cross-reference to Maiden's work. It is not illustrated and this proved to be a short-coming which was solved two decades later in 1952 with the publication of a Forestry and Timber Bureau Leaflet, *Illustrations of buds and fruits of Eucalyptus species*. The Leaflet was in such demand and so useful that it was revised several times.

In 1968 George McCartney Chippendale arranged for a revised fifth edition. This was published as *Eucalyptus Buds and Fruits*. Initially the drawings were mostly pencil sketches executed by Margaret Flockton and Matilda Smith between 1903 and 1928; and Crystal Stahl, June Collins, Jan Kennedy and Marjorie Hall between 1962 and 1967. For Chippendale's book, Marjorie (Hall) May converted all of the drawings into stippled ink drawings.

The story of this publication very forcibly demonstrates that the average user of Floras and Keys has great difficulty in constructing an image from words. Chippendale states, with caution, that the illustrations should be:[24]

> ... helpful to foresters, botanists, students and others interested in natural history. Those using the book should be careful NOT to rely on the illustrations entirely. It is impossible to represent the range of every species with only one drawing for each, for there is much variation in many species.

The range of variation and inability to present it all in a drawing causes scientists to rely heavily on words. It is a pity that this very real problem, and the purist approach to it, inhibits communication. Scientists tend to hide behind the book-publishers' argument about rising costs for publishing illustrated books and escape their responsibility to produce easy-to-use and reliable identification Keys.

John Black prepared a second edition of his *Flora of South Australia*, the first part appearing in 1943. The fourth part was revised by Enid Lucy Robertson following Black's death in 1951. With the expanded knowledge of the flora, Black arranged for additional illustrations to supplement his own, most of which were reprinted. For Part one, over two hundred and sixty illustrations were added, of which 180 were drawn by Maude Priest.

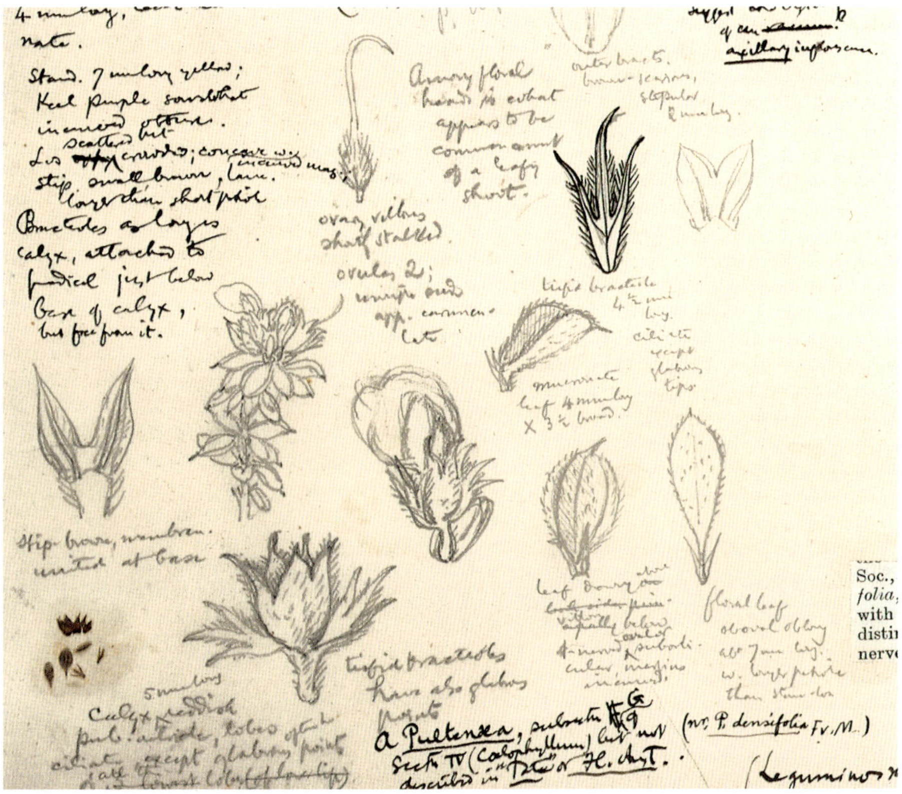

Pultenaea trifida J.Black
John Black, who published illustrated Floras of South Australia, was renowned for his interpretative sketches on herbarium sheets. Here, on part of the sheet, he has analysed the diagnostic parts of the plant with both ink and pencil sketches prior to preparing the type description and illustrating the species in the *Flora of South Australia*, 1922–1929.
[AD]

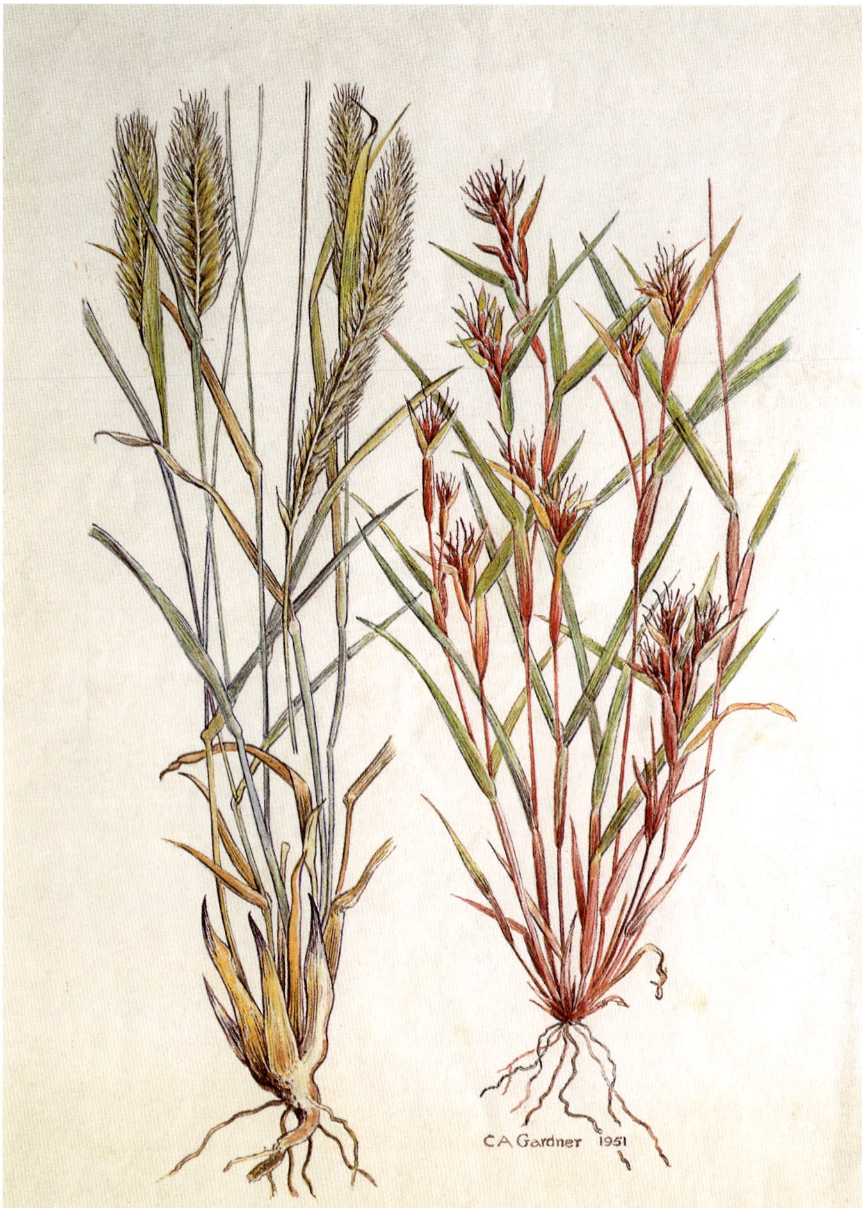

Astrebla pectinata (Lindley) Benth. and *Iseilema fragile* S.T. Blake
Original drawing, using coloured pencil and wash, by Charles Gardner of Mitchell and Flinders grasses for the frontispeice of the first and only volume of the *Flora of Western Australia*, in 1952.
[WA 28.0 x 17.5]

Additional drawings in the later volumes included many drawn by Priest together with Mabel Raymont, Gwen Walsh, and perhaps others. It is not possible to determine which of these illustrators drew which figures. They followed the standard and style of Black, and so their work does not appear markedly different from his.

Charles Austin Gardner[25] began a *Flora of Western Australia* with a volume on grasses in 1952. Only Volume 1 was produced. As with many Australian botanists, Gardner began his botanical career as an amateur. He came to Australia as a boy with his family to settle on a farm at Yorkrakine. In 1920 he was appointed Botanical Collector to the Forests Department of Western Australia. This appointment led not only to rewarding expeditions but also to some publications of his own as Forestry Bulletins. In 1924 he joined the Department of Agriculture and by 1928 found himself appointed as Government Botanist and Curator of the State Herbarium. In the course of his highly productive career he illustrated his own publications.

William Edward Blackall and Brian John Grieve produced *How to Know Western Australian Wildflowers* (1954–1982). Dr Blackall, who died in 1941, had created an illustrated *Key to the Flora* in manuscript form. Grieve revised this manuscript, bringing it up to date with current knowledge. Initially Nathaniel Henry Speck amended Blackall's handwritten script and illustrations for photographic publication. Later Joan Rayner took over and completed the work. The handwritten script is neat and the illustration minimal. It is nevertheless remarkably effective for a dichotomous and diagnostic Key. As an identification tool it is surprising that more Keys do not follow this method of communicating the vital detail.

Part three of the *How to Know...* was published in 1965. Yvonne Chadwick executed the script and illustration for this volume. Part four, published in 1975, abandoned the script and was typeset. C. Gladstone and Claire Muldownie are acknowledged as the major illustrators. Handsome frontispieces were painted for these two volumes by Margaret Stones.

While the scientists were busy, the amateurs were getting on with the job too.

In 1921 Emily Harriet Pelloe had *Wildflowers of Western Australia* published. She stated that the:

> ... non-existence of a simply-written book dealing with the flora of Western Australia is a great handicap to people having no special botanical knowledge, yet wishing to know something of the names and characteristics of the most familiar of the beautiful native flowers for which Western Australia is famous. This little book, written from notes made while painting and classifying a fairly representative set of sketches of West Australian wildflowers, may be found of interest and use.

Pelloe's book is sparingly illustrated with fairly poor pen sketches and eight colour plates, mostly of bunches of flowers. The standard of the painting is adequate, perhaps not matching Pelloe's enthusiasm for her subjects. This book provides a dramatic contrast to that of Diels. Pelloe's *West Australian Orchids* (1930) is very similar in style and content to her first book. There are three colour plates of bunches of orchids.

Emily Pelloe was born and educated in Melbourne. She married Theodore Pelloe, a bank manager, and spent most of her married life in Perth. Here she developed a great love of the native flora and expended much energy in communicating this to others through her painting and writing.

In 1933–1934 Edgar Dell had plates of flowers of Western Australia published as supplements to *The Western Mail*. Charles Gardner prepared a descriptive-text caption for each plate. In 1935 a compilation of these was published as *West Australian Wild Flowers*. This ran to eight editions, later ones including colour photographs of plants. In 1959

Gardner published *Wildflowers of Western Australia*, a revised ninth edition. This features eighty of Edgar Dell's paintings together with some colour photographs of additional species. In contrast to the three-colour process of the previous editions, a four-colour process was used which has provided an enhanced portrayal of Dell's work.

Edgar Dell was born in England in 1901 and came to Western Australia in 1924. He worked on an orchard ostensibly to earn sufficient money to return to England to continue his studies at art school. However, he was captivated by Western Australia's native flowers and began painting them, frequently under difficult conditions. In the 1930s he started his own orchard on a property at Kalamunda. During the Depression of the 1930s he secured a couple of commissions for his work, one of which was that for *The Western Mail*. For botanical art, Dell frequently worked closely with Charles Gardner at the herbarium.

Whilst there are many publications on flowers of Western Australia there still is no complete Flora of Western Australia.

In 1938 Thistle Yolette Harris published *Wildflowers of Australia*. This was aimed at the layman interested in observing nature and it encouraged nature conservation. The book is liberally illustrated with colour paintings by Adam Forster. The plates are reproduced in halftone colour. The poor registration does not do justice to Forster's work. In spite of that, the book apparently made an impression because it was reprinted several times over the following four decades.

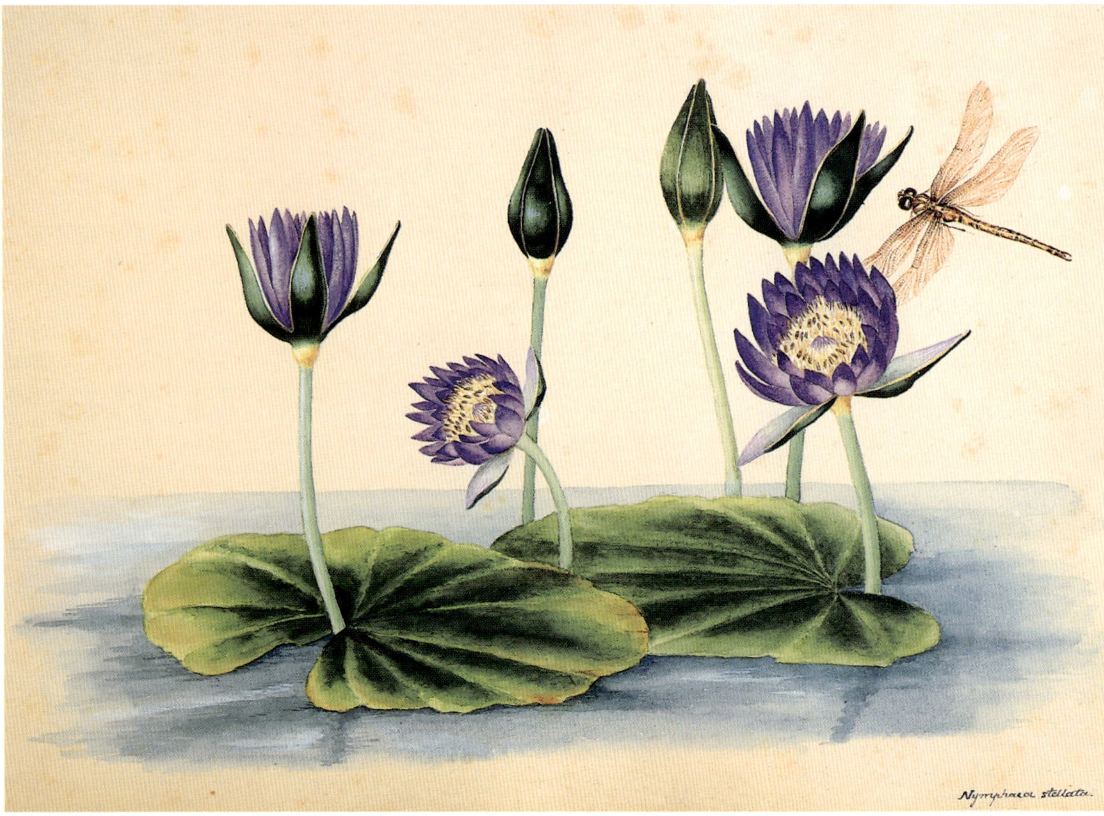

Nymphaea gigantea Hook. (= *Nymphaea stellata* Willd.)
Original watercolour painting by Adam Forster.
[NLA 29.2 x 39.5]

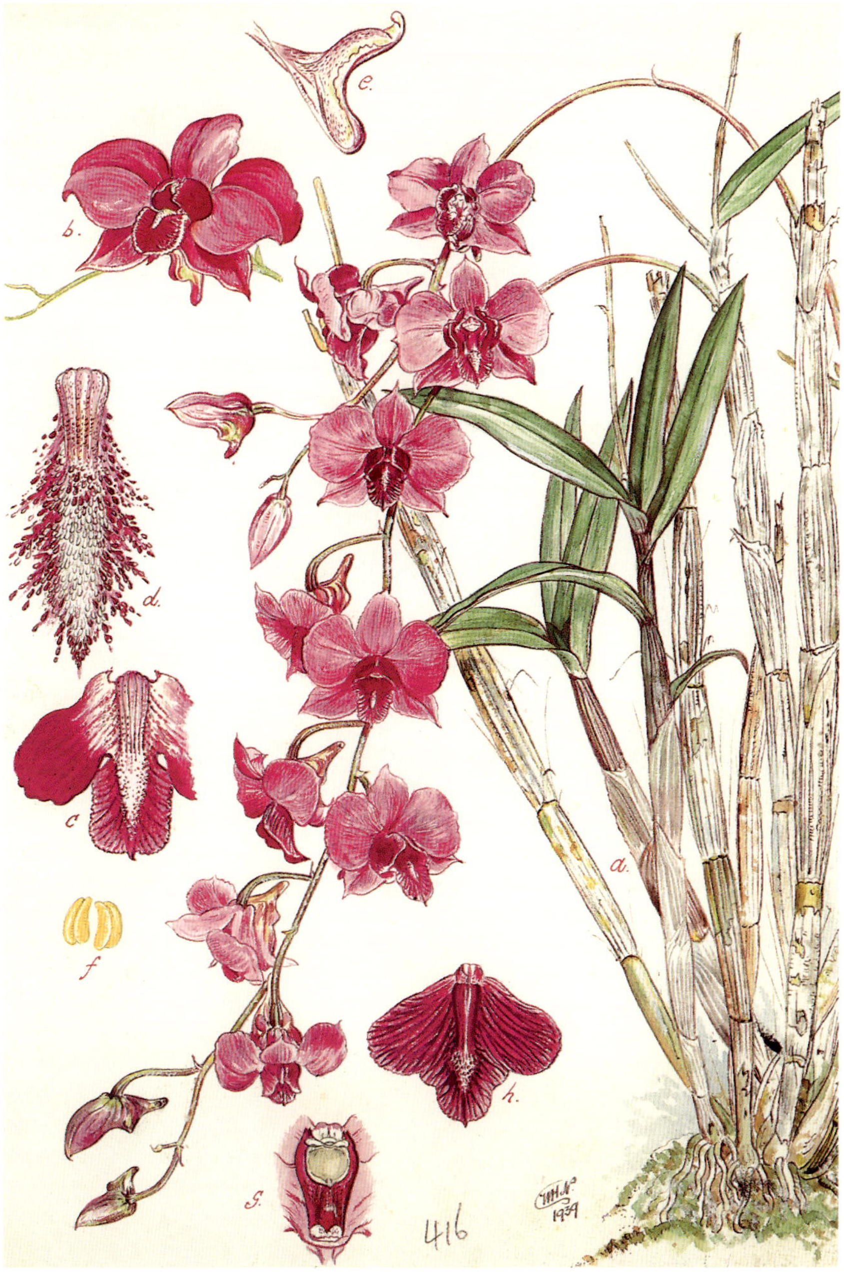

Dendrobium phalaenopsis Fitzg. (= *Dendrobium bigibbum* Lindley & Paxton)
Original watercolour painting by William Nicholls. This was published posthumously for him in *Orchids of Australia* in 1969.
The book was edited by David Jones and Bruce Muir.
[MEL 30.6 x 20.4]

Adam Forster was born in Prussia. He came to Australia via South Africa in the 1870s and practised in Sydney as a pharmacist. He was a competent artist and painted almost a thousand studies of Australian plants. The collection for *Wildflowers of Australia* was surely completed long before Harris' book was published.

While Forster and others had an interest in the broad representation of Australia's flora, others specialised in one group.

William Henry Nicholls, working as an enthusiastic amateur, set out to describe 'by pen and brush, the orchid wealth of our continent'.[26] Starting in the 1920s he worked tirelessly for thirty years to complete the task. In the end, 476 plates were published posthumously in 1969 in *Orchids of Australia*, edited by David Lloyd Jones and T. Bruce Muir.

Nicholls did not have the benefit of formal training in botany and art. Professionally he worked as a bookbinder. Nevertheless he was an amateur naturalist with a gift. His work is not amongst that of the greats, but it has an appealing style, used to good effect in accurate depiction of the subjects. It is fortunate indeed that a publisher was found to issue the compendium.

Meanwhile, taxonomists illustrated their own publications in a range of journals and books. This became a common practice in Australia during the twentieth century. As observers and describers, it is entirely appropriate that taxonomists should execute both modes of description — words and images. Mostly, they achieve a pleasing degree of competency.

10. The Revival

There is a need for new and modern colour-plate treatments about plants, of both the exotic and the mundane, of the indigenous and the cultivated. In our present economic society it can be done, and with satisfaction to all, if through excellence and proper balance of text and colour-plate, there is produced a work that ... serves the needs as well as the interests of a substantial segment of discriminating plantsmen and scientists. The demand does exist.

George Lawrence[1]

The later decades of the twentieth century witnessed a revival in botanical illustration with a wealth of high quality work as had not been seen for over a hundred and fifty years.

An early inspiration from England was provided in 1950 when Wilfred Blunt, with the assistance of William T. Stearn, published *The Art of Botanical Illustration*. This book provides an historic review of botanical illustration and puts some Australian art in context. In addition, it contains a reprint of an essay by Walter Fitch on Botanical Drawing, a 'how to ...' of draftsmanship. Overall, the book provides a gentle revelation of the development of the art-form of botanical illustration without any particular, zealous message. For those interested in botanical art and illustration, and who had not enjoyed the benefit of a specialist library in which to browse, the book surely provided an inspiration and a goal.

Within Australia, Winifred Mary Curtis[2] became a participant in the revival. Curtis prepared *The Student's Flora of Tasmania* and the first Part was published in 1956. The entire Flora was completed in 1994. Illustration in the Flora was sparse and somewhat

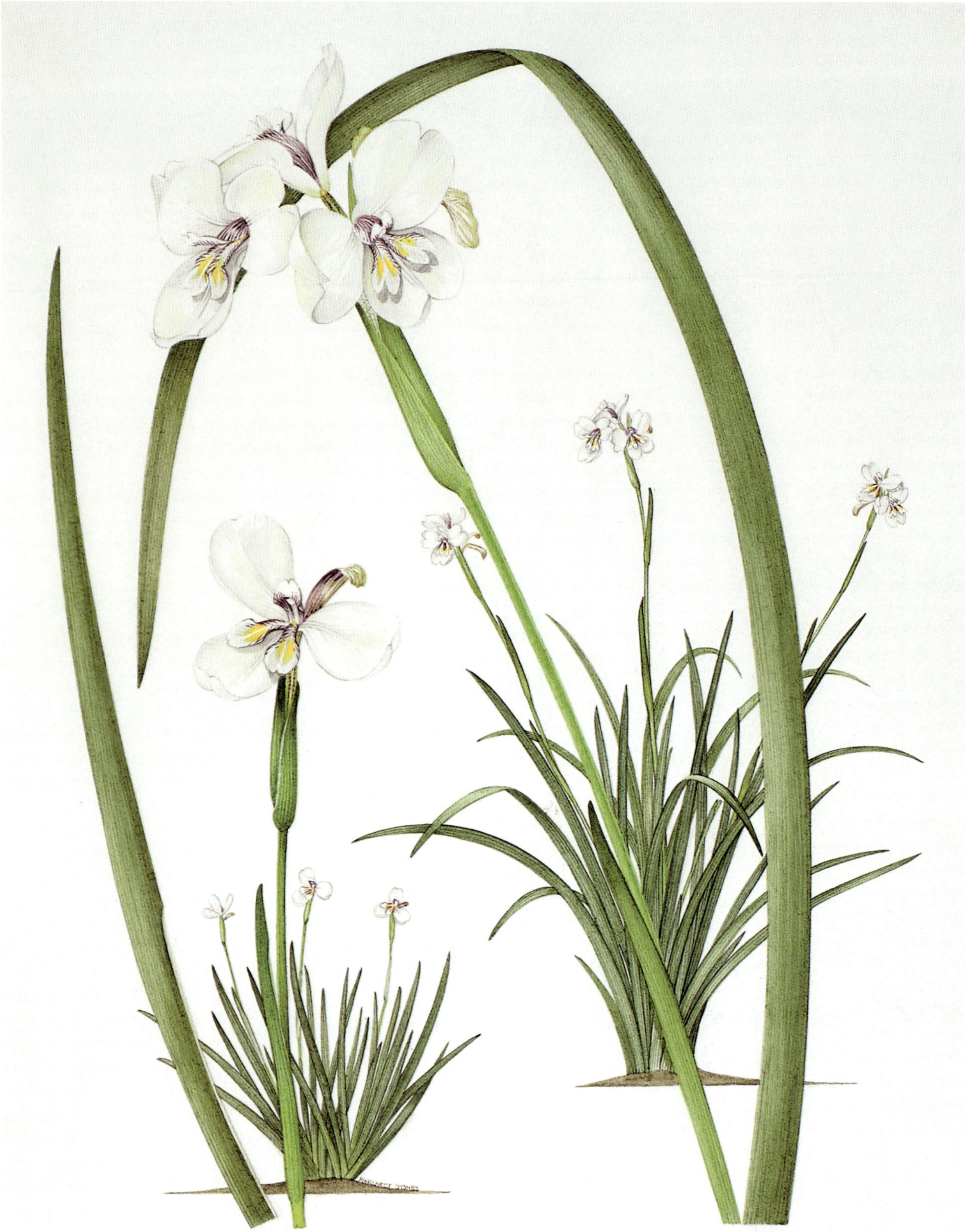

Diplarrena latifolia Benth.
Original watercolour painting by Margaret Stones, 1968, published in *The Endemic Flora of Tasmania* in 1978.
[QVM 36.8 x 28.2]

vignette-like. For the first three Parts, Curtis did the drawings. The style is interesting because it relies on line to give form. While it is reminiscent of John Black's illustration in the *Flora of South Australia*, it is more crisp. However, there was nothing particularly special about it. What was special was that Winifred Curtis was the resident specialist on the Tasmanian flora. On the strength of that she was invited to participate in the writing of the text for a grand, illustrated Flora, *The Endemic Flora of Tasmania* (1967–1978). It became one of the major stimulants to the eventual revival of botanical illustration in Australia.

Winifred Curtis was born in England in 1905 and came to Hobart in 1939 to join the staff of the University of Tasmania. There she practised and taught botany for almost thirty years. Both the *Endemic Flora* and the completion and revision of the *Student's Flora* ultimately became retirement activities.

Milo John Reginald Talbot — Lord Talbot de Malahide — an inveterate and unashamed collector, chose the endemic plants of Tasmania as the objects of his collection. Lord Talbot, though English-born, had ancestral links with Tasmania and was drawn to its unique wilderness. It is most fortunate that he not only chose to have each plant painted, but also to share his collection by having it published with botanical and ecological text. To do this he needed to find, and bring together, a botanist and an artist. Those two people were to be Winifred Curtis and Margaret Stones.

Elsie Margaret Stones was born in Colac, Victoria, in 1920 and later lived in Melbourne while undertaking her secondary and tertiary education. Stones' vocation was to become an artist and she studied industrial and commercial art, and kept body and soul together working piecemeal as a commercial artist. During World War II she trained and worked as a nurse. After the war a series of fortunate coincidences brought her to the attention of Professor John Turner, Professor of Botany at the University of Melbourne. Under the influence of Turner and other staff at the University, she found that her destiny as an artist lay with botanical art. This opportunity led to a study of botany and to important commissions such as one from Sir Russell Grimwade. Sir Russell had Stones paint some Australian wildflowers for his home, Miegunyah. Stones was soon sufficiently motivated to follow her calling to England and to the Royal Botanic Gardens. She left Australia in 1951 and finally settled in London working freelance, mostly for the Botanic Gardens.

In 1961 Stones was 'discovered' by Lord Talbot and he initiated the commission to paint the endemic plants of Tasmania. Winifred Curtis was the obvious choice for his botanist. Stones executed most of the painting for Lord Talbot in England in autumn from fresh material collected in Tasmania's spring and air-freighted to London for immediate work. Curtis participated in the collection and dispatch-work and then prepared the detailed descriptions for the text of the Flora.

Stones' career has been a busy and rewarding one with numerous important commissions and successful exhibitions together with an impressive portfolio of published scientific illustration.[3]

The *Endemic Flora of Tasmania* took over fifteen years to complete, and therefore spanned a significant part of both Curtis' and Stones' careers. The role played by Rose Talbot, Lord Talbot's sister, also needs to be acknowledged. The Flora was published in six Parts and

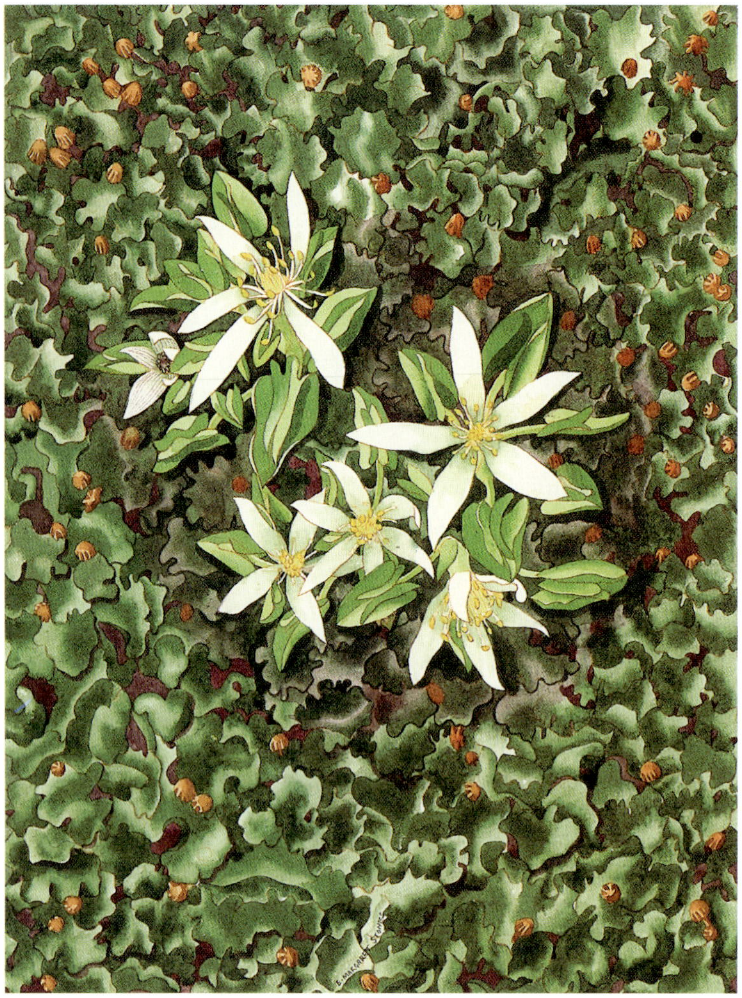

Caltha introloba F.Muell. growing in *Marchantia berteroana* Lehm & Lindenb. Watercolour, painted by Margaret Stones in the late 1940s, for a project documenting the alpine flora of Victoria. It is unusual to have a painting of a subject together with the associated plants. The *Marchantia*, one of the large thalloid liverworts, is rarely painted. [MEL 17.2 x 12.7]

Lord Talbot died just as Part Four was going to press. Rose persevered to honour her brother's commitment to complete the collection.

Sir Wilfred Russell Grimwade[4] was a very successful industrialist and philanthropist with a wide range of interests including science, especially chemistry and to a lesser extent forestry, and the arts. Sir Russell was to have an influence on Stones' career and on botanical illustration. He was a Chairman of the Felton Bequests Committee, set up by a former business partner and philanthropist Alfred Felton, which controlled a trust fund enabling the Melbourne National Gallery to purchase works of art. Not only did he commission paintings, but also the Felton Bequest had a London-based adviser who encouraged Stones and gave her a copy of Wilfred Blunt's book.

Sir Russell was also on the board of the Melbourne Botanic Gardens and was instrumental in setting up the Maud Gibson Trust to commemorate William Gibson, father of Maud Gibson and founder of the Melbourne retailing business, Foy & Gibson Ltd. The Trust was established in 1945 to enhance the Melbourne Botanic Gardens and to aid research carried out in the National Herbarium of Victoria. The Trust has gone on to support and encourage botanical illustration in numerous ways over the years. It provided early commissions for Stones, Celia Rosser and Betty Conabere and supported the appointment of a permanent staff illustrator in 1977.

In South Australia the Herbarium appointed a Botanical Illustrator, Ludwik Dutkiewicz, to the permanent staff in 1953. Dutkiewicz was thus the first permanently appointed artist at an herbarium during the revival period. Being an artist in his own right, Dutkiewicz was able to bring the skills of line, technique and composition to the job at hand. While learning botany and the requirements of botanical observation and scientific illustration were an early challenge, Dutkiewicz set the goal for others.

Dutkiewicz was born in Stara Sol in Poland in 1921 and trained as an artist at the Lwow School of Art. He specialised in abstract and semi-abstract painting. When he came to South Australia after World War II, he counted himself fortunate to have secured the inaugural position as Botanical Illustrator with the Herbarium. Paid employment allowed Dutkiewicz to indulge in his real vocation working as a contemporary modern abstract artist in his free time. He worked at the Herbarium for thirty years until his retirement, this term rivalling that of Margaret Flockton in Sydney at the beginning of the century.

Dutkiewicz was succeeded briefly by Beth Chandler and then later in 1983 by Gilbert Roelof Maria Dashorst, who still holds the position.

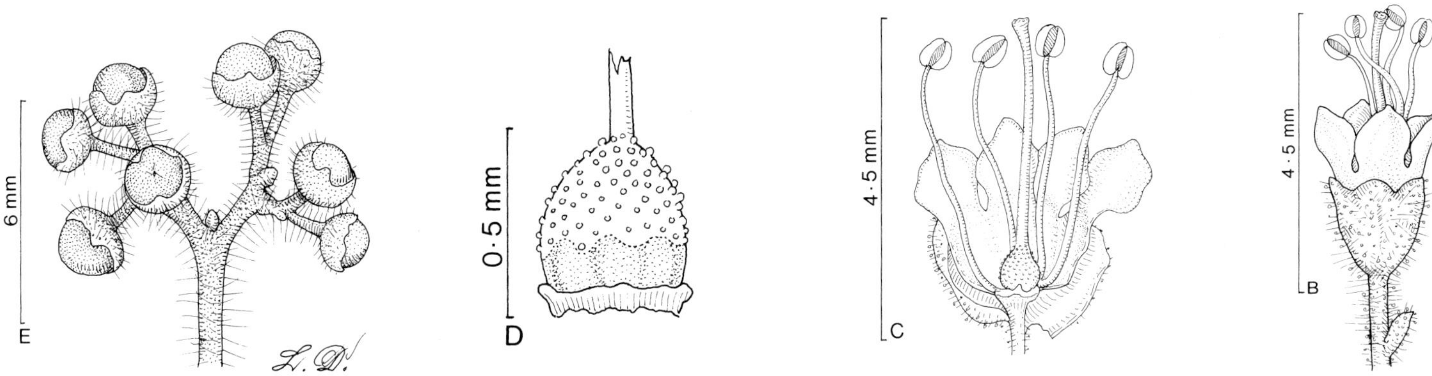

Callicarpa caudata Maxim.
Original pen drawings by Ludwik Dutkiewicz, prepared for publication in the *Journal of the Adelaide Botanic Garden* in 1982.
[AD]

Nancy Tyson Burbidge[5] and Max Gray followed in the revival with the *Flora of the Australian Capital Territory*, published in 1970. This Flora is liberally illustrated with pen drawings, all by Burbidge. At that time 1035 species in 412 genera were known to occur in the Australian Capital Territory. Burbidge illustrated every genus with a drawing, providing a comprehensive pictorial representation. This achievement was very significant at a time when Floras, with the exception of Black's *Flora of South Australia*, had carried far fewer illustations. The standard and level of detail and analysis provided by Burbidge more than matched that of Black and his artists, and of Curtis. Moreover, the standard is equivalent to that of most contemporary research taxonomists. The significance of the artwork in this Flora was that Burbidge was following Black's example and (perhaps passively) challenging other Flora writers to make their texts more effective for readers. Burbidge was influenced by Blunt's book and encouraged other potential artists to consult it.

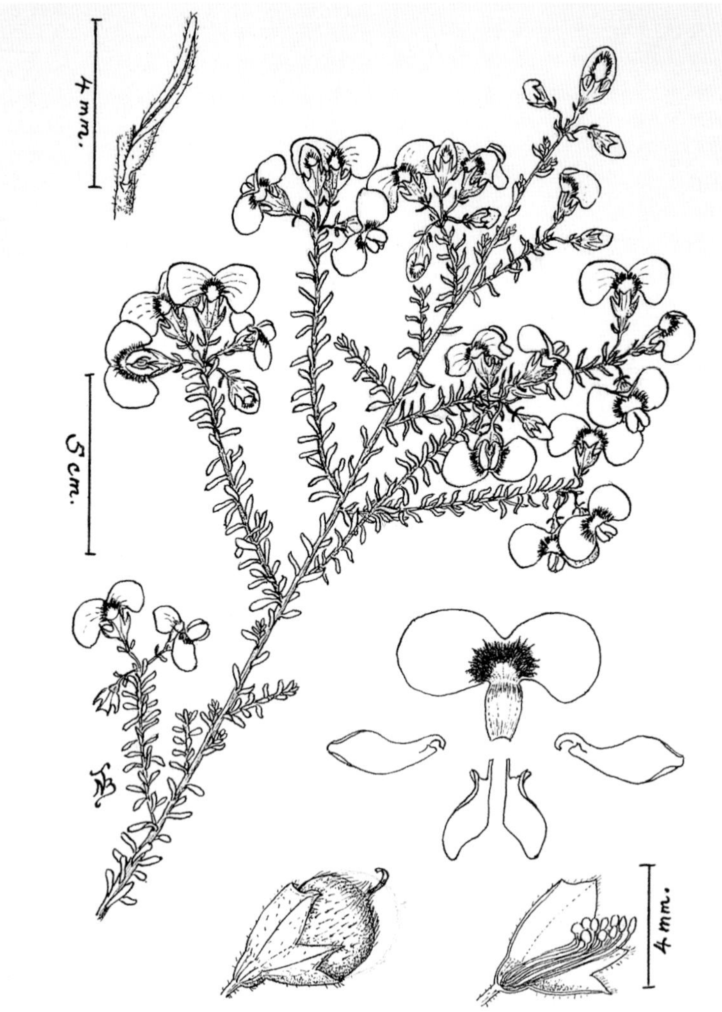

Dillwynia retorta (Wendl.) Druce
Original pen drawing prepared by Nancy Burbidge for publication in the *Flora of the Australian Capital Territory* in 1970.
[CANB 25.0 x 20.0]

Nancy Burbidge was born in England and came to Perth in early life with her family. Soon after graduating from the University of Western Australia, she won a scholarship to study grasses at Kew during 1939–1940. And so began the brilliant career of one of the energetic modern-day botanists. In 1946 Burbidge was appointed Systematic Botanist for CSIRO[6] Division of Plant Industry in Canberra.

Burbidge had a great sense of history, of the bigger picture, and of nurturing the interest of young folk and of enthusiastic amateurs. Her publications reflect this. William Hartley, a colleague, wrote of her as having a 'constant desire to stimulate interest in the Australian flora, and to make information available to people in all walks of life'.[7] Towards the end of her career, Burbidge was working tirelessly to get the latest *Flora of Australia* off the drawing board, the first for over one hundred years.

The year 1970 was the bicentennial year of Cook's landing on the east coast of Australia, an occasion which was to be honoured in a range of ways, but none with a particular emphasis on botanical art. Perhaps Australian botanists overlooked their heritage until it was too late to publish a grand publication during 1970. Sydney Parkinson's moment came in the 1980s with the publication of *Banks' Florilegium* (1981–1988). This is a complete set of Parkinson's work carried out on Cook's first voyage of discovery to the southern hemisphere. However, in 1970 in recognition of Cook and the achievements of the *Endeavour* voyage, the Australian Academy of Science commissioned a painting of *Banksia serrata* by Celia Rosser for presentation to the Royal Society of London.

While Burbidge and Gray were preparing the ACT Flora, Celia Elizabeth Rosser[8] emerged in Victoria as a highly competent botanical artist. Rosser was born in Melbourne in

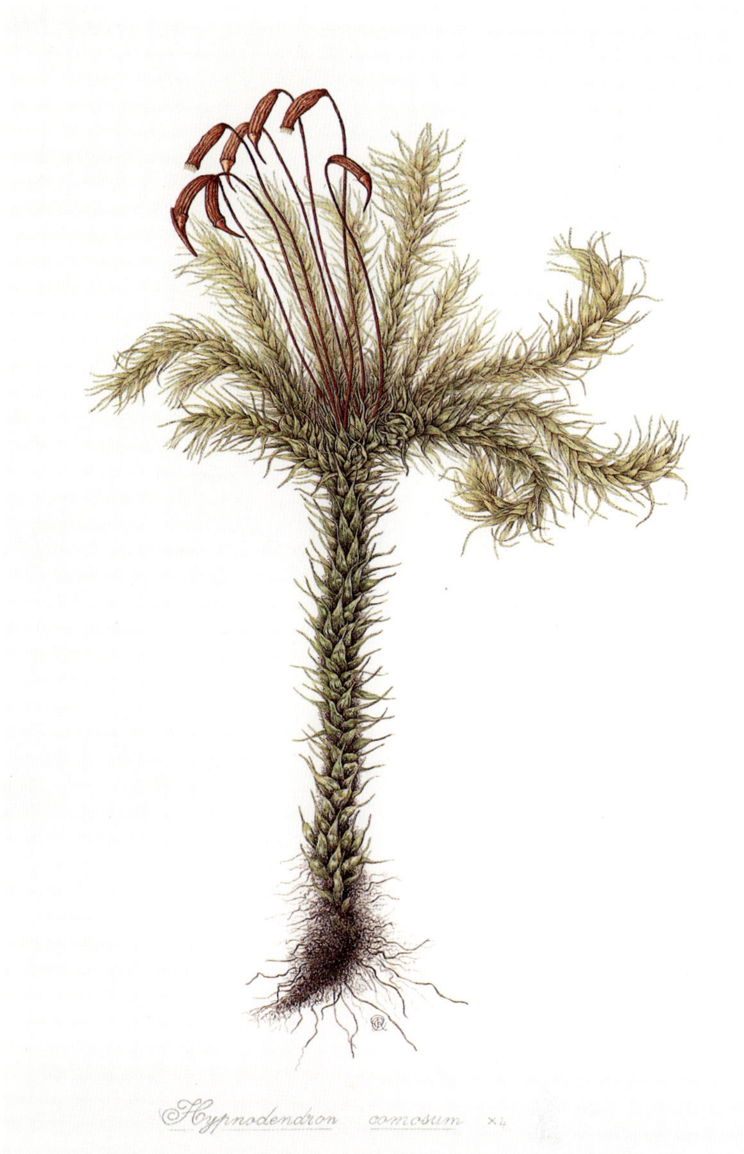

Hypnodendron comosum (Labill.) Mitt. Original watercolour painting by Celia Rosser published as the frontispiece in *The Mosses of Southern Australia* in 1976. [PC 37.0 x 32.0]

1930 and had art training at the Royal Melbourne Institute of Technology. While living in the country and raising a young family she became interested in native plants and began painting them. At an early exhibition of her work she was 'discovered' by the scientific community and, through the enterprise of Thomas Carrick Chambers, Rosser began her formal career as a botanical illustrator.

In 1970 Rosser was appointed Science Faculty Artist at Monash University, Melbourne. Her first major project for the University was to illustrate a book on mosses for George Anderson Macdonald Scott and Ilma Grace Stone, *The Mosses of Southern Australia* (1976). The eighty-six drawings are pencil studies of whole plants and associated diagnostic features, frequently leaf-cell detail. This must have been an extraordinary challenge for Rosser, as she had worked with flowering plants previously. However, her work on complex *Banksia* inflorescences must have helped her manage the intricate detail of mosses. The scale and idiosyncrasies of mosses, however, present a range of different challenges, with all the observation having to be done under a microscope. The size of mosses requires macroscopic study using a dissecting microscope for whole plant studies, and the cellular detail requires the use of a compound microscope. Moreover, the heat of the lights causes the plants to dry and to move — they are very responsive to changing humidity. The results demonstrate that Rosser adapted and helped produce one of the finest of our modern illustrated Floras. The images are reproduced in halftone which portrays the nuances of pencil tone-shading. A frontispiece painting of *Hypnodendron comosum* (Labill.) Mitt. is a crowning glory to the book.

Celia Rosser was appointed as the Monash University Botanical Artist in 1974, a unique position in Australian universities. This appointment took account of her extraordinary

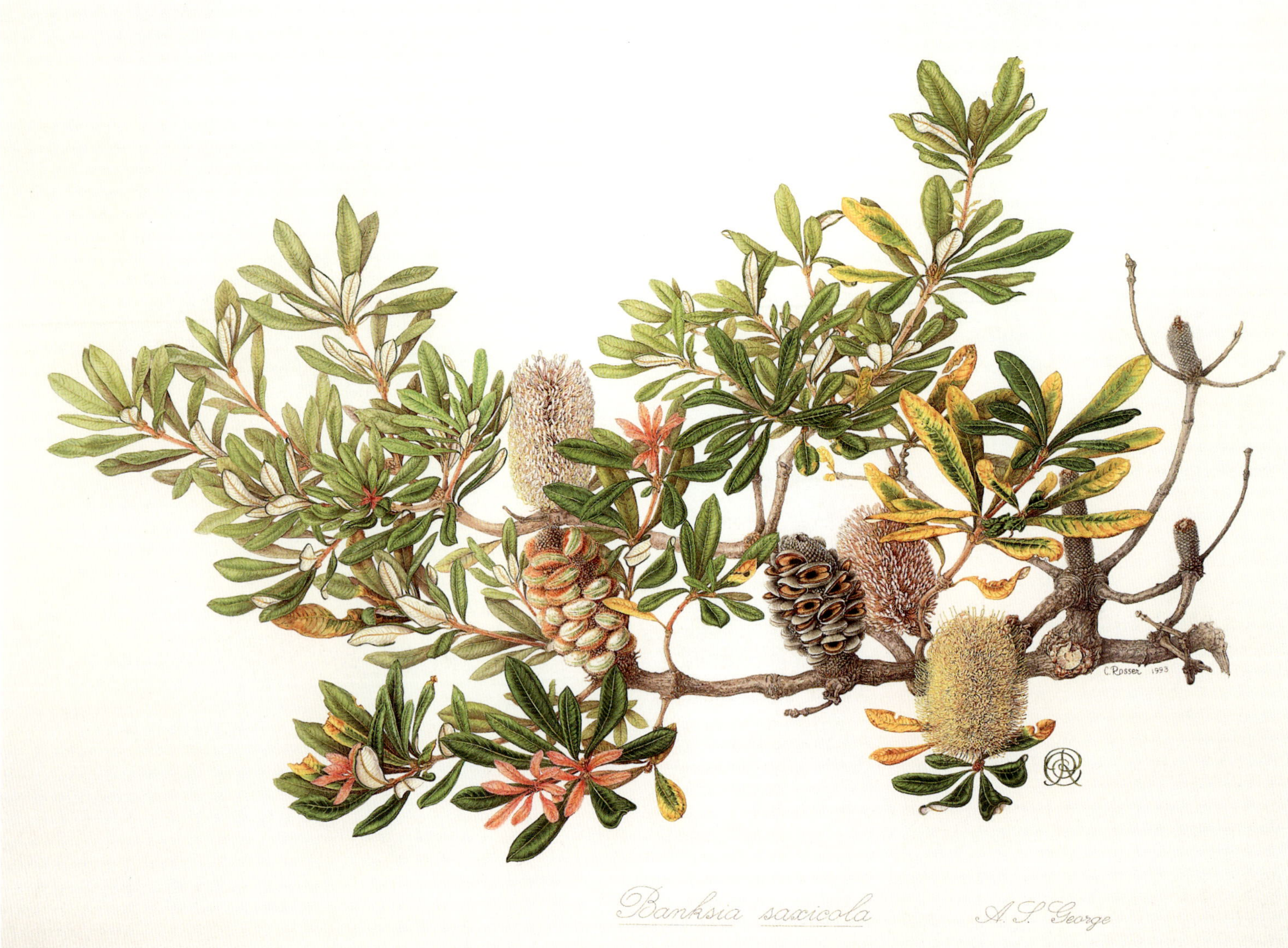

Banksia saxicola A.S. George
Original watercolour painting by Celia Rosser for publication in the third volume of *The Banksias*, a monograph of the genus with every species painted by Rosser together with text by Alexander George.
[MU 57.5 x 76.0]

ability to paint one of the most technically difficult genera, the genus *Banksia*. The University undertook to have all species of *Banksia* painted by Rosser and Alexander Segger George,[9] the world authority on the genus, was chosen to provide the descriptive text. *The Banksias* was destined to appear as an elephant-sized, three-part publication.[10] Rosser's work is meticulous down to each single hair and places her in the realm of the world's greatest botanical illustrators.

Meanwhile, the revival was gaining momentum on all fronts, including lavishly illustrated books.

Eucalyptus rhodantha Blakely & Steedman
Original watercolour painting of Rose Mallee by Stanley Kelly and published in his *Eucalypts*.
[MEL 25.8 x 20.6]

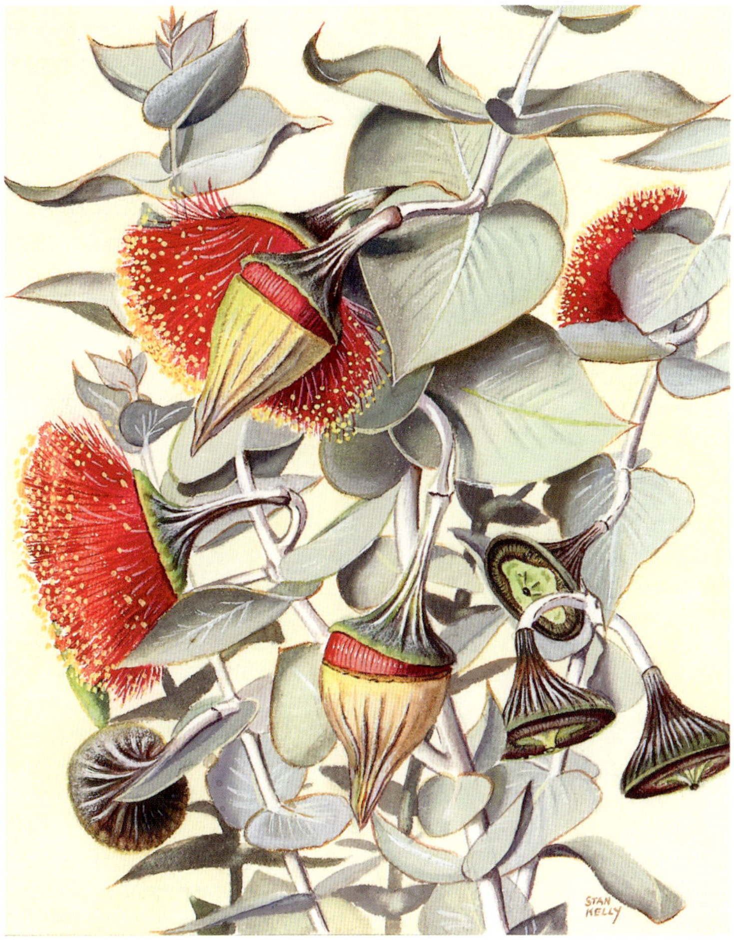

Stanley Kelly undertook a massive task. He wrote:[11]

> I began painting eucalypts as a hobby in 1945, little dreaming that thirty-two years later, having travelled over a wide area of Australia in my quest for specimens ... I would realise my dream of painting the whole of the genus.

His work is published in two volumes of *Eucalypts* (1971, 1978). *Eucalyptus* is a very large genus and Kelly painted 508 species. More species are recognised now. However, a combination of Kelly's work together with Chippendale's publication, *Eucalyptus Buds and Fruits*, and Blakely's Key would take one a long way towards making a reliable identification.

Stanley Kelly was born in 1911 and began his working life as an engine-driver for the Victorian Railways. He had an obvious talent as a botanical artist and capitalised on it in realising his dream.

Helen Isobel Aston, a botanist at the National Herbarium of Victoria, published the *Aquatic Plants of Australia* in 1973. Aston illustrated her work with her own drawings and with borrowed drawings. Included in the borrowed artwork is illustration by Stella Ross-Craig, a staff artist at the Royal Botanic Gardens. Ross-Craig became one of those artists who set an inspiring standard. She had her own style which embraced the use of parallel line of graded width to give form.

In 1974 Elizabeth Vivienne Conabere painted wildflower studies for John Roslyn Garnet for his book, *Wildflowers of South-Eastern Australia*. Conabere was born in Alexandra in Victoria in 1929 and began her artistic career by training herself. Early in her venture she was inspired and encouraged by Jim Willis, a botanist at the National Herbarium of Victoria, and she received support from the Maud Gibson Trust. Staff at the Melbourne University Botany Department, particularly John Turner and Carrick Chambers, were instrumental in helping Conabere gain the botanical knowledge to underpin her inherent ability as a botanical artist. They, and the staff at the National Herbarium of Victoria, helped her

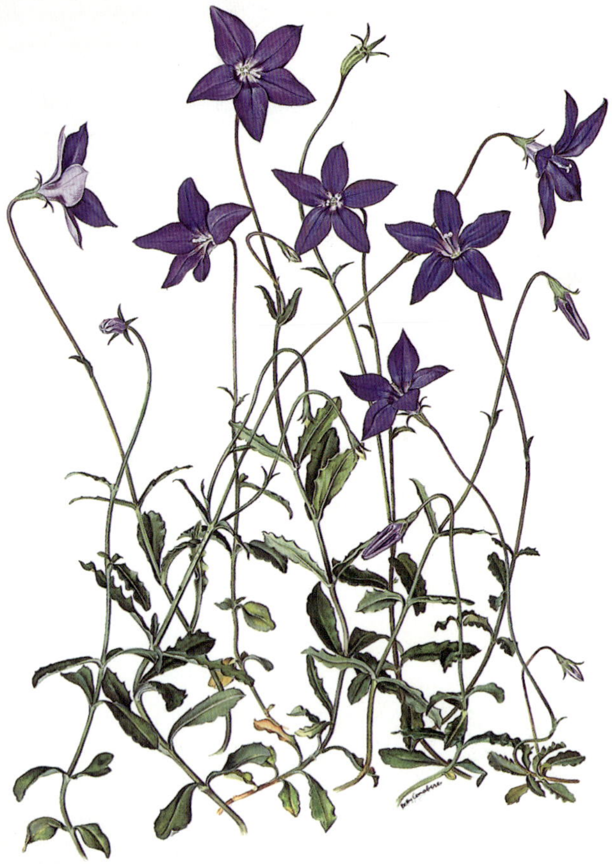

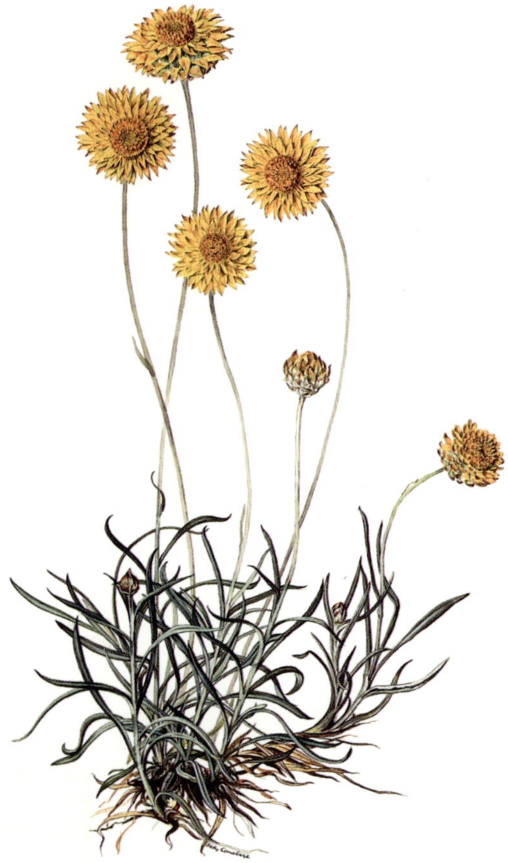

Wahlenbergia gloriosa Lothian (left);
Leucochrysum albicans (Cunn.) Paul G.Wilson (= *Helipterum albicans* (Cunn.) DC.) (right).
Watercolour paintings by Elizabeth Conabere. The Royal Bluebell (left) is the floral emblem for the Australian Capital Territory.
[MEL 56.0 x 38.0] [MEL 37.0 x 27.0]

establish the network necessary to undertake the massive task of painting so many species for Garnet's compendium. Conabere's acknowledgment in the first volume emphasises the amazing extent of the professional and amateur network which operates behind the front doors of herbaria and the front covers of texts and journals.

Ros Garnet was an industrial chemist with the Commonwealth Serum Laboratory. He had a special interest in natural history, national parks, conservation and the protection of the natural environment. His book provided a means of promoting a conservation message together with an exposé of the biology of the plants depicted. These messages were just beginning to be sounded, *en masse*, about thirty years ago and issued more forcibly these days but not always so aesthetically.

In the interim, William Stearn teamed again with Wilfred Blunt to attempt to mark another moment of historical significance for Australia. They revealed another treasure of the past in publishing *The Australian Flower Paintings of Ferdinand Bauer* (1976) to, belatedly, commemorate the bicentenary of the birth of Matthew Flinders in 1774. This grand elephant-folio publication of Ferdinand Bauer's paintings achieved all sorts of wonderful things, perhaps the greatest being that it gave Australian botanical artists a standard of excellence to aim for.

Ferdinand Bauer's Australian flower paintings had not been issued other than as uncoloured or hand-coloured lithographs in rather rare early-nineteenth-century publications. His work was therefore obscure, if not inaccessible. Indeed, through this publication, Stearn and Blunt have drawn attention to the phenomenon that not only are Australia's early herbarium and museum specimens in European custodianship, but also Australia's early botanical art. They opened the window on one of Australia's visiting artists and his works. Twenty years later a collection of Bauer's work travelled for exhibition in Sydney and Canberra and reinforced this realisation.

The growth of the Australian herbaria following World War II saw a need for extra publications for the increasing scientific output. Each herbarium began its own house journal which provided a further stimulus for scientific botanical illustration. Looking at the first volume of each of these journals provides a window on this initiative.

Muelleria was begun in 1955 for the National Herbarium of Victoria, financed by the Maud Gibson Trust. As the title implies, the journal commemorates the enormous contribution to botany made by Ferdinand von Mueller during the latter half of the 1800s. Volume 1 was sparingly illustrated. Of particular interest is illustration of *Stylidium* by Rica Erickson, a Western Australian artist.

Erickson is an amateur botanist whose interest in plants, coupled with her ability to draw and paint them, led to her descriptive work being recognised and used by scientists, and to her fostering a network of budding young botanical illustrators in Western Australia. She had several illustrated books published — *Orchids of the West* (1951), *Triggerplants* (1958), and *Plants of Prey* (1968). *Orchids of the West* is illustrated with line drawings and colour paintings. The colour plates are ruined (from an artistic and design point of view), or enhanced (from a user's point of view) by a series of labelled analytical drawings of the finer structure of the flowers. The latter are printed in blue and, while very useful, detract from the more delicate painting alongside. *Triggerplants* has a colour painting for the frontispiece and eight other colour plates. The majority of the illustration is very creditable line drawing, some of which required dissection microscopy to effect.

Three of the line drawings are by Tarlton Rayment who wrote an entry on pollen studies. Rayment was an entomologist with the National Museum of Victoria. In addition to screen-printed colour and black and white plates of Erickson's paintings, *Plants of Prey* includes two plates of illustration of *Nepenthes mirabilis* (Lour.) Druce, the tropical pitcher plant, painted by Ellis Rowan. Erickson made another valuable contribution to Australian botanical science as an historian and she has published work in this vein. *The Drummonds of Hawthornden* (1969), documenting the work of James Drummond, the collector, is a most important record of early Western Australian settlement and the early Western Australian botanical record.

Nuytsia was initiated in 1970 for the Western Australian Herbarium and named to glorify the Western Australian Christmas Tree.[12] *Nuytsia* was named in honour of Pieter Nuyts, a navigator on the Dutch ships, the *Leeuwin* and the *Gulge Zeepard*. On separate voyages Nuyts explored the Cape Leeuwin region, in 1622, and along the south-west coast of New Holland into the Great Australian Bight, in 1627. Illustration in the first volume was carried out mainly by the scientists themselves with some particularly pleasing illustration of *Acacia* by Bruce Roger Maslin, an Australian specialist on the genus.

Telopea was initiated in 1975 for the National Herbarium of New South Wales, the Waratah[13], and state emblem, being glorified in the title and cover illustration. Volume 1 was generally poorly and sparsely illustrated. An exception is the illustration of *Acacia* by Robyn Griffiths and Christine Lillian Payne.

Christine Payne was born in Newcastle in 1949. She developed an interest in botanical illustration and is self-taught. She was appointed to the staff of the National Herbarium of New South Wales in 1974 together with Robyn Griffiths, an English botanical illustrator. Payne remained on the staff for four years and rejoined again in 1983 for a further seven years. During that time she was the major illustrator for staff scientific papers and for the *Flora of New South Wales*. From time to time Payne had the opportunity to train future illustrators in-house for the Herbarium. During the 1990s Payne worked freelance and taught botanical art in The Happiest Boy in the World, School of Decorative Arts.

Payne had numerous contemporaries while the *Flora of New South Wales* was being prepared and was succeeded by Nicola Oram and David Mackay.

The Adelaide Botanic Garden and State Herbarium of South Australia chose a conventional title for their house journal. The *Journal of the Adelaide Botanic Garden* was initiated in 1976 and the Chairman of the Governors, D. Scott Young, in describing the scope of the Journal stated, in Volume 1, that:

> ... it may also be the means of collecting work by botanical artists showing a range of styles and techniques depicting plants which have hitherto not been illustrated, or have been poorly documented.

The illustration in the first volume is exciting and generally of a high standard, notably that by Ludwik Dutkiewicz.

The Queensland Herbarium initiated its house journal, *Austrobaileya*, in 1977. In Volume 1 Robert Johnson, Director, wrote:

> The association of the name *Austrobaileya* with F.M. Bailey and C.T. White, the doyens of the first century of Queensland botany, the endemic nature of the genus and the challenge it has presented the botanical world make *Austrobaileya* a very apt title for this new journal.

Illustration is almost notable for its absence in the first three parts of Volume 1. Part Four (1981), however, is refreshing with work by three artists, striking among these being Margaret A. Saul.

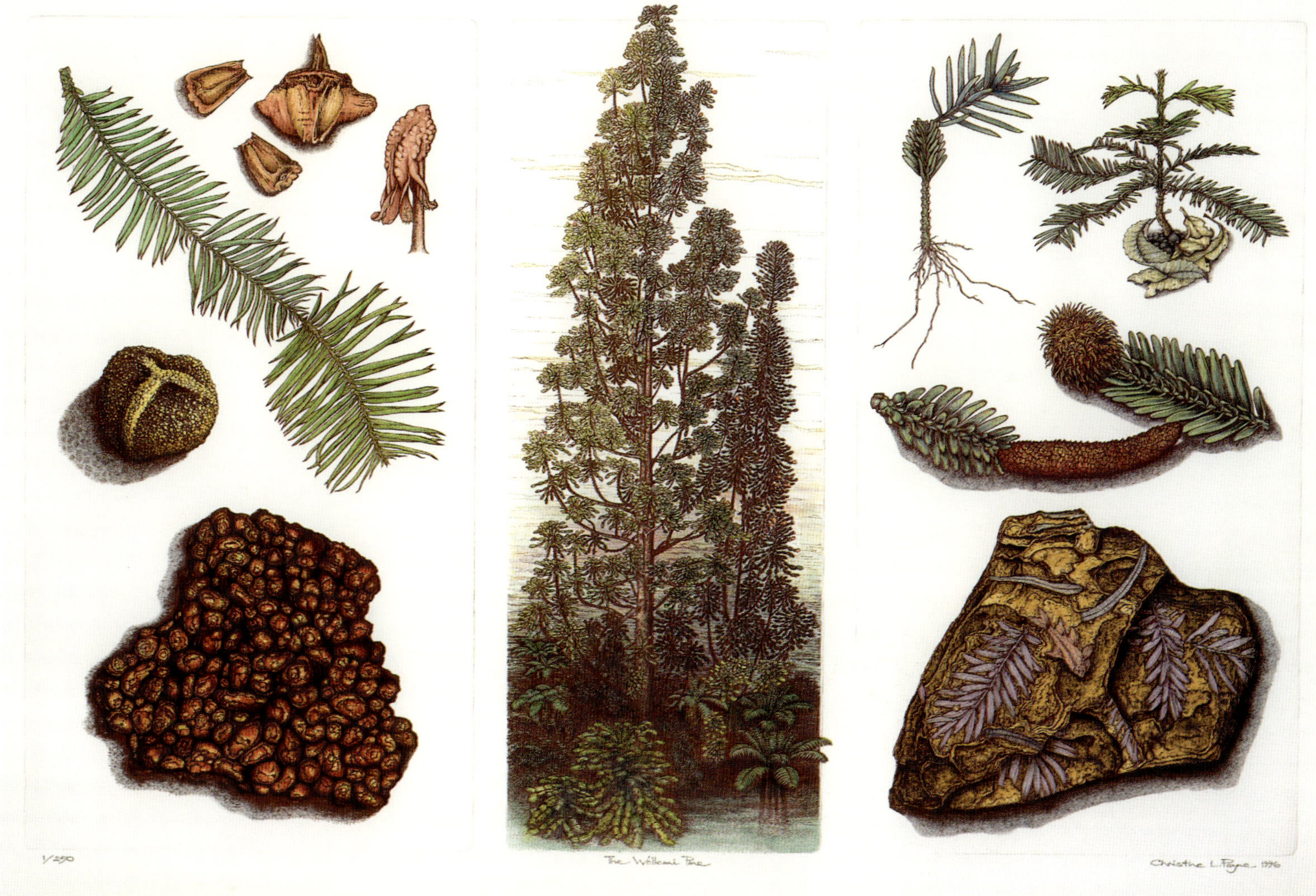

Wollemia nobilis W.G. Jones, K.D. Hill & J.M. Allen
Embossed hand-coloured print by Christine Payne of the recently discovered Wollemi Pine.
[MEL 44.0 x 64.0]

Margaret Saul was born in 1951 and trained at the Brisbane Art School between 1967 and 1971. She was fortunate to secure appointments in natural history with museums for a few years before being appointed as the inaugural Botanical Illustrator for the Queensland Herbarium in 1975. Saul worked in the Herbarium until 1980, when she decided to go free-lance. Her work is of a high standard and her commissions include cover and frontispiece paintings for the *Flora of Australia*. Saul now runs Eeco Illustration Art and Illustration Studio and is a founding member of the Botanic Artists' Group of Queensland.

Saul was succeeded by Gillian Marea Rankin who worked for the Herbarium between 1980 and 1985. She was followed by William Alexander Smith.

Hansjoerg Eichler[14] initiated a house journal for the Herbarium Australiense, the herbarium of CSIRO, Canberra, in 1978. He chose to call this *Brunonia* in honour of Robert Brown, the botanist with Matthew Flinders on board the *Investigator* and after whom the

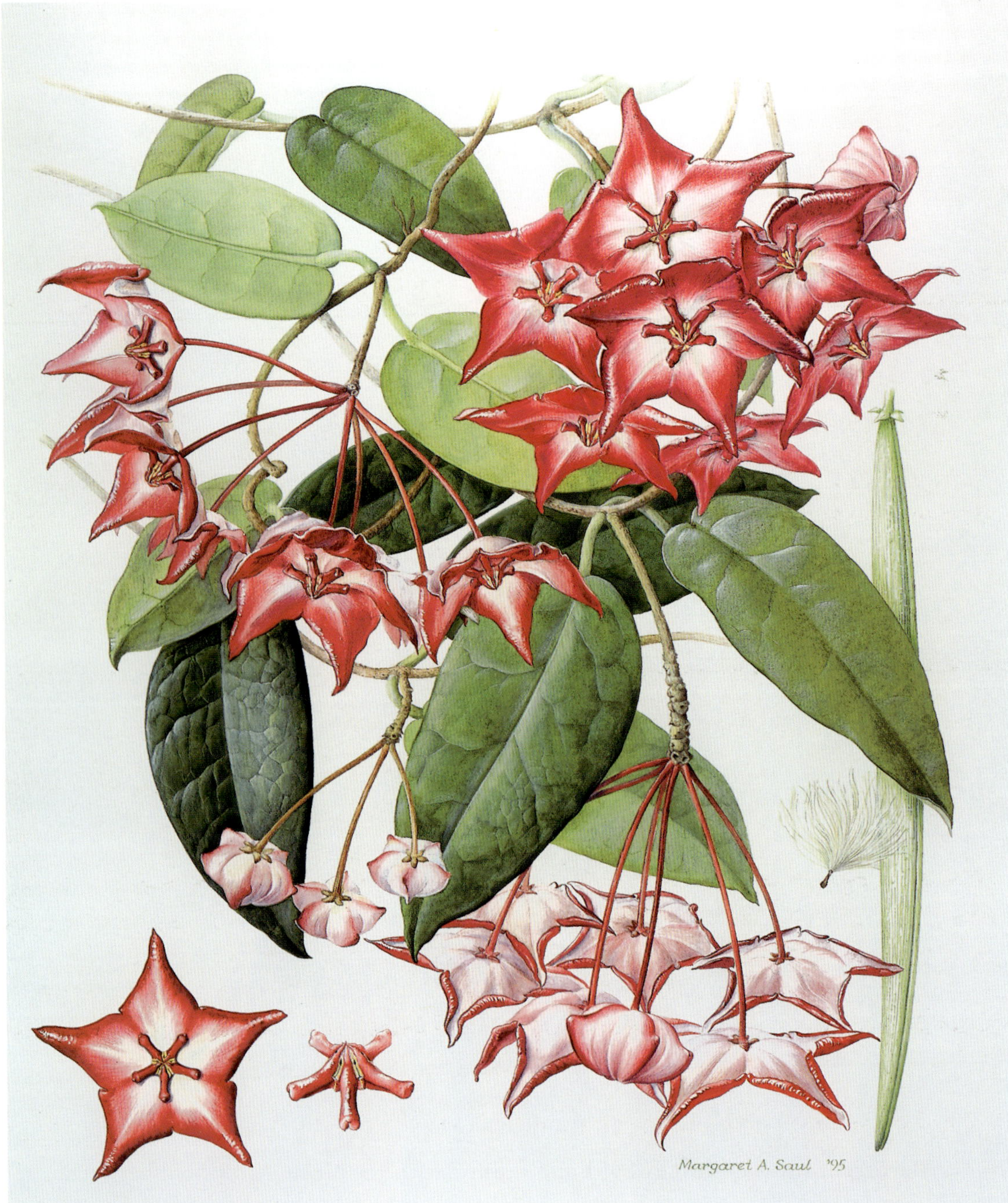

Hoya macgillivrayi Bailey
Original watercolour painting by Margaret Saul and published in the *Flora of Australia* in 1996.
[ABRS 46.0 x 39.0]

genus *Brunonia* (Goodeniaceae/Brunoniaceae) was named. The Ferdinand Bauer painting of *Brunonia australis* R. Br. was reproduced as a frontispiece for the first issue and the whole plant image from that painting reproduced in blue-monochrome on the cover.

This issue also provides an obituary to Nancy Burbidge; a review of *The Mosses of Southern Australia* by G.A.M. Scott and Ilma G. Stone with illustrations by Celia Rosser; and a range

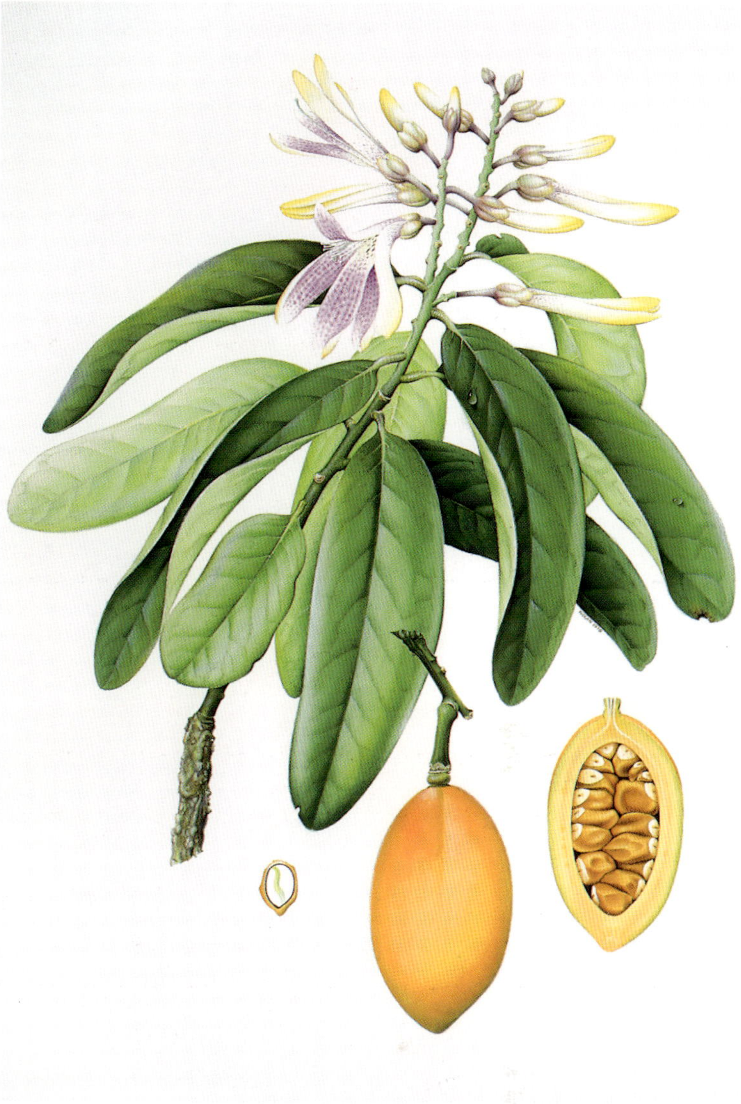

Xanthophyllum fragrans C. White
Watercolour painting by Terrence Nolan.
[CANB 50.0 x 40.0]

of illustration methods in the various papers. In four parts, Volume 1 provides a dramatic example of how botanical illustration in Australia was rapidly moving into revival mode. Eichler was not an artist himself but he fostered an interest in botanical art and illustration and encouraged taxonomists to have their subjects illustrated. He seized appropriate opportunities to make the point that a well-executed botanical illustration was much more effective than a photograph.

Hansjoerg Eichler was born and educated in Germany. During his education he came under the influence of Ludwig Diels. Eichler joined the staff of the State Herbarium of South Australia as its first Keeper in 1955. In 1973 he was appointed Curator of the Herbarium Australiense, Canberra, following the secondment of Nancy Burbidge to the Australian Biological Resources Study. Hansjoerg's contribution to Australian botany was manifold and positive. He helped to establish Australia's recognition in an international context, he played an important role in the politics of the science of plant taxonomy, and he raised standards of collection, curation and documentation in Australia. Students and staff of his have all absorbed aspects of his high ideals.

The influence of Terrence W. Nolan is evident in the first volume of *Brunonia.* Nolan was born in Victoria, trained at the Royal Melbourne Institute of Technology and began to specialise in scientific illustration. Nolan worked for CSIRO on the illustration of *The Insects of Australia* (1970). This led to an appointment at the Lae Herbarium, Papua New Guinea. There he illustrated scientific papers for the staff. At the same time he trained several of the indigenous members of staff, notably T. Iwagu. Nolan maintained the highest of standards and expected the same of others so his students, if they endured, were also doing extremely high quality work.

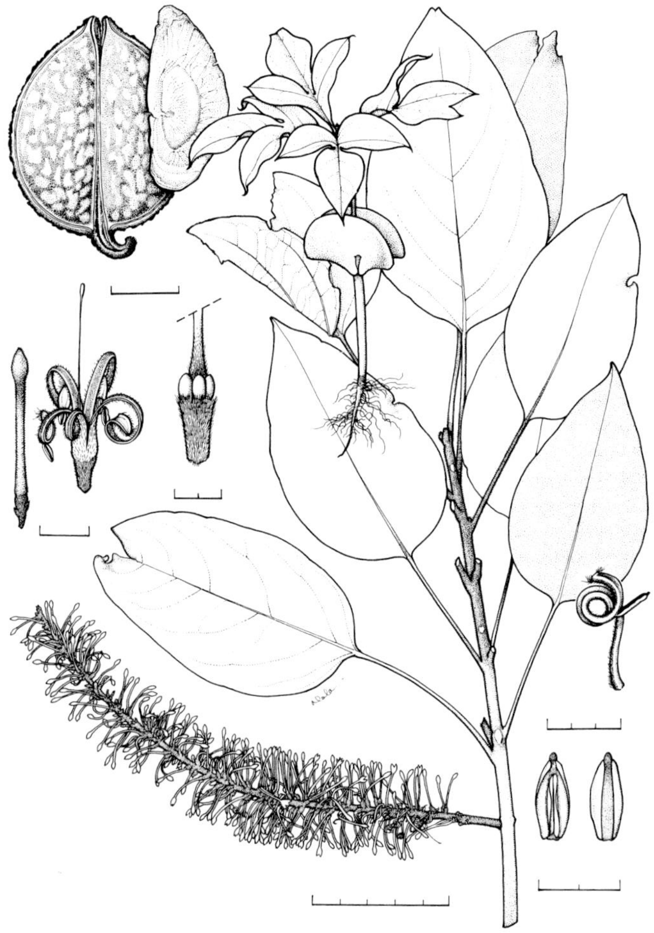

Austromuellera valida B. Hyland
Pen drawing by Terrence Nolan for publication in the *Flora of Australia* in 1999. Nolan's pen work served to raise the standard of pen illustration during the later decades of the 1900s.
[ABRS 36.5 x 26.0]

Nolan returned to Australia and worked again for CSIRO at the Atherton Herbarium from 1976 to 1979. Here he continued to demand and provide excellence in illustration. Nolan's line drawings of plants are among Australia's best.

In these journals many scientists continued to illustrate their own work. Their standard is variable yet some are obviously quite talented including Winifred Curtis, Nancy Burbidge, Helen Aston and Bruce Maslin.

The appointment of botanical illustrators to the staff of several of the herbaria provided perhaps the greatest stimulus of all to the revival. Following Dutkiewicz, Payne and Saul in other states, Anita Miriam (Podwyszynski) Barley was appointed to the staff of the National Herbarium of Victoria in 1977. Barley was born in Melbourne in 1955 and had the benefit of training as a graphic artist at the Royal Melbourne Institute of Technology. Barley remained on the staff until 1992 and so became the major illustrator for the *Flora of Victoria*. She also provided illustrations for the *Flora of Australia* and was commissioned for frontispiece and cover paintings. Barley was succeeded by Mali Moir, Su Pearson and Enid Mayfield.

The other herbaria contracted when required from a bank of highly competent artists. Margaret Anne Menadue has a high profile in this group. Menadue was born in Melbourne in 1942. She is self-taught and is passionate about her art and her subjects. Menadue worked freelance in a number of contexts in Western Australia, including formal botanical illustration. The Western Australian Herbarium arranged for her to illustrate both the cover painting and the line work in the *Flora of the Perth Region* (1987). Menadue has also illustrated for Western Australian contributors to the *Flora of Australia* and provided one frontispiece and cover painting. Her attention to detail and her ability to portray the most complex of subjects is quite outstanding. In addition, she is concerned for the conservation of plants and is applying her art towards their recognition and preservation.

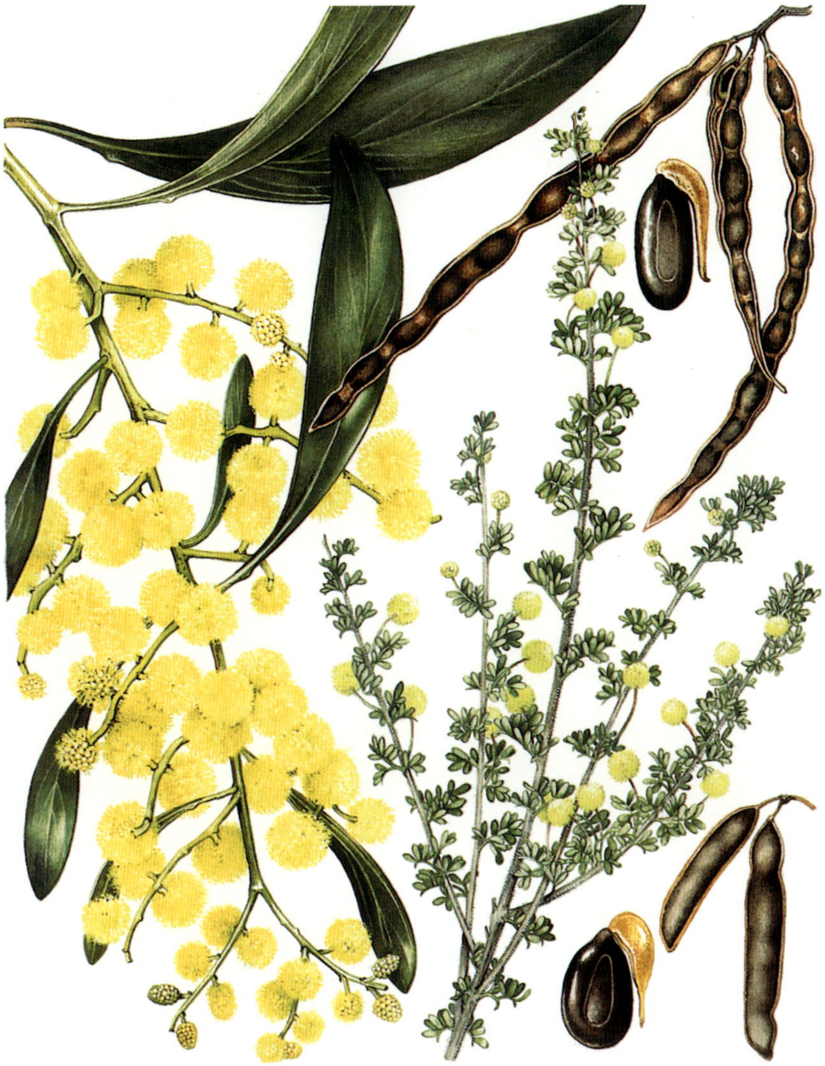

Acacia pycnantha Benth. and *Acacia mitchellii* Benth.
Watercolour painting by Anita Barley, published in the *Flora of Victoria* in 1996.
[MEL 20.0 x 15.1]

Philippa Nikulinsky was born in 1942 in Kalgoorlie and later trained as an art teacher at Claremont Teacher's College and James Street Technical College. Nikulinsky has built a successful freelance career. She has illustrated for the Western Australian Herbarium and the Australian Biological Resources Study. The latter includes two cover and frontispiece paintings for the *Flora of Australia*.

Katryna Syme, who was born in 1947, won a Churchill Fellowship to study botanical art in England in 1988. Syme had three cover and frontispiece paintings published for the *Fungi of Australia*. She teaches natural history illustration. Neale Bougher and Syme teamed to produce the *Fungi of Southern Australia* (1998). This book contains 130 plates of paintings of fungi together with numerous line drawings. In a sense Syme's and Bougher's achievement in this book issues a challenge to science, art and publishing, not to mention mycology (the study of fungi), representing a modern standard to aim for.

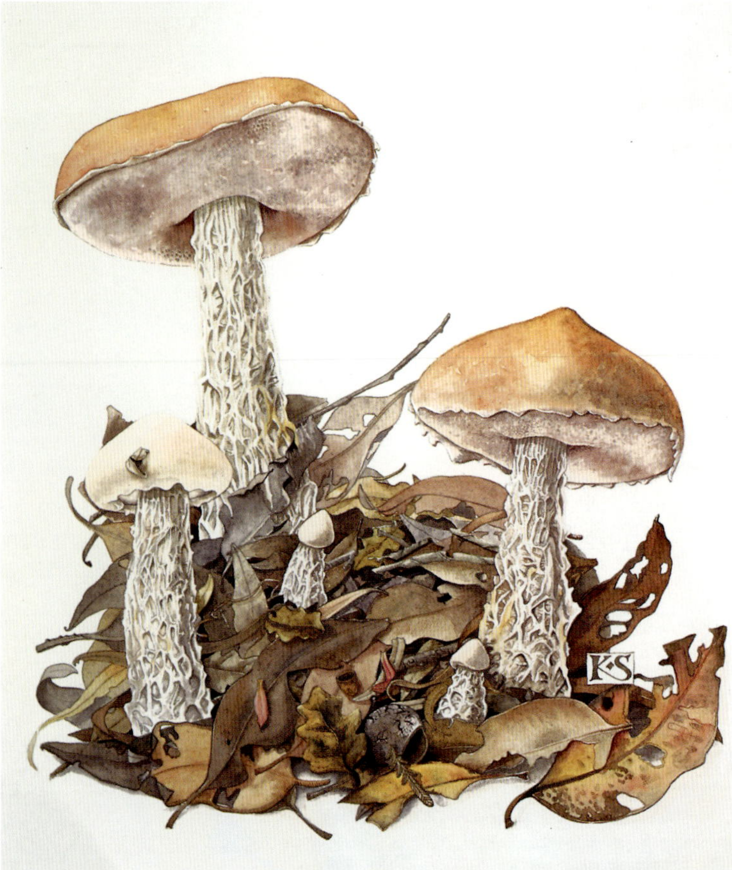

Austroboletus occidentalis Watling & N.M. Greg. Watercolour painting by Katrina Syme and published in the *Fungi of Australia* in 1997. [ABRS 34.0 x 28.5]

In 1973 the Australian Biological Resources Study was set up by the Commonwealth Government of Australia to document all of Australia's plants and animals. On the plant side, the *Flora of Australia* is the publication for the manifold research stimulated by the Study. Alexander George was the inaugural editor from 1980 to 1993. Publication of the Flora began in 1981 with an introductory volume. Volume 29, treating the Solanaceae — the potato, tomato and nightshade family — was next to be published with Betsy Osborne (Green) as the first illustrator. The frontispiece is a painting by Stones of *Solanum semiarmatum* F.Muell.

It is estimated that the knowledge of the flowering plants alone has almost trebled since Bentham recorded them one hundred years earlier, making this Flora a massive undertaking. For mainland Australia, forty-seven volumes and an introductory volume are to be published.[15] Each volume is illustrated with numerous line drawings, colour photographs and a colour painting for the frontispiece/dust jacket. In this way, the Study is encouraging the art and is providing a limited opportunity for the publication of paintings. The Study does not employ a staff artist but maintains a register of botanical illustrators and artists, and contracts artists where the contributing scientist is unable to provide artwork. The register is quite sizeable, indicating the growth of expertise in the last two decades of the twentieth century. To help maintain a reasonable standard of illustration, the Study prepared a *Guide for Illustrators* (1986). The contribution to botanical illustration provided by the Study is somewhat piecemeal and the standard of the line artwork published in the Flora is variable, ranging from adequate through to outstanding. In general the outstanding work is coming from herbarium staff artists, or those who have left the staff and become freelance.

In 1981 the XIIIth International Botanical Congress was held in Sydney. This provided an incentive and a focus for a number of projects to be published before the Congress, or to get on to the drawing board with a realistic (and in some cases, an unrealistic) hope of being

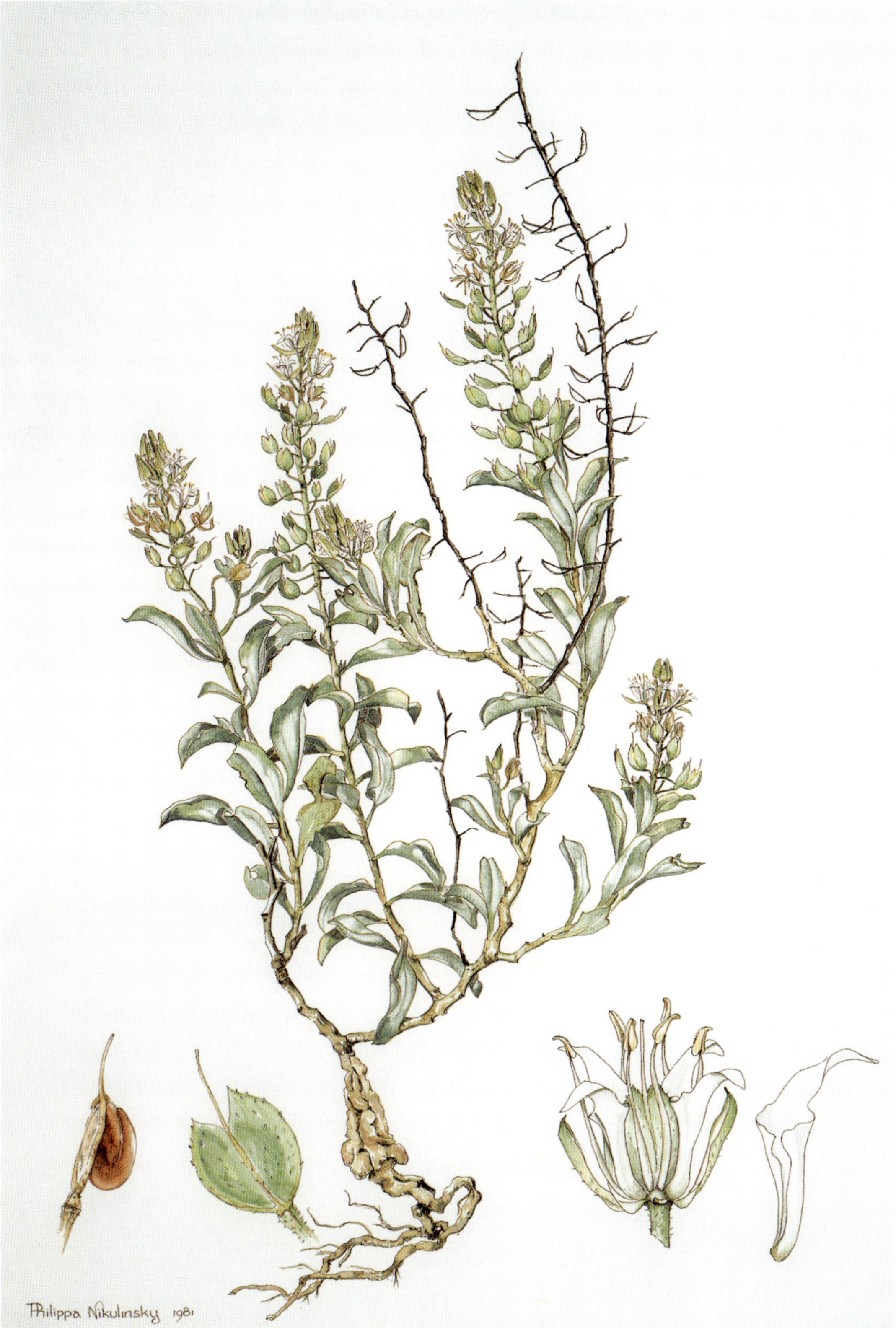

Lepidium pedicellosum F.Muell.
Watercolour painting of *Lepidium*, a native crucifer, by Philippa Nikulinsky published in the *Flora of Australia* in 1982.
[PC 42.5 x 28.5]

published in time for the Congress. It also provided the Sydney organisers, in particular, with a great opportunity to show off the resources and gems otherwise hidden in dark corners. The new books, the stamps, the historical displays and the art exhibitions provided a rich and enjoyable experience and evoked emotions of pride in the depth and quality of the heritage, and disbelief of the volume.

The books included *People and Plants in Australia* (1981) edited by Denis John Carr and Stella Grace Maisie Carr. This book brought together a number of contributors, including the Carrs, with essays on the history of plants and people concerned with them in Australia. The botanical artists emerge in the essays time and again, though only one essay is entirely devoted to *Collectors and Illustrators*. This was written by Ann Moyal who has made a huge contribution in other works towards the documentation of the history of Australian science.

Banks' Florilegium (1981–1988) was a very special case. Alecto Press, together with the Natural History Museum, decided to publish Sydney Parkinson's artwork as prepared for the proposed Joseph Banks' Florilegium which had not eventuated two hundred years earlier. For this work, the engraved plates made during the two decades after Cook's voyage in the *Endeavour* were cleaned and enhanced to effect publication of Parkinson's paintings and drawings. The plates of Australian material had been used by Britten in 1900–1905 and were then printed with black ink and reproduced by process engraving.

In the 1980s printing the plates were printed in full colour using the *à la poupée* method of colouring the plate. Had the *Florilegium* been printed around the 1780s, it is almost certain that the plates would have been printed in black ink and possibly hand-coloured. This would have been a grand publication in the 1780s. It is even grander in the 1980s, so much so that only the wealthy libraries and entrepreneurs own it. While the publication is of great historical interest, and is very fine in quality, it is doubtful that it did much to stimulate modern illustration to greater heights. The engraving style and the reproduction methods have long since been superseded. In addition, botanical illustration methods in detail and analysis have greatly advanced since Parkinson's time.

Denis Carr edited *Sydney Parkinson: Artist of Cook's* Endeavour *Voyage*, which was published in 1983. This book reproduces a selection of Parkinson's paintings, many finished by Frederick Nodder, together with a great deal of associated historical detail about the voyage and the artworks. A chapter by Rodney John Francis Henderson elaborates the plants of Australia. Both of these books, the *Florilegium* and Denis Carr's compilation, are a great, and belated, tribute to Parkinson.

Several modern illustrated books were prepared in Australia in time for the 1981 Botanical Congress, notably the *Native Trees and Shrubs of South-Eastern Australia* (1981) by Leon F. Costermans. It is regional and ecological in its approach to describing the dominant components of the flora in south-eastern Australia and is very usable, even by an amateur. Illustration is provided with diagnostic, vignette line drawings (including a map of distribution) and a selection of colour photographs. Costermans did all of the drawing.

The *Proteaceae of the Sydney Region* (1981) by Alec Morris Blombery and Betty Maloney is in the genre of the grand coffee-table book. Eighty-six plates of paintings of the species of this spectacular family of plants from around Sydney make this book a gem amongst

Eremophila oppositifolia R.Br. and *Eremophila maculata* (Ker Gawler) F.Muell. Coloured pencil drawings by Collin Woolcock published in the *Flora of South Australia* in 1986.
[AD 36.0 x 27.5]

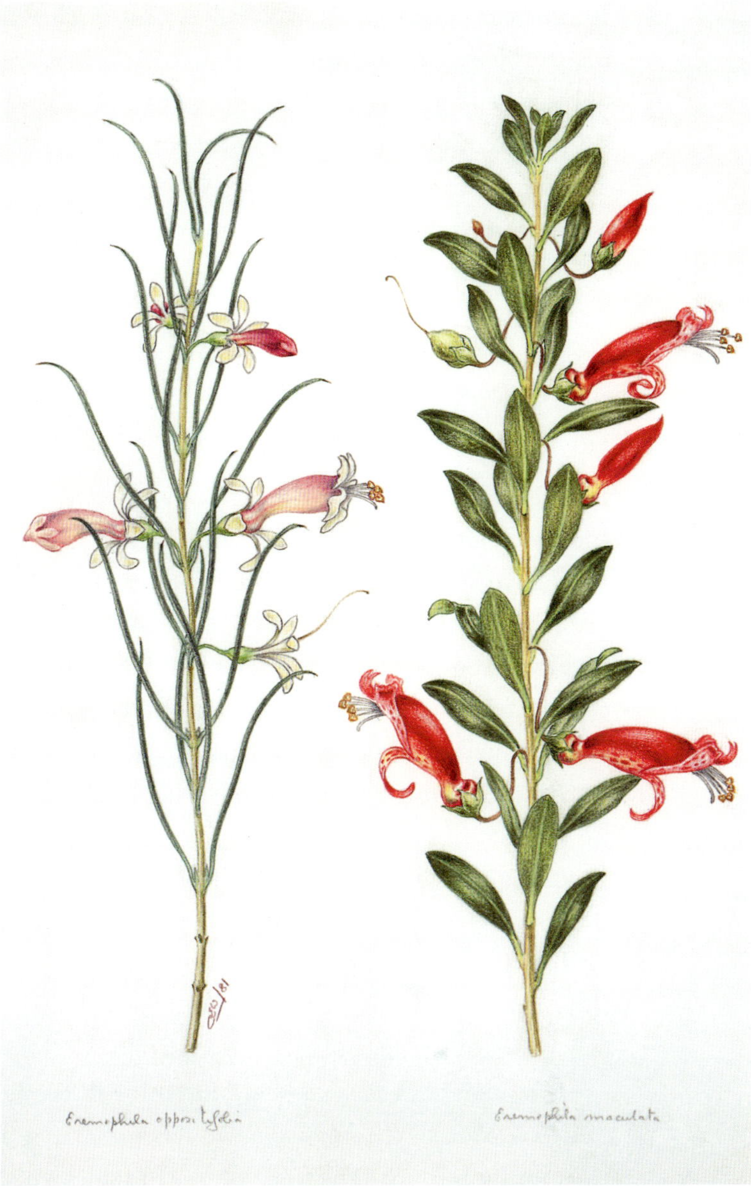

Australia's botanical art book collection. Betty Maloney was born in Colac in 1925, studied art at the Royal Melbourne Institute of Technology and taught art in Melbourne. When she married and moved to Sydney, she teamed with Alec Blombery on his many publications on Australian native plants. She was converted to growing Australian native plants and this is a recurring theme in the books she has illustrated.

In 1983 Brian Derek Morley and Hellmut R.Toelken of the Adelaide Botanic Gardens released *Flowering Plants of Australia* which they had teamed to edit and compile. This is an 'overview of both native and introduced families of cycads, gymnosperms and angiosperms found in Australia' — effectively an overview of all of the seed-bearing plants. Seventy scientists contributed text and three artists illustrated this informative tome which, with its colour plates, doubles as a reference text and coffee-table book. There are 230 plates, both line drawings and colour paintings. The artists are Ludwik Dutkiewicz, Ronald Adams and June Morley. Dutkiewicz executed all of the line drawings and Adams and Morely provided the paintings. Ronald Adams is an artist, graphic designer and gallery owner who had no prior experience as a botanical artist. He adapted remarkably well to the discipline. June Morley, wife of one of the editors, is an agricultural botanist turned artist and gallery owner.

The bicentenary of European settlement in Australia was celebrated in 1988. Botanists, however, did not make best use of this occasion, with one special exception. The Australian Systematic Botany Society held a symposium on the History of Systematic Botany in

Australasia. The organisers were very aware that the history of the taxonomy of Australian plants is inextricably locked into the history of discovery, colonisation, exploration and development of Australia. Botanical art was embraced as one of the themes. Three papers on the subject are published in the proceedings, *History of systematic botany in Australasia* (1990).[16] The cover of the book is greatly enhanced with a painting of *Correa bauerlenii* F.Muell. by Anita Barley.

The momentum was by now increasing so fast that it is impossible to tell all of the story in this book. The end of the millenium, however, deserves some attention with the release of a spate of completed, revised, or new, editions of state Floras during the last two decades.

In 1975 Winifred Curtis led the way again, this time in the publication of revised editions of the state Floras. Curtis teamed with Dennis I. Morris to prepare a revised edition of Part 1 of the *Flora of Tasmania* in which the original drawings were reprinted. In 1979 Curtis published a revised edition of Part 4a *Angiospermae: Orchidaceae*. This is illustrated with twenty-seven very fine pen drawings by Morris. All the while, the first edition had not been completed. Finally, in 1994, Curtis and Morris published Part 4b to complete the Flora, with very fine pen drawings by Morris.

In 1983 Trevor Donald Stanley and Estelle M. Ross (editors) began the process of preparing a Flora of Queensland with the publication of a *Flora of south-eastern Queensland*. This appeared in three volumes with the illustration limited to line drawings. The first volume is illustrated with eighty (mostly full-page) plates of line illustration, mostly by Margaret Saul. Gillian Rankin also provided some work. The second volume (1986) was predominantly illustrated by Rankin in eighty-four plates. The third volume (1989) has work by both these artists, but is predominantly illustrated by Will Smith with sixty-four plates.

In 1986 John P. Jessop and Hellmut R.Toelken (editors) began the publication of the fourth edition of the *Flora of South Australia*. It is described as a Jubilee edition and is dedicated to John Black. The first volume provides a history of botany in South Australia, though it does not pay particular attention to the scientific illustrators. The editors attempted to maintain the high level of illustration for this new edition and had the flora re-illustrated. Ludwik Dutkiewicz, Elizabeth Chandler, Gilbert R.M. Dashorst and Collin Elwyn Woolcock provided the artwork. Sixteen colour plates were drawn, in crayon, by Woolcock. Most of the pen drawings are signed and attributable to Chandler, Dutkiewicz or Dashorst.[17] While the latter two provided most of the artwork for the last three volumes, several additional artists also made contributions.

Some artists include animals in botanical illustration, but this is unusual in formal illustration. In this Flora Dashorst has provided creatures in many of his illustrations, providing light relief in the midst of otherwise heavy-going data.

Western Australia has begun to prepare regional Floras, two being completed. *The Flora of the Perth Region*, by Neville G. Marchant and others,[18] was published in two volumes in 1987. The cover painting and the line drawings are all by Margaret Menadue, there being 317 figures.

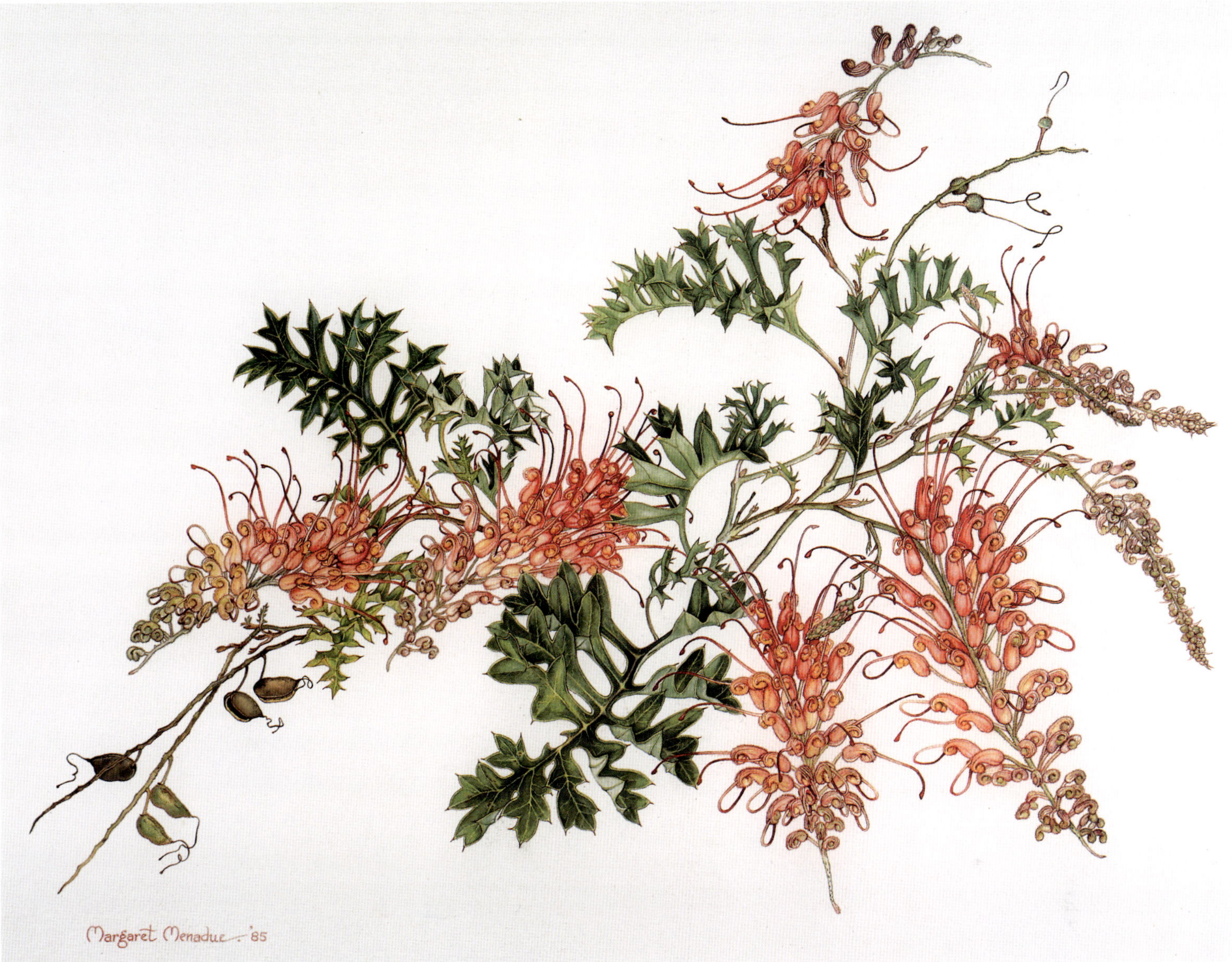

Grevillea bipinnatifida R.Br.
Watercolour painting by Margaret Menadue published as the cover painting of the *Flora of the Perth Region* in 1987.
[WA 48.0 x 60.0]

The *Flora of the Kimberley Region* was produced in 1992 under the editorship of Jocelyn R.Wheeler. It is illustrated with a cover painting by Pat Dundas and 354 line drawings by nineteen artists, together with some reprinted from elsewhere.[19]

New South Wales began the *Flora of New South Wales* in 1990, under the editorship of Gwen Jean Harden. This Flora, at last, achieves the ideal — it illustrates every species. Colour is provided by photographs of a selection of species. All species have a vignette illustration in the liberal margin immediately adjacent to the description, reminiscent of Leon Costerman's style. The illustrations vary in subject to reflect the most diagnostic characters. For the first volume, seven artists provided line drawings with the greatest number by David Mackay and

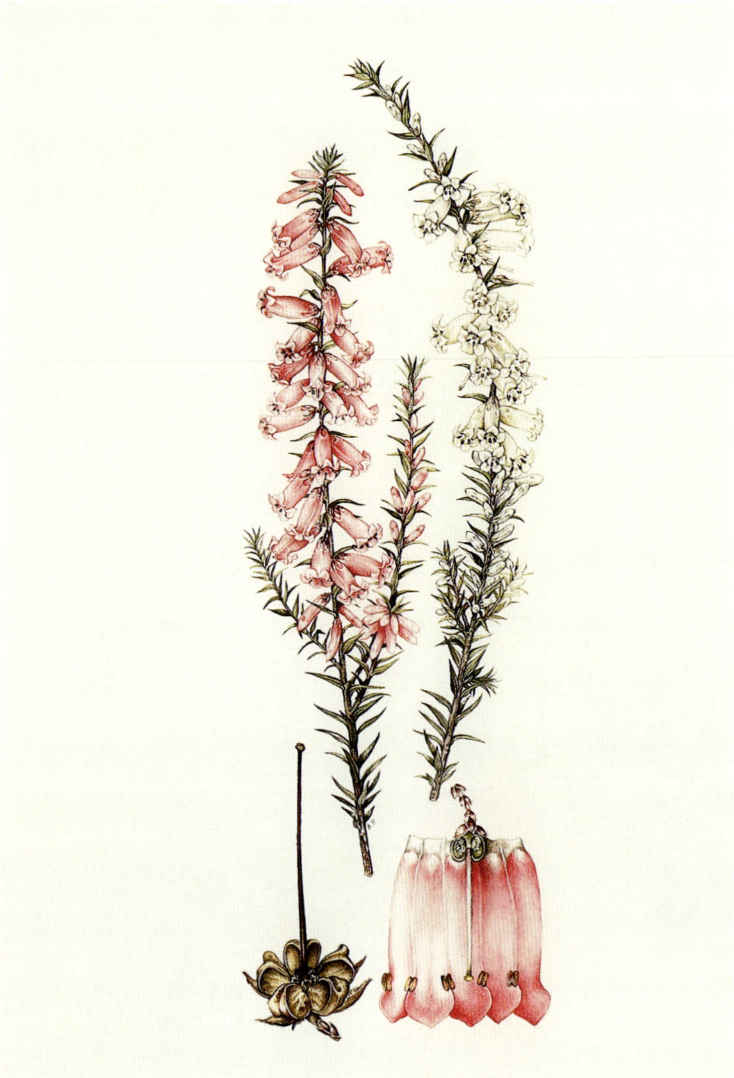

Epacris impressa Labill.
Watercolour painting by Anita Barley published on the back cover of the *Flora of Victoria* in 1993.
[MEL 29.8 x 21.0]

Christine Payne. In later volumes, an additional ten artists contributed with Nicola Oram joining Mackay and Payne as a major contributor. This massive work was completed in four volumes in 1993.

Victoria began the *Flora of Victoria* in 1993 under the editorship of Donald B. Foreman, Neville G. Walsh and Timothy J. Entwisle. This four-volume Flora is illustrated selectively with paintings and line drawings. Anita Barley painted all the studies for the fifty-six colour plates. The line drawings were completed by several artists with Enid Mayfield and Mali Moir making the major contribution. The cover painting for Volume 1 is a study of *Epacris impressa* Labill., the state emblem.

In 1995 Clyde Robert Dunlop, Gregory John Leach and Ian D. Cowie began a Northern Territory Flora with a *Flora of the Darwin Region for the Northern Territory*. The cover painting is by Bee Fong Gunn. The line drawings are by Milton Andrews and Monica Osterkamp Madsen.

Staff at the Darwin Herbarium have been liaising with Aboriginal communities to record the botany of Aboriginal plant usage. This work has resulted in several *Northern Territory Botanical Bulletin*'s featuring regional ethnobotany. The cover plates are Aboriginal paintings and the species are illustrated with pencil or ink drawings by Milton Andrews. Patricia Marrfurra is among the artists contributing cover paintings.

Training for botanical illustrators is an aspect which seems to have ranged from very focused instruction, through apprenticeship and adaptation, to being completely self-taught. The Bauer brothers had a mixture of a very focused training and an apprenticeship. Most successful illustrators either learned on the job or adapted skills from graphic- or commercial-art training. Friederich Schoenfeld, Emil Todt, Eirene Mort, Rosa Fiveash, Ludwik Dutkiewicz, Terrence

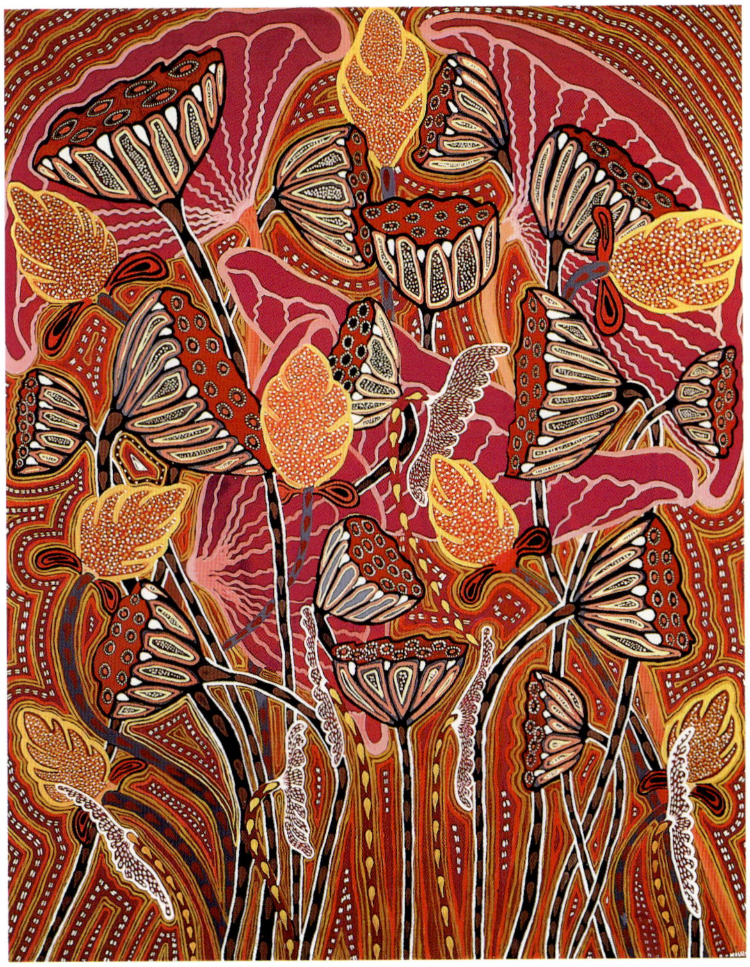

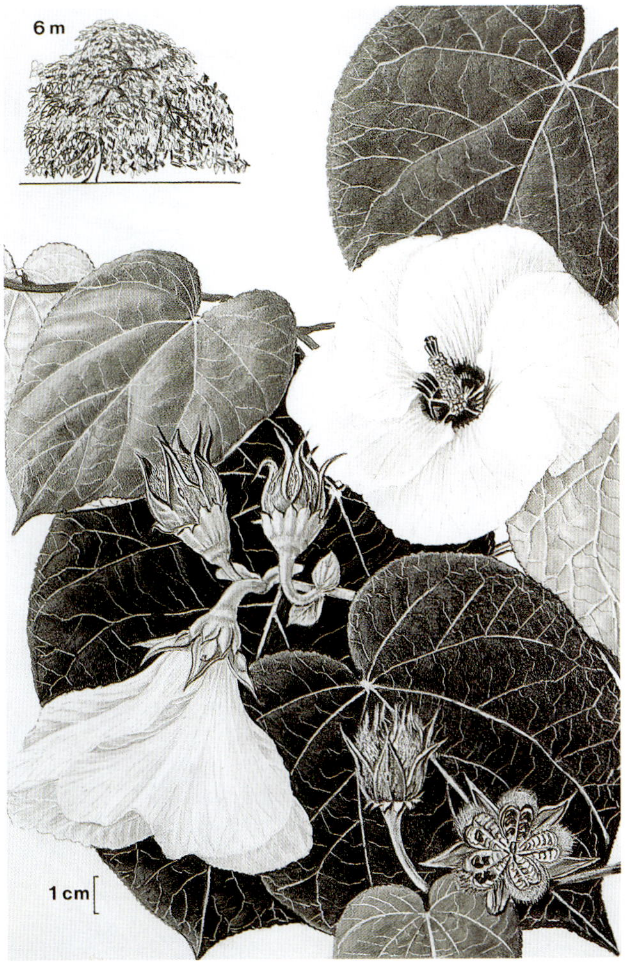

Nelumbo nucifera Gaertner (= *Lotus extravaganza*); *Hibiscus tiliaceus* L.
Painting in acrylic by Patricia Marrfurra (left) and pencil drawing by Milton Andrews (right), each published on the cover of a *Northern Territory Botanical Bulletin* during 1995. The *Bulletin* documents Aboriginal plant use in the Northern Territory.
[DNA 130.0 x 102.0] [DNA 26.1 x 17.2]

Nolan, Anita Barley and Margaret Saul adapted other training-experience to the discipline. Margaret Menadue, Christine Payne and many others were completely self-taught.

Generally professional training for botanical illustrators was largely overlooked until the latter decades of the twentieth century. The employment niche is very small and highly specialised, so the demand for such training is small. The Royal Melbourne Institute of Technology has been responsible for training several important botanical artists. Almost all believe that they underwent an unofficial apprenticeship on the job to develop their skills in observation and portrayal. In 1978 the University of Newcastle began a course providing training in wildlife illustration which includes botanical illustration. Graduates gain a Bachelor of Arts in Visual Arts and, for most, a career outside formal science.

William T. Cooper was a lecturer in this course in the early 1980s. Cooper is now famous for his ornithological painting, but from time to time ventures into painting plants. An outstanding example of his work is published in *Fruits of the Rainforest* (1994).

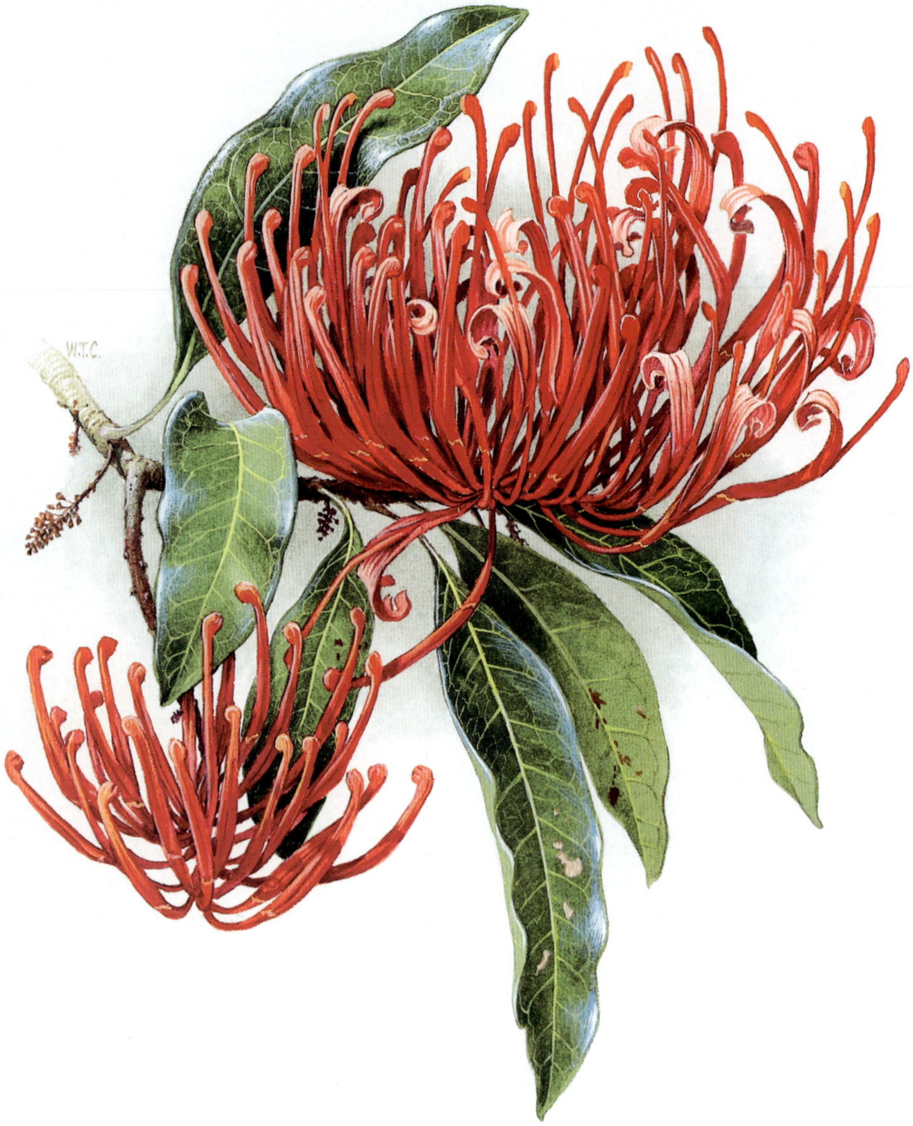

Alloxylon flammeum P.H. Weston & Crisp
Watercolour painting by William Cooper used for the promotion of the interactive key to *Australian Tropical Rainforest Trees* which was published in 1993.
[PC 31.0 x 24.0]

This is a collection of paintings of about a third of the fruits of Australian rainforests together with synoptic notes and a line drawing of a typical leaf of the plant.

The Newcastle course is unique in Australia, indeed few countries offer such specialised training in illustration at the tertiary level. Nicola Oram is one of its graduates.

The artists who have played their role in documenting Australian flora have done so by chance, choice, or default. Those who have illustrated by chance are generally those who have had a more cosmopolitan role to play in botanical illustration. Artists for the grand periodicals of eighteenth- and ninteenth-century Europe fall into this category — James Sowerby, Sydenham Teast Edwards, Walter Hood Fitch and many others.

Those who have illustrated by choice include those who came to Australia on voyages of discovery and exploration, such as Sydney Parkinson and Ferdinand Bauer. Artists living, or born, in Australia also fit into this category. Rosa Fiveash, Margaret Flockton and Celia Rosser represent a sizeable bank of these artists. Some of these people became zealots, painting with a message (either for themselves, or for particular authors) — Ellis Rowan, Rosa Fiveash, Elizabeth Conabere and Stanley Kelly having education, conservation, or the aim to paint every species as their driving force.

Some have had the benefit of a full-time career in botanical art such as Margaret Flockton, Celia Rosser, Ludwik Dutkiewicz, Anita Barley, Margaret Saul and Christine Payne. Others retained their independence, including Eirene Mort, John Black, Adam Forster, Margaret Stones, Rica Erickson and Margaret Menadue.

Those who illustrated by default are, in the main, scientists who illustrated their own works. Winifred Curtis, Nancy Burbidge and many, many others (most not recorded here), found that lack of funds forced them into executing their own illustration.

Some have endured great privation, even death, for the cause. Sydney Parkinson lost his life in the course of duty. For lack of botanical work, many performed other work rather than starve. Ludwig Becker kept body and soul together with miniature portrait painting. He later died on the Burke and Wills Expedition. Friederich Schoenfeld tried to keep himself going with teaching, but succumbed to despair and committed suicide. Even today, it is very hard to maintain botanical illustration as a full-time career. Most artists have to have several additional sources of income, including teaching. Very few have been able to use a career in botanical illustration to support another of their artistic capacities. Ludwik Dutkiewicz did this very successfully.

Private training schools are now available in a few capital cities. Jennifer Phillips began teaching botanical art at the National Herbarium of Victoria. More recently Phillips has set up her own school, the Botanical Art School of Melbourne, and has had considerable influence on the quantity and quality of work produced in Victoria. Christine Payne runs classes in Canberra and Margaret Saul teaches in Brisbane.

In 1998 and 1999 Dr Shirley Sherwood ran a Master Botanical Art Course in Sydney in conjunction with an exhibition of her Contemporary Botanical Art Collection. Jennifer Phillips was the teacher. Dr Sherwood has become a promoter of the world's botanical artists and has written a book on the subject based on her own private collection — *Contemporary Botanical Artists* (1996).

Botanical artists and illustrators have united in some states to form societies. In Melbourne in 1990 the Friends of the Royal Botanic Gardens began Botanical Art Workshops from where a Botanical Illustrators group developed. This group now runs a biennial exhibition of botanical illustration. Western Australia, Queensland, New South Wales and the Australian Capital Territory all now have similar groups and activities.

Formal training schools, botanical artist and illustrator societies and associated exhibitions will inevitably result in rising standards. Already some of our artists are

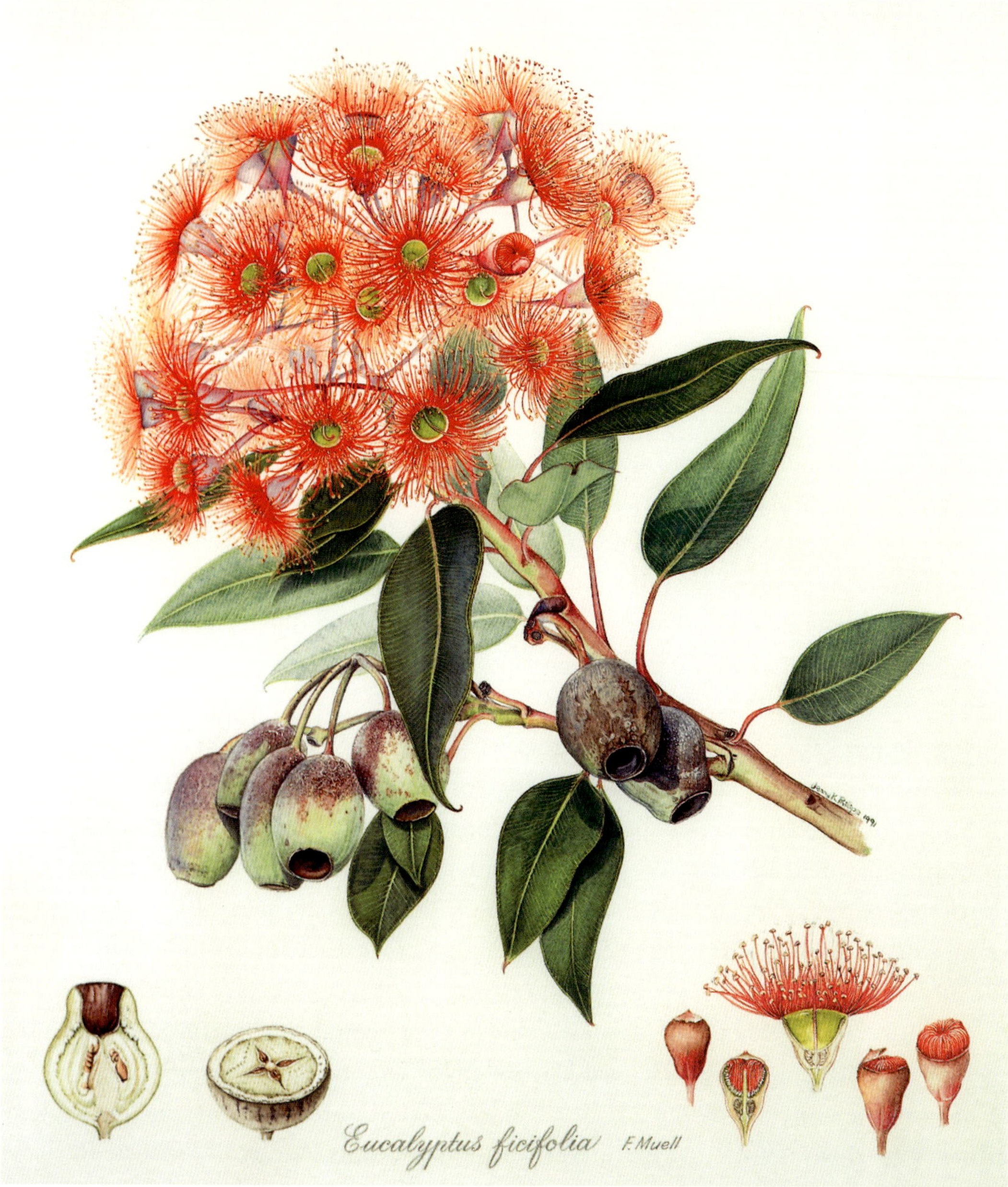

Eucalyptus ficifolia F.Muell.
Watercolour by Jennifer Phillips, painted in 1991.
[PC 36.0 x 29.8]

competing successfully at the international level for work and for awards. Others are challenging the illustration tradition developed two hundred years ago by deliberately moving illustration back into the realms of art. Where this revitalisation and enthusiasm will lead the art-form in the future can only be speculated on. One hopes that in both senses of speculation — contemplation and investment — the outcomes will be rewarding.

Endnotes

Preface

1 Frans Stafleu, *Linnaeus and the Linnaeans*, IAPT, Utrecht, 1971.

Chapter 1: The Art of Illustration

1 Carl Linnaeus in *Philosophia Botanica*, 1751, translated by Frans Stafleu in *Linnaeus and the Linnaeans*, IAPT, Utrecht, 1971.

2 W. Blunt and W.T. Stearn, *The Art of Botanical Illustration*, Antique Collectors Club and The Royal Botanic Gardens, Woodbridge, 1994.

3 Ferdinand von Mueller, *Transactions of the Philosophical Institute of Victoria* 2: plate between 72–73, 1858. The plate was supervised by Ludwig Becker and the imprint taken by H. Friend.

Chapter 2: Discovery

1 Sir Rutherford Robertson, Foreword in D.J. Carr (ed.), *Sydney Parkinson*, BM(NH) & ANU Press, Canberra, 1983.

2 Abel Tasman (1603–1659) was commemorated with the genus *Tasmannia* (Winteraceae) by Robert Brown ex A. P. de Candolle in 1817.

3 Anthony van Diemen (fl. 1636–1645) was commemorated with the family and genus *Vandiemenia* (Vandiemenaceae) by Helen Hewson in 1982.

4 William Dampier, *A Voyage to New Holland, &c In the Year, 1699,* James Knapton, London, 1703.

5 The bulk of the plant collection, which is now in the Sheradian Herbarium at Oxford, is of great historic interest. William Dampier (1652–1715) collected *Dampiera incana* R.Br., the type of the genus named in his honour by Robert Brown in 1810. This specimen is the only one with significant scientific value. The drawing of *Diplolaena dampieri* Desf. (= *D. grandiflora* Desf.) is the type of that species and the drawings of *Beaufortia dampieria* Cunn. ex Hook. (= *Beaufortia sprengelioides* (DC.) Craven) and *Clianthus dampieri* (= *Swainsona formosa* (G.Don) J. Thompson – Sturt's Desert Pea) form part of the type-description (protologue) of those species and all, therefore, have added scientific value.

6 A.S. George, 'A Tercentenary', *Australian Systematic Botany Society Newsletter* 89:20–21, 1996.
7 Frans Stafleu, op. cit.
8 George Ehret (1708–1770) was made a Fellow of the Royal Society in 1757.
9 James Cook (1728–1779) was commemorated with *Cookia* (Thymeliaceae) by J.F. Gmelin in 1792. This name is redundant due to the same name for a non-Australian member of the Rutaceae (named by P. Sonnerat in 1782) having priority. Sonnerat's name also honours James Cook. Cook was elected a Fellow of the Royal Society in 1776.
10 Joseph Banks was commemorated with the genus *Banksia* (nom. cons.) (Proteaceae) in 1782 by Carl Linnaeus jnr and with numerous specific epithets including some for species of algae.
11 Daniel Solander was commemorated with several species including *Banksia solandri* (Proteaceae) by Robert Brown in 1830.
12 Sydney Parkinson was commemorated with *Ficus parkinsonii* (Moraceae) by W.P. Hiern in 1901.
13 Daniel Solander was appointed Keeper, Natural History Department of the British Museum, in 1773.
14 Joseph Dalton Hooker (ed.), *Journal of the Right Hon. Sir Joseph Banks*, Macmillan, New York, 1896.
15 Sydney Parkinson, *A Journal of a Voyage to the South Seas*, Stanfield Parkinson, London, 1773.
16 James Cook lost forty-one men in total, thirty-three to tropical disease contracted in Batavia.
17 Sir Rutherford Robertson, loc. cit.
18 Now known as *Allocasuarina torulosa* (Aiton) L. Johnson, and *Allocasuarina verticillata* (Lam.) L. Johnson.
19 The prolific engravers were Daniel MacKenzie (fl. 1775–1800) – 251 plates, Gerard Sibelius (fl. 1759–1785) – 195 plates, Gabriel Smith (1724–1783) – 118 plates. Charles White (1751–1785), Robert Blyth (1750–1784), Frederick Nodder, William Tringham, Jabez Goldar, van Drazowa, Thomas Scratchley, John Lee, Jean-Baptiste Michell, William Smith, Edward Walker, John Roberts, Thomas Morris, Bannerman Chesham and Francis Chesham (all fl. 1770s), all made smaller contributions.
20 Johann Reinhold Forster (1729–1798) and Johann George Forster (1764–1794) were both Fellows of the Royal Society and were commemorated with the genera *Forstera* by Carl Linnaeus jnr in 1780 and *Forsteropsis* by O.W. Sonder in 1845, both in the Stylidiaceae.
21 William Anderson was commemorated by *Andersonia* (Epacridaceae) by Robert Brown in 1810.
22 David Nelson was commemorated by *Nelsonia* (Acanthaceae) by Robert Brown in 1810.
23 Archibald Menzies (1754–1842) became a Fellow of the Linnean Society in 1790 and is commemorated by several Western Australian species, e.g. *Drosera menziesii* (Droseraceae), in this case by Alphonse de Candolle in 1824.
24 C. Koenig & J. Sims, *Annals of Botany* 2:366,1806.
25 Charles-Louis L'Héritier was commemorated with *Heritiera* (Sterculiaceae) by William Aiton in 1789.
26 Jonas Dryander was commemorated with the genus *Dryandra* (Proteaceae) by Robert Brown in 1810.

Chapter 3: Colonisation

1 Joseph Banks, 'Some account of that part of New Holland now called New South Wales', *Journal*, 1770.
2 John White, *Journal of a Voyage to New South Wales*, Debrett, London, 1790. John White (c.1756–1832) was made a Fellow of the Linnean Society in 1796.
3 James Smith was founder of the Linnean Society, a Fellow of the Royal Society and knighted in 1814. Smith was commemorated with the genus *Smithia* (Fabaceae) by William Aiton in 1789.
4 J. Calaby in B. Smith and A. Wheeler, *The Art of the First Fleet & other early Australian drawings*, OUP, Oxford, 1988.
5 Thomas Watling, quoted from J. Hackforth-Jones, *The Convict Artists*, Macmillan, Melbourne, 1977.
6 James Sowerby (1757–1822) was a Fellow of the Linnean Society and was commemorated with the genus *Sowerbaea* (Liliaceae) by James Smith in 1798.

7 *The Hunter Sketchbook* (ed. J. Calaby) was published for the first time in 1988 by the National Library of Australia, which now owns the original.

8 Commemorations arising from Flinders' voyage are many. For Matthew Flinders (1774–1814) the genus *Flindersia* (Rutaceae) was named by Robert Brown in 1814.

9 Thomas Watling, quoted from C.M. Finney, *To Sail Beyond the Sunset*, Rigby, Adelaide, 1984.

10 For Robert Brown the genus *Brunonia* (Brunoniaceae) was named by J.E. Smith ex R. Brown in 1810, and *Brunoniella* (Acanthaceae) by C. Bremekamp in 1964. Brown also had academic recognition by being made a Fellow of both the Linnean Society and Royal Society.

11 Ferdinand Bauer has numerous commemorations. Among them is the genus *Bauera* (Cunoniaceae) by H.C. Andrews in 1801 (also for Franz Bauer). *Bauerella* (Rutaceae) by A. Borzi in 1897, and *Bauerella* (Fabaceae) by A.K. Schindler in 1926, are both illegitimate names, the latter being 'legitimised' with *Baueropsis* by J. Hutchinson in 1964.

12 Peter Good (fl.1801–1803) had the genus *Goodia* (Fabaceae) named in his honour by Richard Anthony Salisbury in 1806.

13 Robert Brown letter to Joseph Banks, 6 August 1803, quoted from Phyllis Edwards, 'Botany of the Flinders voyage' in D.J. Carr and S.G.M. Carr, *People and Plants in Australia*, Academic Press, Sydney, 1981.

14 William Paterson (1755–1810) had *Patersonia* (Iridaceae) named for him by Robert Brown in 1807. William Paterson also had academic recognition by being made a Fellow of both the Linnean Society and Royal Society.

15 Stephan Ladislaus Endlicher (1804–1849) published *Prodromus Florae Norfolkicae* (1833) largely based on Bauer's collections and drawings. The drawings are cited and given plate numbers (117 in all), but were not published with the *Prodromus*. A few of the plates were published in other works, but the bulk of them remain archived in Vienna.

16 John Lindley, *Gardener's Chronicle*, 1844.

17 Most of Ferdinand Bauer's original drawings are now at the Naturhistorisches Museum, Vienna, and most of the original paintings are at the Natural History Museum in London.

18 In John Sibthorp's *Flora Graeca*, 1806; A.B. Lambert's *A description of the genus Cinchona*, 1797, and *A Description of the genus Pinus*, 1803, 1824; John Lindley's *Digitalium Monographia*, 1821; and a few other minor contributions as reported by C. Nissen. See Claus Nissen, *Die Botanische Buchillustration*, Hiersemann Verlags-Gesellschaft, Stuttgart, 1951, 1966.

19 William Withering (1741–1799) was a physician with an interest in natural history and botany.

20 Frans Stafleu, op. cit.

21 Walter Lack, 'Recording Form in Early Nineteenth Century Botanical Drawing. Ferdinand Bauer's "Cameras"', *Curtis's Botanical Magazine* 15 (4): 254–274, 1998.

Chapter 4: The French Impression

1 Jacques J.H. Labillardière, *Voyage in search of la Pérouse*, John Stockdale, London, 1800.

2 Étienne Ventenat honoured Boissieu de la Martinière (fl.1788) with *Bossiaea* (Fabaceae) in 1800. La Martinière accompanied La Pérouse as a naturalist on his voyage.

3 Jacques J.H. Labillardière was commemorated with *Billardiera* (Pittosporaceae) by James Smith in 1793.

4 Louis Deschamps (1765–1845) was honoured with *Deschampsia* (Poaceae) by P. Beauvois in 1812.

5 Claude Riche (1762–1779) was honoured with *Richea* (Epacridaceae) by Robert Brown in 1810.

6 Jacques J.H. Labillardière, op. cit.

7 Frans Stafleu, *Adanson, Labillardière, De Candolle*, Verlag Von J. Cramer, Lehre, 1967.

8 Louis Anemaat, *Dare to Know*, State Library of New South Wales, Sydney, 1998.

9 The original drawings for Jacques Labillardière's *Sertum Austro-Caledonicum* are bound plate-by-plate in the Royal Horticultural Society's Lindley Library copy.

10 Jean Leschenault (1773–1826) was honoured with *Latouria* (Goodeniaceae) by Stephan Endlicher (1838) and with *Leschenaultia* by Robert Brown (1810).

11 Louis de Freycinet (1779–1842) was honoured with *Freycinetia* (Pandanaceae) by C. Gaudichaud-Beaupré (1824).

12 Nicolas Baudin, *The Journal of Post Captain Nicolas Baudin*, translated by C. Cornell, Libraries Board of South Australia, Adelaide, 1974.
13 Étienne Ventenat (1757–1808) was honoured with *Ventenatia* (Stylidiaceae) by James Smith in 1804.
14 Antonio Cavanilles honoured Sohn Sellier (1757–1835) with *Selliera* (Goodeniaceae) in 1799.
15 Étienne Ventenat honoured Empress Joséphine with *Josephinia* (Pedaliaceae) in 1804.
16 Aimé Bonpland (1773–1858), who had met Pierre-Jean-François Turpin in America, teamed with Alexander Humbolt (1769–1859) in several publications for which Turpin provided illustrations.
17 The engravers for Australian plants in Aimé Bonpland's, *Description des plantes rares...* were R. Bessin (fl. 1812–1817) (2), Louis Bouquet (fl. 1800–1817) (8), Jean Louis Denis Coutant (1776–1831) (4), de Gouy (fl. 1800–1820) (1), Antoine Maxime Monsaldi (1768–1816) (3), and François Plée (fl. 1812–1864, son of Auguste) (1); two are unsigned.
18 Claus Nissen, in *Die Botanische Buchillustration* (1966), reported that Wendland did both the drawing and the etching.

Chapter 5: The Banksian Collectors

1 Ann Moyal, *Scientists in Nineteenth Australia: a documentary history*, Cassell Australia, Sydney, 1976.
2 Franz Bauer was made a fellow of both the Royal and Linnean Societies. *Bauera* (Cunoniaceae) was named in honour of both the Bauer brothers, in 1801, by Henry Charles Andrews based on a manuscript name given by Joseph Banks.
3 The Royal Horticultural Society was founded in 1804 and took a strong and active role in having plants recorded by commissioning artists to paint them. *Curtis's Botanical Magazine* was therefore entirely appropriate as an acquisition by the Society.
4 Henry Andrews (1794–1830) was honoured with *Andreusia* (Myoporaceae) by Étienne Ventenat in 1805.
5 Denis Considen (fl.1788–1815) had *Eucalyptus considenіana* (Myrtaceae) named in his honour by Joseph Henry Maiden in 1904.
6 Quoted from J.H. Maiden, *Sir Joseph Banks*, Government Printer, Sydney, 1909.
7 David Burton (fl.1791–1792) had *Burtonia* (Dilleniaceae) and *Burtonia* (Fabaceae) named in his honour by Richard Anthony Salisbury (1807) and Robert Brown (1811), respectively. Neither name is in current use, the former being an *Hibbertia* and the latter a *Gompholobium*.
8 George Suttor was recognised by the scientific community by being made a Fellow of the Linnean Society in 1843.
9 Quoted in J.H. Maiden, op. cit.
10 George Caley was honoured with *Caleana* (Orchidaceae) by Robert Brown in 1810.
11 F.A. Stafleu and R.S. Cowan, *Taxonomic Literature*, Bohn, Scheltema & Holkema, Utrecht, 1976.
12 Sydenham Edwards was not a botanist, and did not keep an herbarium, but he succeeded in becoming a Fellow of the Linnean Society of London in 1804.
13 John Bellenden Ker(-Gawler) was honoured with *Bellendena* (Proteaceae) by Robert Brown in 1810.
14 George Loddiges (1784–1846) was made a Fellow of the Linnean Society in 1821.
15 Allan Cunningham and his brother Richard were honoured with *Cunninghamia* (Cupressaceae) by Robert Brown in 1826, now regarded as *Athrotaxis, Alania* (Liliaceae) by S.L. Endlicher in 1836, and with several species — *Casuarina cunninghamiana* and *Araucaria cunninghamii* — being well-known trees.
16 John Oxley (1783–1828) was honoured with *Oxleya* (Rutaceae) by William Hooker in 1830.
17 Charles Fraser (1788–1831) was honoured with numerous species including *Casuarina fraseriana* (Casuarinaceae) by F.A.W. Miquel in 1848. This is now known as *Allocasuarina fraseriana* (Miq.) L. Johnson.
18 Phillip Parker King (1791–1856) and his father were honoured with *Kingia* (Xanthorrhoeaceae) by Robert Brown in 1826. Philip Parker was a Fellow of both the Linnean and Royal Societies.
19 Louis Duperry (1786–1865) was honoured with *Duperrya* (Convolvulaceae) by C. Gaudichaud-Beaupré in 1828.
20 James E. Smith in *Exotic Botany*, Vol. 1, R. Taylor and Co., London, 1804.

Chapter 6: Hooker, Lindley and the Private Collectors

1 Allan Cunningham, 1827. Quoted from K. Fitzpatrick, *Australian Explorers*, OUP, London, 1965.

2 William Hooker became a Fellow of the Linnean Society in 1806, a Fellow of the Royal Society in 1812, and knighted in 1836.

3 Robert Greville became a Fellow of the Linnean Society and was honoured with the genus *Kayea* (Clusiaceae) by N. Wallich in 1831. *Grevillea* (Proteaceae) was named in honour of Charles Francis Greville by J. Knight in 1809, for his being a founder of the Horticultural Society of London and for his work in introducing and cultivating rare plants – not for Robert, as erroneously reported by R. Desmond in *Dictionary of British and Irish Botanists and Horticulturists* (1977, 1994).

4 John Lindley became a Fellow of both the Linnean Society and the Royal Society, and was honoured with many Australian species. An early example is *Eucalyptus lindleyi* (Myrtaceae) named for him by Alphonse de Candolle in 1829.

5 Walter Fitch was made a Fellow of the Linnean Society in 1857. William Hooker honoured him with a non-Australian genus in 1845. However, in 1855, C.F. Meissner honoured him with *Fitchia* (Proteaceae). This is now known as a species of *Grevillea*. In 1992 Jan Lewis had a beautifully illustrated book, *Walter Hood Fitch: a Celebration*, published to honour her great-grandfather.

6 Robert Sweet (?1782–1835) was elected a Fellow of the Linnean Society in 1812.

7 F.A. Stafleu and R.S. Cowan, *Taxonomic Literature*, Bohn, Scheltema & Holkema, Utrecht, and W. Junk, The Hague, 1983, Volume 4.

8 William Baxter (fl.1823–1830) was honoured with *Baxteria* (Xanthorrhoeaceae) by William Hooker in 1843.

9 Franz Sieber (1789–1844) was honoured with *Siebera* (Apiaceae) by H.G.L. Reichenbach in 1828.

10 From a letter quoted in J.H. Maiden, op. cit.

11 Thomas Mitchell became a Fellow of the Royal Society in 1839. *Capparis mitchellii* (Capparaceae), named by John Lindley in 1838, is one of several species named in his honour. Perhaps, however, Mitchell's name is kept more to the fore with Mitchell Grass (*Astrebla pectinata* F. Muell.), the Major Mitchell Cockatoo which he very ably illustrated in his journal, and with the township of Mitchell in Queensland.

12 *Nepenthes kennedyanum* (Nepentheaceae) was named for Edmund Kennedy (1818–1848) by Ferdinand von Mueller in 1866. It is now regarded as being embraced in the variation of *Nepenthes mutabilis*.

13 Charles Sturt (1795–1869) was elected a Fellow of the Linnean Society in 1833. The genus *Sturtia* (Malvaceae) was named for Sturt by Robert Brown in 1849 in the Botanical Appendix to Sturt's *Narrative*, the type having been collected by Sturt.

14 Charles Sturt published *Two expeditions into the interior of Southern Australia* (1833) based on his journals. He published on the later expedition in a *Narrative of an expedition into Central Australia* (1849). The latter has a Botanical Appendix which was written by Robert Brown.

15 John Bidwill (1815–1853) has the special honour of having one of Australia's finest trees, the Bunya Pine, *Araucaria bidwillii* (Araucariaceae), named for him by William Hooker in 1843. Bidwill observed that this tree is different from *Araucaria cunninghamii*, collected the type and, together with copious notes, sent his collection to Hooker. The genus *Bidwillia* (Liliaceae) was also named for him by W. Herbert in 1844.

16 Ludwig Leichhardt (1813–1848) was honoured with *Leichhardtia* (Asclepiadaceae) by Robert Brown in 1849 in the Botanical Appendix to Sturt's *Narrative; Leichhardtia* (Cupressaceae) by T.W. Shepherd in 1851; and *Leichhardtia* (Euphorbiaceae) by von Mueller in 1876.

17 John Gilbert is honoured with a few species including *Acacia gilbertii* (Mimosaceae) by C.F. Meissner in 1848.

18 Quoted from Ann Moyal, op. cit.

19 Paul Strzelecki (?1796–1873) had *Strzeleckya* (Rutaceae) named in his honour by von Mueller in 1857.

20 John Eyre (1815–1901) had *Eyrea* (Asteraceae) named in his honour by von Mueller in 1853.

CHAPTER 7: ESTABLISHMENT

1 Joseph Hooker, *Flora Tasmaniae*, Lovell Reeve, London, 1860.
2 Robert Lawrence (1807–1833) had *Lawrencia* (Malvaceae) named in his honour by William Hooker in 1836.
3 Ronald Gunn (1808–1881) was commemorated with *Gunnia* (Orchidaceae) by John Lindley in 1834. *Gunnia* of F. Muell. (1859), *Gunniopsis* of F.A. Pax (1889) and *Neogunnia* of F.A. Pax and K. Hoffman (1934), all in the Aizoaceae, were named for him too.
4 Joseph Hooker was a Fellow of both the Linnean Society and the Royal Society. In 1877 he was knighted. He was honoured with *Hookerochloa* (Poaceae) by E.B. Alexeev in 1985, and with numerous species.
5 Joseph Milligan (1807–1884) was honoured with *Milligania* (Liliaceae) by Joseph Hooker in 1853.
6 James Backhouse (1794–1869) was honoured with *Backhousia* (Myrtaceae) by William Hooker in 1845.
7 Johann Lhotsky (1795–1866) was honoured with *Lhotskya* (Myrtaceae) by J.C. Schauer in 1836.
8 Joseph Hooker, op. cit., Introductory Essay.
9 Augustus Oldfield (1820–1887) was commemorated with several species including *Eucalyptus oldfieldii* (Myrtaceae) by von Mueller in 1860.
10 Joseph Hooker, loc. cit.
11 William Archer was a Fellow of the Linnean Society and was honoured with *Archeria* (Epacridaceae) by Joseph Hooker in 1857.
12 Joseph Hooker, op. cit., Preface.
13 James Stirling (1791)–1865) had *Stirlingia* name in his honour by Stephan Endlicher in 1837.
14 *Drummondita* (Rutaceae) was named for James Drummond and his brother Thomas by William Harvey in 1855. In naming this genus Harvey stated:

This genus is dedicated to the brothers Thomas and James Drummond, two of the ablest and most indefatigable of botanical collectors and explorers; the one in North America, the other in Western Australia. That such men deserve all the honours our science can bestow, all will agree; but as yet no universally adopted genus has been assigned to either. ... Lest however there should be any confusion, I have adopted the termination 'ita,' – being an I for James, and a T for Thomas!

It transpires that, somewhat egotistically, James Drummond especially asked that this genus be named for himself, a request which Harvey honoured. Ironically, this is now embraced within *Philotheca*.

15 James Mangles (1786–1867) was a Fellow of the Royal Society. *Manglesia* (Proteaceae) was named for both James and Robert Mangles (fl.1830s–1860) by Stephan Endlicher in 1839 (now embraced within *Grevillea*), however Mangles is probably better known through *Anigozanthos manglesii* (Haemodoraceae), the Mangles Kangaroo Paw or Red and Green Kangaroo Paw, the emblem of Western Australia, named for James by D. Don in 1834.
16 R. Roberts and M. Whitten, Biographies of Australian Botanists, manuscript.
17 John Lindley in an Appendix to *Edwards's Botanical Register* 23 (1839).
18 *Molloya* (Proteaceae) was named for Georgiana Molloy by C.F. Meissner in 1855 as a replacement name for *Fitchia* of C.F. Meissner (this having been originally named for William Fitch). It is now embraced within *Grevillea*.
19 Quoted from A. Hasluck, *Portrait without background*, Fremantle Arts Centre Press, Perth, 1990.
20 John Roe (1797–1878) was honoured with the genus *Roea* (Fabaceae) by George Bentham in 1837. In an attempt to correct a nomenclatural problem, *Roea* was renamed *Hugelroea* for both Hügel and Roe by E.G. Steudel in 1840. This is now known as *Sphaerolobium*.
21 Carl Hügel (1795–1870) was honoured with *Huegelia* (Apiaceae) by H.G.L. Reichenbach in 1830 and by Stephan Endlicher in 1840, and with *Hugelroea* (Fabaceae) by E.G. Steudel in 1840. None of these names is in current use. However, *Hardenbergia* (Fabaceae) was named in honour of Hügel's sister, Francisca Hardenberg von Hügel, by George Bentham in 1837.
22 Ludwig Preiss (1811–1883) was honoured with many species and with the genus *Neopreissia* (Chenopodiaceae) named for him by E. Ulbrich in 1934.

23 Charles Moore (1820–1905) was made a Fellow of the Linnean Society in 1863 and was honoured with several species including *Eucryphia moorei* (Eucryphiaceae) by von Mueller in 1863.

24 John Dallachy was honoured with *Dallachya* (Rhamnaceae) by von Mueller in 1875.

25 Louis B. Van Houtte (ed.), C. Lemaire, & M. Scheidweiler, *Flore des Serres et des Jardins de l'Europe*, ou descriptions et figures des plantes les plus rares et les plus mértantes, nouvellement introduites sur le continent ou en Angleterre, et extraites notament des Botanical Magazine, Botanical Register, Paxton's Magazine of Botany, etc., etc., etc.; ouvrage publié en allemand, en français et en anglais, enrichi de notices historiques, scientifiques, étymologiques, synonymiques, horticulturales, etc.1845–1883.

Chapter 8: Ferdinand von Mueller

1 Joseph Hooker in a letter to Ferdinand von Mueller, 1857, quoted from A.M. Moyal, op. cit.

2 Several species of both algae and flowering plants were named in William Harvey's honour, the latter by George Bentham and Ferdinand von Mueller, for example *Verticbordia harveyi* Benth. (Myrtaceae) in 1867.

3 By 1858 von Mueller had forty-nine papers published.

4 George Bentham (1800–1884) was honoured with *Benthamina* (Loranthaceae) by P.E.L. van Tieghem in 1895. He is also honoured with several species named for him.

5 G. Bentham, *Flora Australiensis*, Preface, Volume I, Lovell Reeve, London, 1863.

6 Euphemia Henderson was honoured with *Nematolepis euphemiae* (Rutaceae) by von Mueller in 1863. His honorific stated:
Speciem perrarum singularum et cultura dignum dicavi ingeniosae et praecomi Euphermiae Henderson, quae nuperius plures algarum species in insula Phillip Island detexit, et arte nunc ad illustrandam floram indigenum impendenda complurium plantarum eleganter pinxit imagines.
The description included a lithograph executed by Friederich Schoenfeld. The species is now treated as a *Rhadinothamnus*.

7 In *The Tasmanian Journal of Natural Science, Agriculture. Statistics*, &c. 3: plate opp. 72 (1849). The drawing of *Richea*, a lithograph, was done by James Grant (fl. 1840s) for Ronald Gunn and printed by Henry Dowling, Launceston.

8 B.V. McCleary, *National Geographic*, March 1995. Thiaminate in nardoo destroys vitamin B1 (thiamine), the resulting deficiency causing beriberi.

9 Von Mueller named *Eremophila beckeri* (Myoporaceae) for Ludwig Becker in 1859, stating 'Speciem designavi nomine amici Ludovici Becker, qui plures observationes in regnum animalium et vegetalium terrae australis institut, et nostrae faunae et florae cognitionem delineationibus pilchris et fidelissimis auxit.'

10 *Antennaria nubigena* var. *meredithae* (Asteraceae) was named for Louisa Meredith by von Mueller in 1871. Meredith (1812–1895) was also made an honorary member of the Royal Society of Tasmania for achievements in art, literature and science.

11 F. von Mueller, *Fragmenta Phytographiae Australiae*, Volume 8, Government Printer, Melbourne, 1872–1874.

12 Helene (Scott) Forde (1832–1910) was honoured with *Poa fordeana* (Poaceae) by von Mueller in 1873.

13 Louisa Atkinson (1834–1872) was honoured with the mistletoe, *Atkinsonia* (Loranthaceae), by von Mueller in 1865. Both Louisa and John Calvert were honoured with *Epacris calvertiana* by von Mueller in 1873.

14 *Phaius grandiflorus* var. *rowanae* (Orchidaceae) was named in honour of Ellis Rowan (1848–1922) by Frederick Bailey in 1912. Not only did Bailey name the variety after Rowan but he also used her painting as the type.

15 See E.L. Robertson, *Alison Ashby's Wildflowers of Southern Australia*, South Australian Museum Board, Adelaide, 1981.

16 Quoted from B. Archer and S. Maroske in *The Mueller Issue, The Victorian Naturalist* 113 (4):190,1996.

17 Amalie Dietrich (?1821–1891) was honoured by having several Australian species named for her. One of these is *Acacia dietrichiana* (Mimosaceae) named for her by von Mueller in 1882. Several species of Cyperaceae were named for her by J.O. Boeckler, a German botanist.

18 *Pimelea husseyana* (Thymelaeaceae) was named in Jessie Hussey's (1862–1899) honour by Alfred Ewart in 1921. Two seaweeds are also named in her honour.

19 *Erysimum richardsii* (Brassicaceae) was named for Annie Richards (1873–1894) in 1877 by von Mueller.

20 The genus *Wehlia* (Myrtaceae) was named in honour of Clara Wehl (1873–1901) and her husband in 1876 by von Mueller.

21 Sarah Brooks (1850–1928) had a species of *Hakea* (Proteaceae) and a species of *Scaevola* (Goodeniaceae) named in her honour by von Mueller. Like Jessie Hussey, Brooks also collected seaweed and had a species named in her honour. Brooks also drew plants.

22 William Woolls (1814–1893) was commemorated with *Woollsia* (Epacridaceae) by von Mueller in 1873.

23 Quoted from L.A. Gilbert in D.J. Carr and S.G.M. Carr, op.cit.

24 Emil Todt (c.1810–1900) had *Eucalyptus todtiana* (Myrtaceae) named for him by von Mueller in 1882.

25 F. von Mueller, *Eucalyptographia*, Government Printer, Melbourne, 1879–1884.

26 Robert Graff (?1841–1914) had *Acacia graffiana* (Mimosaceae) named for him by von Mueller in 1887. This species is now regarded as being synonymous with *Acacia hemetiles* Benth.

27 F. von Mueller, *Iconography of Australian Species of Acacia*, Government Printer, Melbourne, 1887–1888.

28 A woodcut published in *The Victorian Naturalist* (10:195–196 (1894)) was attributed to Robert Graff and confirms Graff's capability. Johan Paul Eckert (1861–1924) was also a competent woodcut engraver. He had work published in the *Agriculture Journal of Victoria* (1: 716–722 (1902)).

29 F. von Mueller, *The Native Plants of Victoria*, Preface, Government Printer, Melbourne, 1879.

30 The woodcuts are not usually acknowledged in any of these works so most engravers will remain anonymous.

31 F. von Mueller, ibid.

32 Robert Fitzgerald (1830–1892) was a Fellow of the Linnean Society and is honoured with *Dracophyllum fitzgeraldii* (Epacridaceae) by von Mueller in 1869. Von Mueller later (1882) named the genus *Fitzgeraldia* (Orchidaceae) in his honour. The name is illegitimate and the plant a *Leperanthus*. Several orchid species have been named in Fitzgerald's honour by other authors.

33 Frederick Bailey became a Fellow of the Linnean Society in 1878. *Austrobaileya* (Austrobaileyaceae) was named for him in 1933 by C.T. White, a grandson of Bailey.

34 Karl Staiger, honoured with *Eucalyptus staigeriana* (Myrtaceae) by Frederick Bailey in 1883, was Government Analytical Chemist and had particular research interests in plant poisons and essential oils. *E. staigeriana* has a particularly high oil content and has a strong 'fragrance of lemons'.

35 F.M. Bailey, *Lithograms of the Ferns of Queensland*, Government Printer, Brisbane, 1892.

36 F.M. Bailey and P.R. Gordon, *Plants Reputed Poisonous and Injurious to Stock*, Government Printer, Brisbane, 1887.

37 John Brown honoured Emanuel Spiller (?1824–1888) with *Acacia spilleriana* (Mimosaceae) in 1886. Why Fiveash and Barrett were not honoured, when their skill is what made the publication so very special, remains a mystery. Of Barrett is written, 'J.E. Brown *direxit*, H. Barrett *docet*, and does it well.'

38 Gertrude Lovegrove does not seem to have been honoured but Bauerlen has several taxa named in his honour, one being *Elaeocarpus baeuerlenii* (Elaeocarpaceae) by Joseph Maiden and Ernst Betche in 1896.

39 K.L. Wilson in P.S. Short (ed.), *History of Systematic Botany in Australia*, Australian Systematic Botany Society, Melbourne, 1990.

CHAPTER 9: JOSEPH MAIDEN AND THE FLORAS

1 Joseph Maiden in J.H. Maiden & E. Betche, *A Census of New South Wales Plants*, Government Printer, Sydney, 1916.

2 Maiden was a Fellow of both the Linnean Society and the Royal Society. He was honoured with *Maidenia* (Hydrocharitaceae) by A.B. Rendle in 1916 and with *Maidenia* (Apiaceae) by Karel Domin in 1922. He is also honoured with numerous species names.

3 J.H. Maiden, *A Critical Revision of the Genus Eucalyptus*, Government Printer, Sydney, 1909–1933.

4 *Geijera helmsiae* (Rutaceae) was named in honour of Mrs R. Helms (fl.1890s) by Frederick Bailey.

5 Numerous taxa have been named for Leonard Rodway, including *Eucalyptus rodwayi* (Myrtaceae), by R. Baker and H.G. Smith in 1912.

6 L. Rodway, *The Tasmanian Flora*, Preface, Government Printer, Hobart, 1903.

7 J.M. Black, *The Naturalised Flora of South Australia*, Preface, J.M. Black, Adelaide, 1909.

8 John Black has been honoured with several plant names including *Goodenia blackiana* (Goodeniaceae), named for him by R. Carolin in 1986.

9 Richard Baker (1854–1941) had *Acacia bakeri* (Mimosaceae) named in his honour by Joseph Maiden in 1896.

10 Several species are named in honour of Max Koch (1854–1925), including *Zygophyllum kochii* (Zygophyllaceae) by Ralph Tate in 1899.

11 Ludwig Diels (1874–1945) was honoured with *Dielsia* (Restionaceae) by E.F. Gilg in 1904 and *Dieltzia* (Asteraceae) by Philip Short in 1989.

12 Quoted from D.J. Carr's translation of Ludwig Diels, *Plant World of Western Australia*, 1906, in D.J. Carr and S.G.M. Carr, op. cit.

13 Karel Domin (1882–1953) had *Dominia* (Apiaceae) named in his honour by F. Fedde in 1929. This is now regarded as being *Uldinia*. Numerous species have been named in Domin's honour.

14 J.H. Maiden, Introductory, *The flowering plants and ferns of New South Wales*, Government Printer, Sydney.

15 William Blakely (1875–1941) had *Eucalyptus blakelyi* (Myrtaceae) named in his honour by Joseph Maiden in 1917.

16 Joseph Maiden honoured Margaret Flockton (1861–1953) with *Eucalyptus flocktoniae* (Myrtaceae) and *Acacia flocktoniae* (Mimosaceae) in 1911 and 1916 respectively.

17 N. Hall, *Botanists of the Eucalypts*, Commonwealth Scientific and Industrial Research Organisation, Melbourne, 1978.

18 *Cyrilwhitea* (Chenopodiaceae) was named for Cyril White (1890–1950) by E.H. Ising in 1964. This is now regarded as a *Sclerolaena*.

19 Mrs R.W. Garraway (fl. early 1900s) was honoured by Frederick Bailey with *Citrus garrawayi* (Rutaceae) in 1904.

20 Alfred Ewart (1872–1937) was a Fellow of both the Linnean Society and the Royal Society. He had *Ewartia* (Asteraceae) named in his honour by G. Beauverd in 1911.

21 Alfred Ewart appears to have honoured both Olive Davies and her husband with *Rossittia* (Rutaceae) in the Northern Territory Flora (1917), though Ewart did not state that this was his intent – perhaps it was a wedding present.

22 J. Isaacs, *The Gentle Arts: 200 years of Australian Women's Domestic & Decorative Arts*, Lansdowne Press, Willoughby, 1987.

23 A.J. Ewart, *Flora of Victoria*, Government Printer and Melbourne University Press, Melbourne, 1930.

24 G.M. Chippendale, *Eucalyptus Buds and Fruits*, Forestry and Timber Bureau, Canberra, 1968.

25 Charles Gardner (1896–1970) had *Eucalyptus gardneri* (Myrtaceae) named in his honour by Joseph Maiden in 1924.

26 W.H. Nicholls in D.L. Jones and T.B. Muir (eds), *Orchids of Australia*, Thomas Nelson (Australia), Melbourne, 1969.

CHAPTER 10: THE REVIVAL

1 George Lawrence (Director of the Hunt Botanical Library), in W. Curtis and M. Stones, *The Endemic Flora of Tasmania*, Part 1, The Ariel Press, London, 1967.

2 Winifred Curtis was honoured with *Winifredia* (Restionaceae) by L.A.S. Johnson and B.G. Briggs in 1986.

3 Margaret Stones was awarded an honorary Doctorate of Science by Melbourne University in 1989 for contributions to botany through botanical art and was awarded an MBE in Britain in 1977 and an OAM in the Order of Australia in 1988.

4 Sir Russell Grimwade (1879–1955) was honoured with *Prasophyllum grimwadeanum* (Orchidaceae) by W.H. Nicholls in 1947.

5 Nancy Burbidge (1912–1977) was honoured with several species names including, *Boerhavia burbidgeana* (Nyctaginaceae), named for her by Helen Hewson in 1984.

6 Commonwealth Scientific and Industrial Research Organisation.

7 W. Hartley, *Brunonia* 1: 123–129 (1977).

8 Celia Rosser was granted an honorary Masters Degree in Science in 1981 by Monash University and an OAM in the Order of Australia in 1995.

9 Alexander George was honoured with *Alexgeorgea* (Restionaceae) by S. Carlquist in 1976.

10 The first and second parts of *The Banksias* were published in 1981 and 1988. The third part was published in 1999. In 1993 Monash University published a booklet, also titled *The Banksias*, documenting the project and how the final artwork is achieved.

11 S. Kelly, *Eucalypts*, Volume 2, Preface, Thomas Nelson, West Melbourne, 1978. Stanley Kelley was awarded an Order of Australia Medal in 1981, mainly for services to the community and conservation rather than to botanical illustration.

12 *Nuytsia floribunda* (Labill.) R.Br. ex Fenzl.

13 *Telopea speciosissima* (Smith) R.Br.

14 Hansjoerg Eichler (1916–1992) was honoured with *Eichlerago* (Lamiaceae) by J. Carrick in 1977.

15 In addition to the forty-seven volumes treating the flowering plants, the *Flora of Australia* will have one volume treating the ferns and gymnosperms, three volumes treating the mosses and liverworts, and six volumes treating the lichens. Then there are the fungi and algae. Two additional volumes treat the flora of Australia's offshore island Territories.

16 P.S. Short (ed.), op. cit.

17 The *Flora of South Australia* (1986) ran to four volumes illustrated with sixty-four colour drawings and 1009 line drawings.

18 N.G. Marchant, J.R. Wheeler, B.L. Rye, E.M. Bennett, N.S. Lander and T.D. Macfarlane, *Flora of the Perth Region*, Western Australian Herbarium, Perth, 1987.

19 J.R. Wheeler (ed.), B.L. Rye, B.L. Koch and A.J.G. Wilson, *Flora of the Kimberley Region*, Western Australian Herbarium, Perth, 1992.

Botanical Nomenclature

The names given to plants are closely linked to their classification and how they are related to each other. A so-called binomial system is used. This means that each species has two names: its genus (a noun with a capital) followed by its species (an adjective without a capital), e.g. *Telopea speciosissima*, the Waratah of New South Wales. Currently there are five species recognised in the genus *Telopea*. The best known of the other four is *Telopea truncata*, the Tasmanian Waratah. Species of *Telopea* are more closely related to each other than they are to species of *Banksia* or *Grevillea*. These three genera (*Telopea, Banksia* and *Grevillea*), nevertheless, are all related and belong in the same family, the Proteaceae. They are not closely related to plants in other families, e.g. species of *Eucalyptus* in the family Myrtaceae.

The name of a plant is frequently written with the name of the person who first described it for science. Thus *Telopea oreades* F. Muell., the Gippsland Waratah, was first described by Ferdinand von Mueller.

The first given name has priority, but if the author did not have the relationship correct, then the name has to be changed. The problem of changing names relates to the naming rules and accidents of history.

Take *Telopea speciosissima* (Sm.) R.Br. as an example. The Waratah was first named *Embothrium speciosissimum* by James Edward Smith in 1793, shortly after the establishment of the colony at Port Jackson, i.e. *Embothrium speciosissimum* Sm. Smith described it from a specimen sent back by John White. However, after further study Robert Brown realised that the Australian Waratah is not in the South American genus *Embothrium*. In fact Brown had to describe the new genus in which our Waratah belongs. He called the genus *Telopea* and the Waratah became known as *Telopea speciosissima* (Sm.) R.Br. Smith, as the first to name the species, is remembered with his name in brackets and Brown, who got the relationship correct, has his name last. Both scientists' names are abbreviated, largely to save space.

In 1792 Luis Née, the botanist on the Spanish Malaspina Expedition to the southern hemisphere, collected around Port Jackson. Née collected a Waratah and it was taken back to Antonio Cavanilles in

Spain. Cavanilles had not seen Smith's publication and, thinking it to be new, named it *Embothrium spathulatum* Cav. in 1798. He missed out by five years and his name disappears as a synonym, being an accident of history.

The story for *Telopea truncata* (Labill.) R.Br. is similar. Labillardière described the Tasmanian Waratah in 1805, using specimens he had collected himself in 1792 and 1793, and named it *Embothrium truncatum* Labill. Robert Brown also corrected this and it became known as a *Telopea*. However, in 1835 J. Ross decided to name it *Telopea tasmaniana* Ross. Ross's name too is an accident of history and is a synonym and not used.

The names chosen for plants are Latinised and frequently reflect a quality of the plant. *Telopea* means 'seen from afar'; *speciosissima* means 'spectacular'; *spathulatum* describes the bracts at the base of the inflorescence of head; *truncatum* means 'ending abruptly' or 'cut-off' and describes the shape of the wing of the seed; and *tasmaniana* means 'belonging to Tasmania'.

Sometimes plants are named in honour of a person. For example, *Banksia* is named in honour of Sir Joseph Banks who was the naturalist on the James Cook voyage of discovery and exploration of the eastern coast of Australia. *Eucalyptus considenia*na Maiden, Yertchuk, was named by Joseph Maiden in honour of Denis Considen who was an assistant surgeon with the First Fleet and sent out to help set up the Penal Colony at Botany Bay. Considen was interested in the medicinal use of the local plants. He later gained a Doctorate of Medicine based on his discovery of the medicinal qualities of eucalyptus oil. Many collectors, botanists and artists have been honoured in this way. In this book selected attributions have been recorded for Australian plant honours only.

The concept of a *type* deserves some explanation. When a plant species is described for the first time, the plant specimen(s) used to make the description becomes the type (*holotype* if there was one collection, *syntype* if more than one collection was used). The type provides a 'standard' for the naming of species and as such is quite valuable. For taxonomy there is a set of rules governing the designation and recognition of types. These rules include what happens when the type-specimen is lost. Sometimes an illustration (published at the same time as the description) can be designated as a type, a *lectotype*. If all is lost, a new type (a *neotype*) will need to be chosen. Sometimes this can be an illustration. Such an example is *Dendrobium stuartii* which was described from horticultural material by Frederick Manson Bailey. No specimen was kept. However, Bailey sent a sample of the plant to Robert Fitzgerald. Fitzgerald, in turn, illustrated the plant. This illustration has been chosen as the new type of the species.

More rarely, a description is based on an illustration to start with. Bailey has also provided an example by naming an orchid variety, *Phaius grandiflora* var. *rowanae*, based on a painting by Ellis Rowan. Bailey had not seen any material of the plant, and so the painting became the type.

When a plant is described for the first time, the rules require that it be described in Latin and that the type be clearly specified. In the past, before the rules were established, scientists were less uniform in their approach. The word description, the plant specimens listed as having been seen, and any accompanying illustrations, all become part of the protologue (i.e., 'everything associated with a name at its first publication'.) This is how an illustration can become a type. Where all plant material of the type of a species is lost (common for herbaria, such as Berlin, destroyed during World War II), a drawing from the protologue may be selected as a type and become a lectotype.

Where a species has been too broadly defined, the botanist has to decide which species most ideally fits the original scientist's concept in order to define which species carries the original name and which plant is the type for that name. A drawing can sometimes help the resolution. For *Drakea elastica*, a woodcut in J. Lindley's *Appendix* was used to help make the decision.

Glossary

ad nat: natural or like nature. The illustration is drawn from fresh plant material.

botanical art: the art of portraying plants for a range of purposes from fine art through to science.

botanical illustration: the art of describing plants with images, usually for scientific or educational purposes, and usually highlighting diagnostic (defining) features.

chromolithograph: an image printed in inks of several colours by lithographic process.

classification: the method of categorising similar plants; originally into classes sharing a few like qualities or characters, later based on genetic relationships.

copper engraving: an impression taken from an engraved copper plate.

del., delin., delineavit: Latin for he drew it; refers to the artist who made the original drawing or painting.

dessiné: French for designed; refers to the draftsman who designed or drew the plate.

direxit: Latin for he supervised; refers to the botanist who supervised the artwork.

draftsman (also **draughtsman**): the artist who drew the plant for the plate.

electro-type: a method of preparing a block for printing by metal-coating or metal-plating a mould of an image by electrolysis.

elephant (or **elephant folio**): a paper size used by librarians to refer to large books of 28 x 23 inches or larger.

endemic: of plants which grow naturally only in a specified region or country.

engrave: to draw or write on a block or plate by means of incised lines.

engraver: the artisan who executed the engraving of a block or plate.

engraving: an impression taken from an engraved block or plate. In botanical illustration engraving usually refers to *intaglio engraving* (printed from ink imbedded in the incisions) in the narrow sense. Woodcut engraving (printed from ink deposited on the surface) is *relief engraving*. In its broadest usage the term embraces etching where the incision is made by means of acid rather than a tool. In its modern usage the term embraces photographic incision processes.

etch: to draw on a metal plate by means of acid incision of the surface.

etching: an impression taken from an etched plate.

exotic: of plants which are growing away from their natural locality.

***F., Fig.*:** Latin *figura*; figure, in reference to an illustration or a component of an illustration.

fascicle: a block of text and/or illustrations, published sequentially, and finally bound into one book.

***fecit*:** Latin for he made it; refers to the artisan who executed the engraving.

***fils*:** from Latin *filius*, son; for the son of a man (usually an engraver in this context) who has made a reputation for himself.

***fl.*:** Latin *floruit*, he flourished; used to indicate the period of a person's activity, where dates of birth and death are not known.

flora: the vegetation and its component species for a particular region or country.

Flora: a book documenting the species of plants (usually flowering plants) of a particular region.

Florilegium: a Flora, or selection of flowers, illustrated, usually extravagantly.

folio: a sheet of paper (unfolded) where a sheet is approximately 20 x 12½, 18 x 11½, or 15 x 10 inches; a page in a book; a book size.

form: shape and dimension of an object — artists need to employ a range of methods to give the illusion of *form* in a two-dimensional medium.

genus: Latin for a race; applied to the smallest natural group of distinct species, e.g. *Banksia.*

***gravé*:** French for engraved; refers to the artisan who executed the engraving.

halftone print(ing): a printing process which results in reproduction of tone by means of a series of different sized dots, created by interposing a screen between the original image and a photosensitive plate.

***ic., icone(s)*:** Latin for illustration(s).

imp., imprt.: to imprint, or to create an impression; the image; the artisan who executed the impression.

***intaglio* print(ing):** from the Italian *intagliare*, to engrave; the incised lines are printed in the positive, taking the ink (under heavy pressure) from the grooves of the incisions.

lectotype: the specimen chosen by a taxonomist when the botanist who made the first description of a species did not specify one specimen as the type of the species (see type).

letterpress: text printed from letters or types A direct printing process; a form of relief printing.

***lith.*:** from the Latin *lithographit* meaning he lithographed it. Usually a lithographer (an artisan) transposed the artwork of an artist onto the block or plate. Some artists drew directly on the block themselves.

lithograph: An impression taken from a block or plate, sensitised and printed by lithographic technique. It is planographic. The art of preparing a plate to produce a lithographic print (i.e., sensitising the surface of a block to selectively accept, or repel, an oil-based ink). Originally the block was stone, later it was metal. Originally, too, the block or plate was hand-sensitised, later it was sensitised with light and chemical processes (photo-sensitised).

lithography: the art of printing lithographic prints. The term has been adopted by the printing industry for photographic-sensitisation of a metal plate.

mezzotint: an engraving technique employed to aid the illusion of form or tonal difference. The surface is finely abraded and then the high points rubbed back to take less or no ink.

morphology: the external appearance (shape and form) of features of a plant or other subject.

native: of plants which grow naturally in more than one specified region or country.

nature printing: printing from a piece of plant, usually a leaf, which has been inked. The venation (arrangement of veins) and other raised features provide the impression.

nomenclature: the naming of plants following the rules of the *International Code of Botanical Nomenclature*.

offset printing: images printed from a transferred imprint. The plate is inked and the image transferred to a rubber blanket and this image, in turn, printed onto paper. It is an indirect process, equates to a lithographic process, and is planographic. The method is capable of very large print runs while protecting the original plate.

***perfecit*:** Latin for he accomplished it. Refers to the artist or artisan who perfects or completes another's work.

photoengraving: a printing process made possible with the development of photography in the mid 1800s. A (light sensitive) photographic negative is used to etch an acid-resisting positive on metal, upon exposure to light. The image is prepared in stages by progressively waxing the relief to protect it from the acid and to prevent undermining the high points.

photolithography: a printing process in which the plate is created by photographic sensitisation.

pinx.: Latin *pinxit* for he painted it, the illustration.

planographic: of a plate; the inked and uninked regions are in the same plane as for a lithograph.

Pl., Plate: A sheet/page on which an ink impression is printed; usually numbered in a bound book and titled as Pl., T., or Fig. A smooth polished sheet of metal for engraving or etching or similar reproduction technique.

poupée, à la poupée: French 'by the doll'; inking an engraving or etching with a cloth-tipped pencil-like wooden tool, used to work ink into localised areas of the image.

printing press: a machine which makes repeated impressions from an inked plate; originally a modified wine press to replace hand copying of manuscripts; invented by Johan Gutenberg c. 1455.

Prodromus: (Greek) a treatise which is a precursor to a larger or more detailed treatment of a subject.

protologue: for naming and describing a plant; all of the published information, including illustrations, when a botanist describes and names a new species for the first time.

relief printing: from the Italian *rilievo*, projection; printing in the negative, taking the ink from the surface of the block under limited pressure, leaving the incisions uninked.

screen print: an image created by screen printing.

screen printing: printing method perfected by Max Levy in 1886; used for halftone work. The ink is forced through a screen which has been totally, or partially, blocked to create the image.

sculp., sc.: Latin *sculpsit* for he engraved it, referring to the artisan.

sicc.: Latin *siccatus*, dried; of the specimen from which the drawing was taken; drawn from an herbarium specimen.

stipple engraving: engraving by means of incised dots, the dots being cut through a 'ground' and selectively acid etched; an impression taken from a stipple-engraved block or plate.

stochastic screening: a relatively new method for creating halftone images using randomly placed dots rather than a conventional lined screen.

T., Tab.: Latin *tabula*; a plate or page of illustration bound in a book.

taxonomy: from the Greek *taxis*, order, and the Latin *nomen*, name; the science of describing, naming and classifying plants, the practitioner being a taxonomist or taxonomic botanist.

tone: effect of light and shade achieved in printing through changing the intensity, tint, or shade of colour.

type: the specimen(s) used by a taxonomist to make the first description of a species, and to provide the link to the name given.

wood-block: a carved (engraved) wooden block from which a relief print is taken.

woodcut: a print or impression from the surface of a carved wooden block (relief printing). In its narrow sense it refers to the old process whereby the wood surface used was the long-grained plank (tangential) surface. A very ancient method of printing originating in Japan and China. (See **wood-engraving**.)

wood-engraving: the art of carving an image on a wooden block, using the cross-grain (end-grain or cross-section) surface. The cross grain is stronger and capable of working with a wider range of tools. This method was perfected by Thomas Bewick towards the end of the eighteenth century.

xylography: the art of printing from a wooden block which has been cut (carved) to provide a relief image.

zincography: of lithography; the stone block is replaced with a zinc sheet or block and sensitised in the same way for printing.

References

Adams, B. 1986, *The Flowering of the Pacific,* Collins & BM(NH), Sydney.

Anon. 1993, *The Banksias: watercolours by Celia Rosser*, Monash University, Melbourne.

Arber, A. 1938, *Herbals, their Origin and Evolution*, 2nd edn, Cambridge University Press, Cambridge.

Baudin, N., *The Journal of Post Captain Nicolas Baudin*, transl. C. Cornell, Libraries Board of South Australia, Adelaide, 1974.

Bentham, G. 1863–1878, Preface and Concluding Preface in *Flora Austaliensis,* Volumes I and VII, Lovell Reeve, London.

Blunt, W. 1950, *The Art of Botanical Illustration*, Collins, London.

Blunt, W. & Raphael, S. 1979, *The Illustrated Herbal*, Frances Lincoln/Weidenfeld & Nicholson, London.

Blunt, W. & Stearn, W.T. 1994, *The Art of Botanical Illustration*, 2nd edn, Antique Collectors Club, Woodridge.

Bray, L. de 1989, *The Art of Botanical Illustration,* The Wellfleet Press, New Jersey.

Brosse, J. 1983, *Great Voyages of Exploration*, Doubleday, Lane Cove.

Burbidge, N.T. 1978, *Plant Taxonomic Literature in Australian Libraries*, Australian Biological Resources Study, Canberra.

Calaby, H.J. in Smith, B. & Wheeler, A. 1988, *The Art of the First Fleet & other early Australian drawings*, Oxford University Press, Oxford.

Calaby, H.J. 1989, *The Hunter Sketchbook*, National Library of Australia, Canberra.

Carr, D.J. & Carr, S.G.M. 1981, *People and Plants in Australia*, Academic Press, Sydney.

Carr, D.J. 1983, *Sydney Parkinson. Artist of Cook's* Endeavour *Voyage,* BM(NH) & ANU Press, Canberra.

Chapman, A.D. 1991, *Australian Plant Name Index*, Australian Government Publishing Service, Canberra.

Clark, D. 1994, *New Holland Journal: Baron Charles von Hügel, November1833–October 1834,* MUP, Melbourne.

Clarke, I. (and others) 1996, 'The Role of the Botanical Artist' (and other articles), *Australian Garden History* 7(6):4–5(6–19).

Coats, A.M. 1975, *The Treasury of Flowers*, Phaidon, London.

Cohn, H.M. 1995, 'Australian Plants, the Garden and Botany in the Nineteenth Century Periodical', *Naturae* 5:1–24.

Colsey, D. 1970, *Hand and Eye*, Art Gallery of New South Wales, Sydney.

Dampier, W. 1703, *A Voyage to New Holland, &c In the Year, 1699,* James Knapton, London.

Darragh, T.A. 1997, 'Ludwig Becker, a Scientific Dilettante: His Correspondence with J.J.Kaup and Others', *Historical Records of Australian Science* 11(4):501–522.

Deleuze, M. 1805, 'Some Account of the Life and Writings of the late Dr. Joseph Gaertner', (translated from the French in *Annales de Museum national d'hist. naturelle* 1:207(1803). *Annals Botany* 1:73–99.

Desmond, R. 1977, *Dictionary of British and Irish Botanists and Horticulturalists*, revised edition 1994, Taylor & Francis, London.

Dryander, J. 1806, 'Chloris Novae Hollandiae', *Annals Botany* 2:504–532.

Duchess of Hamilton, J. & Bruce, J. 1998, *The Flower Chain*, Kangaroo Press, East Roseville.

Ducker, S.C. 1988, *The Contented Botanist*, MUP, Melbourne.

Dunthorne, G. 1938, *Flower and Fruit Prints*, repr. 1970, Da Capo, New York.

Duyker, E. & Tingbrand, P. (eds and transls), 1995, *Daniel Solander: collected correspondence 1753–1782*, The Miegunyah Press, Melbourne.

Eisler, W. & Smith, B. 1988, *Terra Australis. The Furthest Shore*, Art Gallery of New South Wales & Int. Cult. Corp. of Australia, Sydney.

Ellis, V.R. 1979, *Louisa Ann Meredith: A Tigress in Exile*, Blubber Head Press, Sandy Bay.

Elliot, B. 1994, *Treasures of The Royal Horticultural Society*, Sagapress/Timber Press, Portland.

Erikson, R. 1969, *The Drummonds of Hawthornden*, Lamb Peterson, Osborne Park.

Finney, C.M. 1984, *To Sail Beyond the Sunset*, Rigby, Adelaide.

Fitzpatrick, K. 1958, *Australian Explorers*, Oxford University Press, London.

Flinders, M. 1814, *A voyage to Terra Australis*, G. & W. Nicol, London.

Gage, A.T. & Stearn, W.T. 1988, *A Bicentenary History of the Linnean Society of London*, Linnean Society of London & Academic Press, London.

George, A.S. 1971, 'The plants seen and collected in North-western Australia by William Dampier', *Western Australian Naturalist* 11(8):173–178.

George, A.S. 1981, 'The Background to the Flora of Australia', *Flora of Australia* 1: 3–23.

George, A.S. 1996, 'A Tercentenary', *Aust. Syst. Bot. Soc. Newsletter* 89:20-21.

George, A.S. & Jordan, C. 1993, 'A Heritage of Botanical Art', *Voices* 3(1):92–107.

Gooding, J. 1991, *Wildflowers in Art*, Art Gallery of Western Australia, Perth.

Gray, P. 1996, Mueller articles by several authors, *Victorian Naturalist* 113(4):128–231.

Hackforth-Jones, J. 1977, *The Convict Artists*, Macmillan, Melbourne.

Hall, N. 1978, *Botanists of the Eucalypts*, CSIRO, Melbourne.

Hall, N. 1984, *Botanists of Australian Acacias*, CSIRO, Melbourne.

Hasluck, A. 1955, *Portrait with Background*, 2nd edn 1990, Fremantle Arts Centre, Fremantle.

Henty, C. 1988, *For the People's Pleasure*, Greenhouse Publications, Richmond.

Hind, Arthur, M. 1923, *A History of Engraving & Etching,* Houghton Mifflin; repr. 1963, Dover Publications, New York.

Hoare, M.E. in D.J. & S.G.M. Carr 1981, *People and Plants in Australia*, Academic Press, Sydney.

Hoare, M.E. 1976, *The Tactless Philosopher. Johann Reinhold Forster 1729–1798,* Hawthorn Press, Melbourne.

Hooker, J.D. 1860, *Flora Tasmaniae*, Lovell Reeve, London.

Horner, F. 1987, *The French Reconnaissance*, MUP, Melbourne.

Karg, A.T. 1974, 'Hunt Institute Collection of Biographical Sketches relating to Australian Botany', *Taxon* 23(4):613–618.

Kerr, J. (ed.) 1992, *The Dictionary of Australian Artists*, Oxford University Press, Melbourne.

Koenig, C. & Sims, J. 1806, 'Review of Novae Hollandiae Plantarum Specimen', *Annals of Botany* 2:365–375.

Kramer, J. 1996, *Women of Flowers*, Stewart, Tabori & Chang, New York.

Kynaston, E. 1981, *A Man on the Edge*, Allen Lane, Ringwood.

Johnson, L.A.S. *et al.* 1983, *Drawn for Life – The Development of Botanical Illustration of Australian Plants*, Royal Botanic Gardens, Sydney.

La Billardière. J.-J.H. 1800, *Voyage in search of La Pérouse,* Stockdale, Piccadilly. (Translation from the French edition, *Relation du voyage à la recherche de La Pérouse*, 1800).

Lack, H.W. 1998, 'Recording Form in Early Ninteenth Century Botanical Drawing, Ferdinand Bauer's 'Cameras", *Curtis's Botanical Magazine* 15(4):254–274.

Lack, H.W., & Ibáñez, V. 1997, 'Recording Colour in Late Eighteenth Century Botanical drawings: Sydney Parkinson, Ferdinand Bauer and Thaddäus Haenke', *Curtis's Botanical Magazine* 14(2):87–100.

Lawrence, G.M.H. (ed.) 1963, *A Catalogue of Redoutéana*, Hunt Botanical Library, Pittsburg.

Lawrence, G.H.M. (ed.) 1963, *Sertum Anglicum 1788 – facsimile with critical studies and a translation*, Hunt Botanical Library, Pittsburg.

Lawson, E. 1995, *The Natural Art of Louisa Atkinson*, State Library of NSW Press, Sydney.

Lewis, J. 1992, *Walter Hood Fitch: A Celebration*, HMSO, London.

Linnaeus, C. 1751, *Philosophia Botanica*, Stockholm, Amsterdam.

Mabberley, D.J. 1985, *Jupitor Botanicus, Robert Brown of the British Museum*, Cramer, Braunschweig.

Mackaness, G. 1936, *Sir Joseph Banks: his relations with Australia*, Angus & Robertson, Sydney.

Maiden, J.H. 1909, *Sir Joseph Banks. The Father of Australia*, Kegan Paul, Trench, Trübner & Co., London.

Mansfield, C. 1998, 'Edgar Dell: A Singular Talent', *Australian Garden History* 9(6):6–8.

McCormick, A. 1992, 'Australian Flowers: the first illustrations', *Australian Antique Collector* July–Dec.:31–35.

McKay, J. 1990, *Ellis Rowan A Flower-Hunter in Queensland*, Queensland Museum, Brisbane.

McMinn, W.G. 1970, *Allan Cunningham*, MUP, Melbourne.

Mills, C. 1998, *Images from Nature*, The Natural History Museum, London.

Mitchell, T.L. 1838, *Three expeditions into the interior of eastern Australia*, 2 Vols, T. & W.Boone, London.

Mitchell, T.L. 1848, *Journal of an expedition into the interior of tropical Australia*, Longman, Brown, Green and Longmans, London.

Moyal, A. 1976, *Scientists in nineteenth century Australia*, Cassell, Sydney.

Moyal, A. 1986, *A Bright and Savage Land*, Collins, Sydney.

Nissen, C. 1951, *Die Botanische Buch-illustration*, revsd edn 1966, Hiersemann Verlags-Gesellschaft, Stuttgart.

Norst, M.J. 1989, *Ferdinand Bauer. The Natural History Drawings*, British Museum (Natural History), London.

North, M. 1980, *A vision of Eden*, 3rd impr. 1999, The Stationery Office, London.

Orchard, A.E. 1999, 'A History of Systematic Botany in Australia', *Flora of Australia*, 2nd edn 1:11–103.

Ord, M. 1988, *Historical Drawings of Native Flowers {by} Harriet & Helena Scott*, Craftsman House, Roseville.

Osborn, T.G.B & Gardner, C.A. 1939, 'Dampier's Australian Plants', *Proc. Linn. Soc. London* 151: 44–50.

Parkinson, S. 1773, *Journal of a Voyage to the South Seas in His Majesty's Ship the Endeavour*, Stanfield Parkinson, London.

Perry, T.M. 1982, *Discovery of Australia*, Mead & Beckett, Sydney.

Rix, M. 1981, *The Art of the Plant World,* The Overlook Press, Woodstock, New York.

Robb, G. & Smith, E. 1993, *Concise Dictionary of Australian Artists*, MUP, Melbourne.

Roberts, R. & Whitten, M. n.d. *Biographies of Australian Botanists*, ms (see Karg, above).

Robertson, E.L. 1981, *Alison Ashby's Wildflowers of Southern Australia*, South Australian Museum Board & Botanic Gardens Board, Adelaide.

Rowan, E.M. 1898, *A Flower-Hunter in Queensland and New Zealand*, Angus & Robertson, Sydney.

Sampson, F.B. 1985, *Early New Zealand Botanical Art*, Reed Methuen, Auckland.

Sherwood, S. 1996, *Contemporary Botanical Artists*, Weidenfeld & Nicholson, London.

Short, P.S. (ed.) 1990, *History of systematic botany in Australasia*, Australian Systematic Botany Society, Melbourne.

Stafleu, F.A. 1967, *Adanson, La Billardière, De Candolle*, Verlag von J. Cramer, Lehre.

Stafleu, F. 1971, *Linnaeus and the Linnaeans*, IAPT, Utrecht.

Stafleu, F.A. & Cowan, R.S. 1976–1988, *Taxonomic Literature*, 2nd edn, Bohn, Scheltema & Holkema, *et al.*, The Hague.

Stearn, W.T. 1960, Introduction in Robert Brown, *Prodromus Florae Novae Hollandiae*, Facs, Engelmann (Cramer) and Wheldon & Wesley, Weinheim.

Stearn, W.T. 1966, *Botanical Latin,* David & Charles, Newton Abbot.

Stearn, W.T. 1990, *Flower Artists of Kew*, The Herbert Press, London.

Stearn, W.T. (ed.) 1999, *John Lindley*, Antique Collectors Club and The Royal Horticultural Society, London.

Stearn, W.T. & Blunt, W. (eds) 1976, *The Australian Flower Paintings of Ferdinand Bauer*, Basilisk Press, London.

Steenis, C.G.G.J. van (ed.) 1950, 'Alphabetical list of Collectors', *Flora Malesiana* 1:5–606.

Stewart, J. & Stearn, W.T. 1993, *The Orchid Paintings of Franz Bauer*, Timber Press, Portland.

Stretton, D. & Doust, R.F. 1981, *A Fine Field for Botanising*, The Library Council of New South Wales, Sydney.

Sumner, R. 1993, *A woman in the Wilderness*, NSW University Press, Kensington.

Tipping, M. 1979, *Ludwig Becker*, MUP, Melbourne.

Turrill, W.B. 1963, *Joseph Dalton Hooker*, Thomas Nelson & Sons, London.

Watts, P., Pomfrett, J.A. & Mabberley, D. 1997, *An Exquisite Eye*, Historic Houses Trust of NSW, Sydney.

Webb, J. 1995, *George Caley*, Surrey Beatty, Sydney.

White, J. 1790, *Journal of a Voyage to New South Wales*, J. Debrett, London.

Willis, J.H. *et al.* 1986, *Australian Plants: Collectors and Illustrators*, Western Australian Herbarium Research Notes 12:1–111.

Willis, J.H. & Cohn, H.M. 1993, 'Botanical Exploration of Victoria' in *Flora of Victoria* 1:61–78.

Willis, M. 1949, *By Their Fruits*, Angus & Robertson, Sydney.

Zdanowicz, I. 1996, *Beauty in Truth: the Botanical Art of Margaret Stones*, National Gallery of Victoria, Melbourne.

INDEX